HAESE MATHEMATICS

Specialists in mathematics education

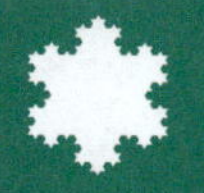

Mathematics 2

PYP 2 Write-on Workbook

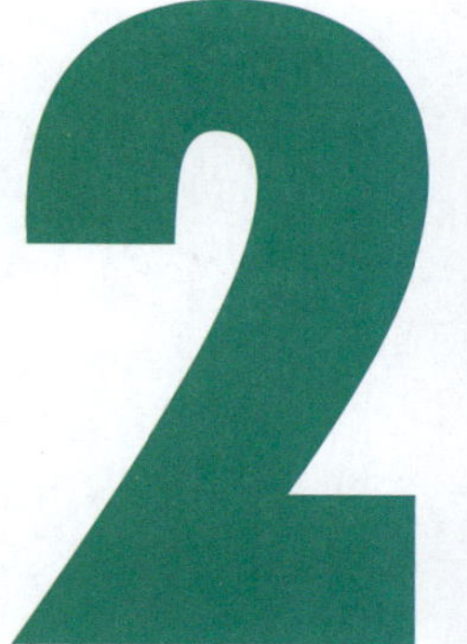

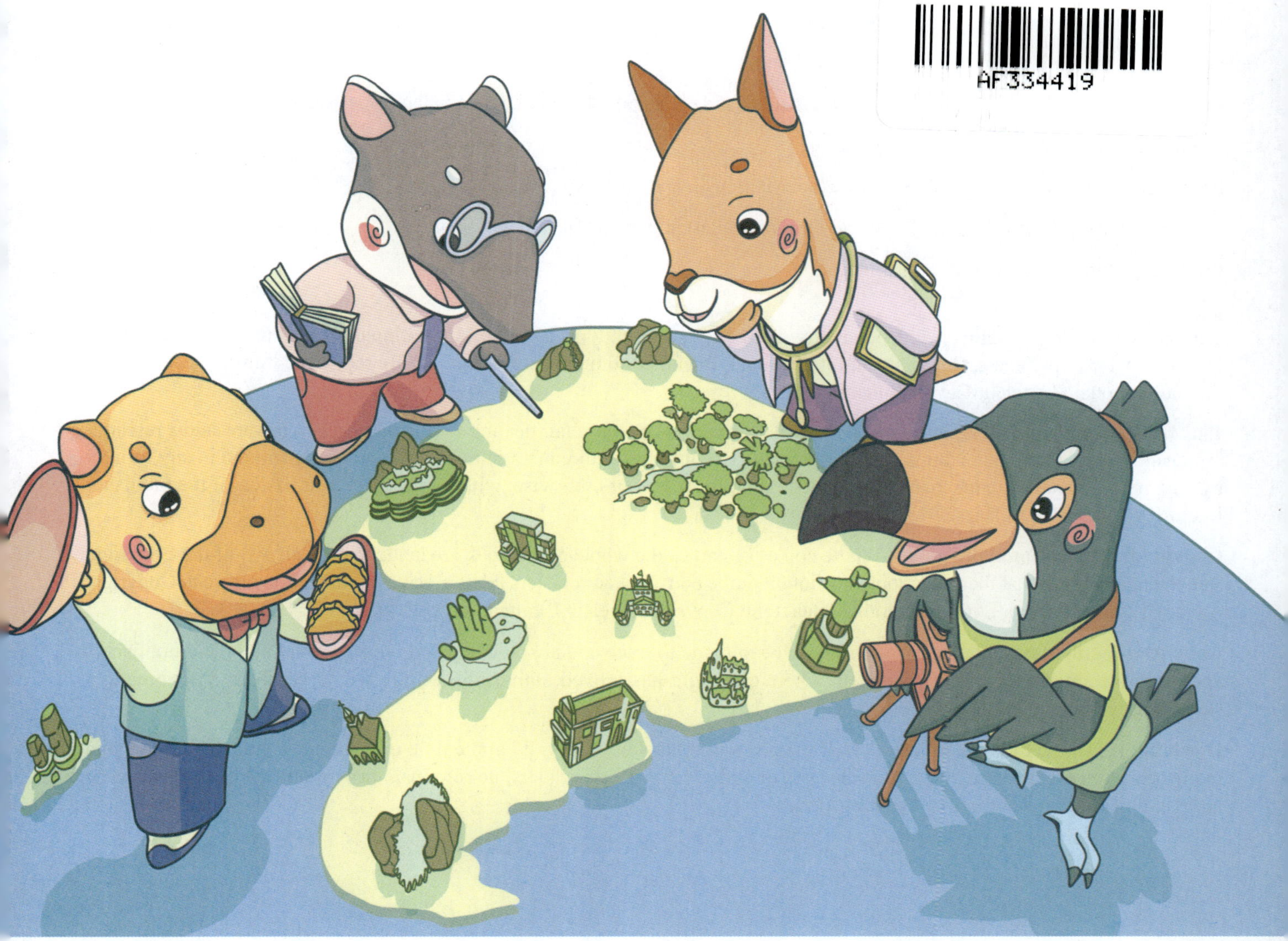

for use with

IB Primary Years Programme

Michael Haese Mark Humphries Philippa Sawyers

MATHEMATICS PYP 2

Michael Haese   B.Sc.(Hons.), Ph.D.
Mark Humphries   B.Sc.(Hons.)
Philippa Sawyers   B.Ed.(Hons.)

Published by Haese Mathematics
152 Richmond Road, Marleston, SA 5033, AUSTRALIA
Telephone: +61 8 8210 4666
Email: info@haesemathematics.com
Web: www.haesemathematics.com

National Library of Australia Card Number & ISBN 978-1-922416-75-9

© Haese & Harris Publications 2024

First Edition   2024

Cartoon artwork by James Hobbs, Yi-Tung Huang, Le Quynh Chi Ta, and Ngoc Vo.

Artwork by Hannah Coleman and Brian Houston.

Cover art by Le Quynh Chi Ta.

Computer software by Patrick French, Ben Hensley, Huda Kharrufa, Rachel Lee, and Han Zong Ng.

Audio recorded by Eloise Quinn-Valentine.

Production work by Hannah Coleman, Michael Mampusti, and Ngoc Vo.

Typeset in Australia by Charlotte Frost and Deanne Gallasch. Typeset in Times Roman 13.

Printed in China by Prolong Press Limited.

This book has been developed independently from and is not endorsed by the International Baccalaureate Organization. International Baccalaureate, Baccalauréat International, Bachillerato Internacional, and IB are registered trademarks owned by the International Baccalaureate Organization.

FOREWORD

Mathematics PYP 2 has been designed and written for the International Baccalaureate Primary Years Programme (IB PYP). The workbook and online interactive material provide an engaging and structured package, allowing students to explore and develop their confidence in Mathematics.

Our aim with this spiral bound write-on workbook is to provide the students and teachers with an alternative to photocopied worksheets, which in turn will give the students an organised record of their work throughout the year.

The book contains a variety of exercises ranging from basic to advanced, to cater for a range of student abilities and interests. The material is presented in a clear, easy-to-follow style to aid comprehension and retention, especially for English Language Learners. A Revision set for each chapter is provided at the end of the book.

Important information and key notes are highlighted, and worked examples provide clear instruction and relevant explanations. Discussions, Activities, and Puzzles are used throughout the chapters to develop understanding. All of these features can be viewed through Snowflake, our online digital textbook platform.

We have endeavoured to provide a stimulating workbook and accompanying interactive material. Our aim is to develop and encourage student understanding and to grow an appreciation and love for Mathematics.

We welcome your feedback:

Email: info@haesemathematics.com
Web: www.haesemathematics.com

ONLINE FEATURES

Each workbook comes with a 12 month subscription to the online edition and its range of interactive features. This can be accessed through the **SNOWFLAKE** online learning platform via a web browser or our offline viewer.

To activate your electronic workbook, please contact Haese Mathematics by emailling info@haesemathematics.com with your proof of purchase such as a copy of your receipt.

For general queries regarding **SNOWFLAKE** and online subscriptions:

- Visit our help page: https://snowflake.haesemathematics.com.au/help
- Contact Haese Mathematics: info@haesemathematics.com

WATCH LISTEN LEARN

Watch Listen Learn is an exciting feature of this book.

This icon in a worked example or exercise denotes an active online link.

Simply click the icon (or anywhere in the example or exercise box) to access the Watch Listen Learn with a teacher's voice explaining each concept.

Play any line as often as you like. See how the basic processes come alive using movement and colour on the screen.

CLICKABLE ICONS

You will see these icons throughout the workbook. Click the icon to access online content.

Printable Worksheets containing extra questions or support material.

Listening Activities.

Video Demonstrations providing instruction.

Activities and Games.

TABLE OF CONTENTS

CHAPTER 1: NUMBER

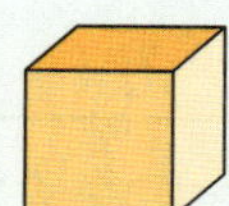

This block is a **unit**.

The smallest counting numbers can be drawn using unit blocks.

Exercise 1

Practise your writing. Copy each number.

1 one ________

2 two ________

3 three ________

4 four ________

5 five ________

6 six ________

7 seven ________

8 eight ________

9 nine ________

10 ten ________

Activity

Count the penguins with:

a a red bow tie ________

b a yellow bow tie ________

c a blue hat ________

d green shoes ________

e a bow tie *and* a hat ________

This block is a **ten**.   It represents 10 units.

Exercise 2

Practise your writing. Copy each number.

11   eleven

____

12   twelve

____

13   thirteen

____

14   fourteen

____

15   fifteen

____

16   sixteen

____

17   seventeen

____

18   eighteen

____

19   nineteen

____

20   twenty

____

Listening Activity

Click and follow the instructions.

1 _______________   2 _______________   3 _______________

4 _______________   5 _______________   6 _______________

Exercise 3

Complete the chart:

Diagram	*Numeral*	*Word*
	3	three
	7	
	2	two
		five

	8	
		six
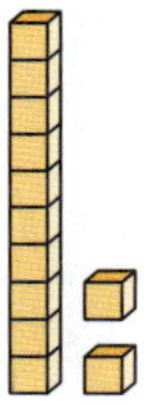		
	10	
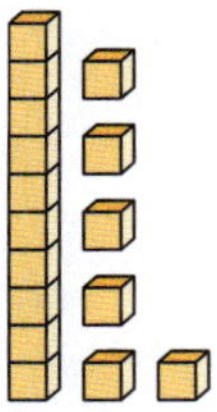		
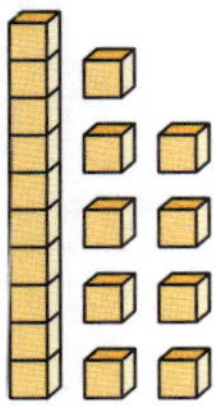		

Exercise 4

Practise your writing. Copy each number.

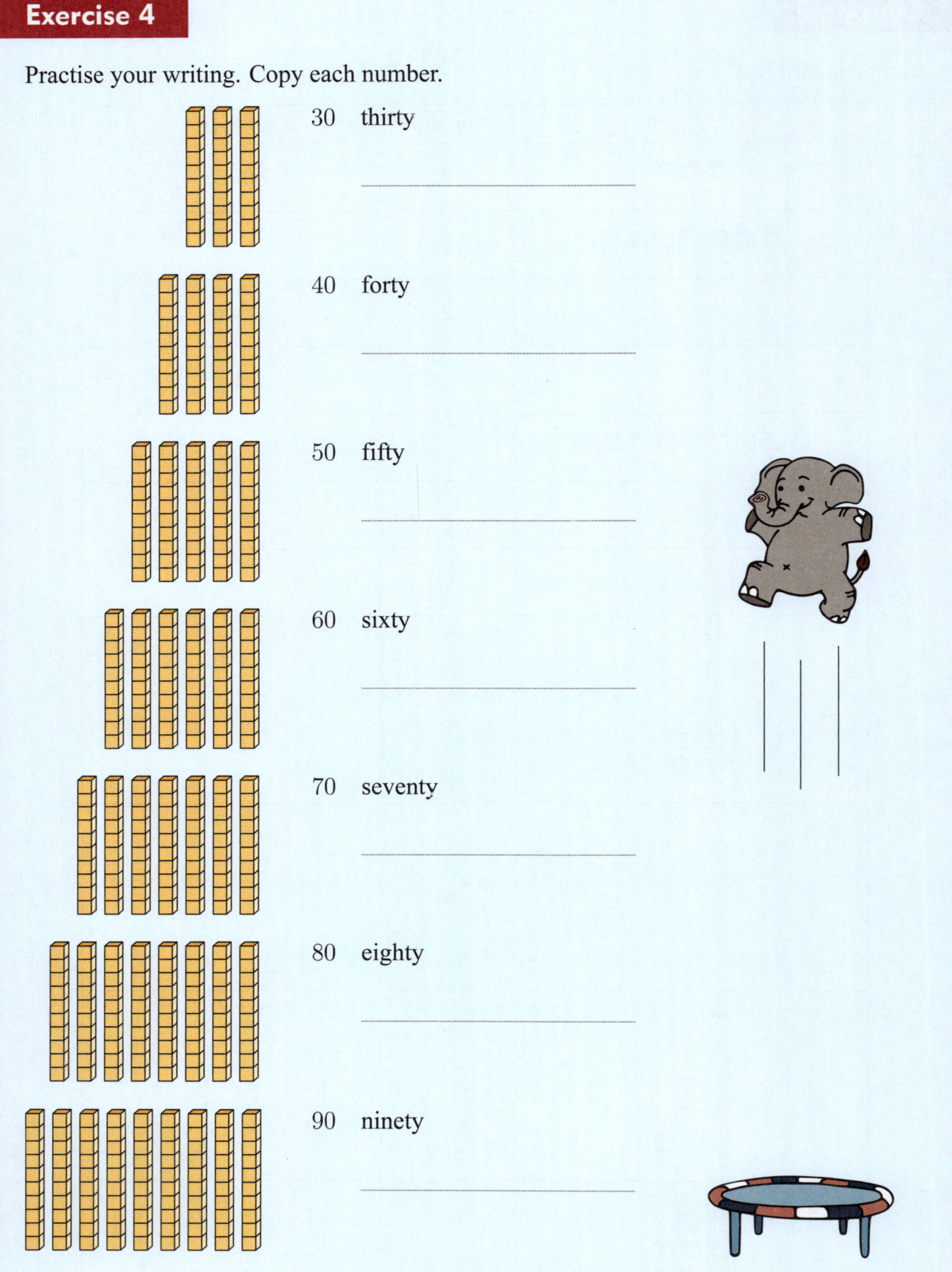

30   thirty

40   forty

50   fifty

60   sixty

70   seventy

80   eighty

90   ninety

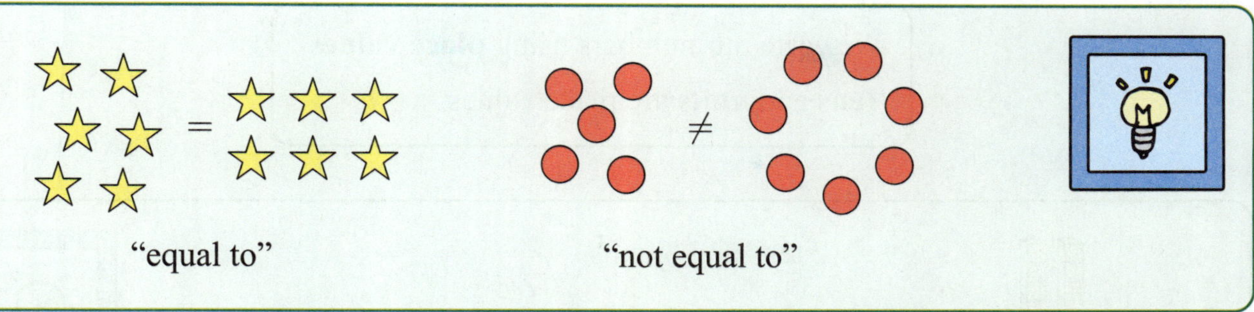

Exercise 5

Practise your writing.

= equal to ≠ not equal to

___ ________________ ___ ________________

___ ________________ ___ ________________

Exercise 6

Use = or ≠.

a

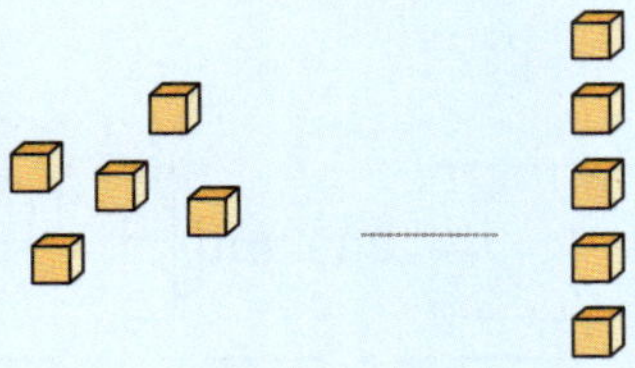

b

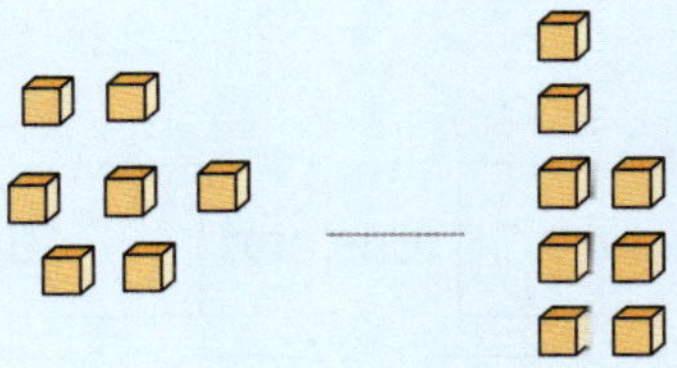

c

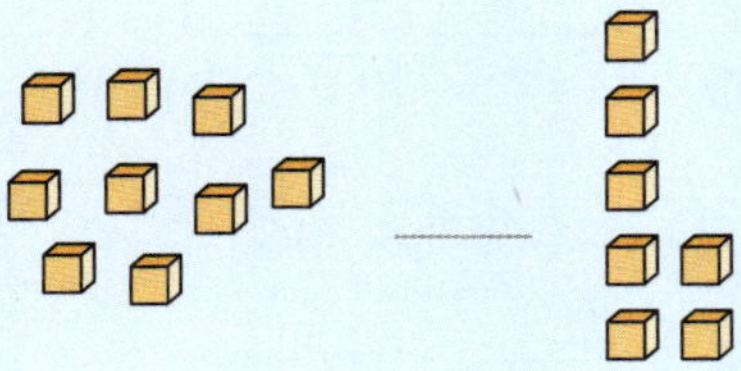

d

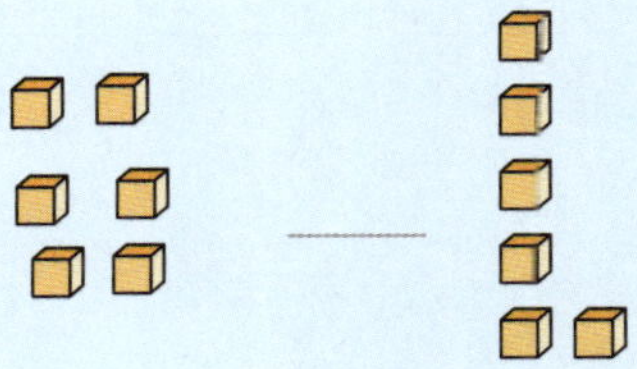

e

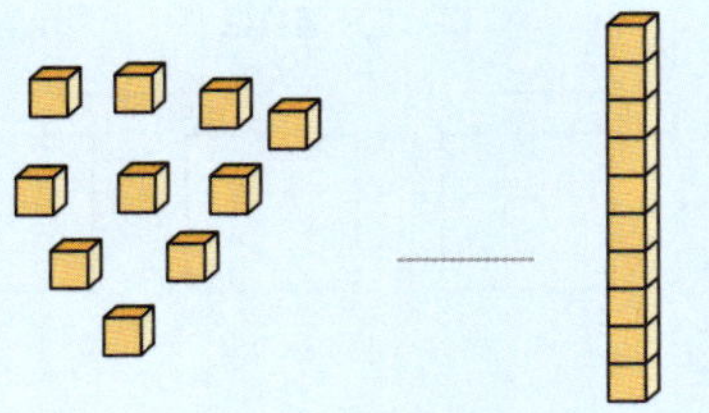

f

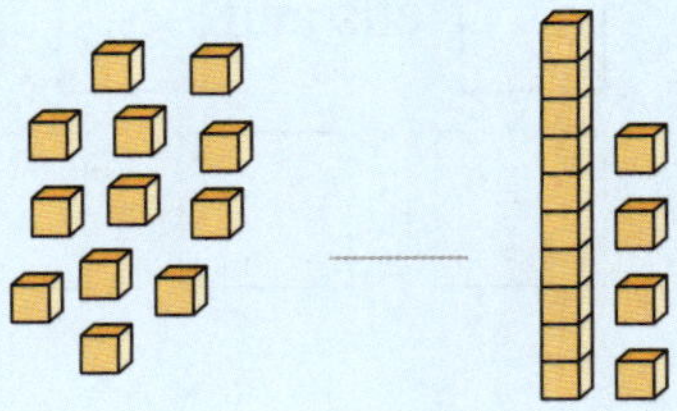

g

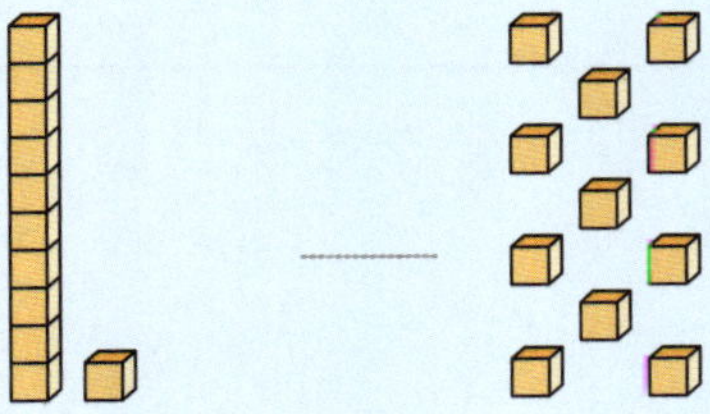

h

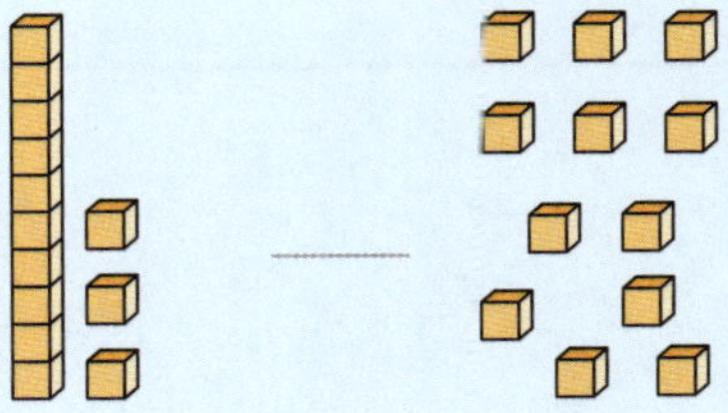

We write our numbers using **place values**.
Tens and **units** are place values.

2 tens and 3 units

20 + 3 = 23

twenty three

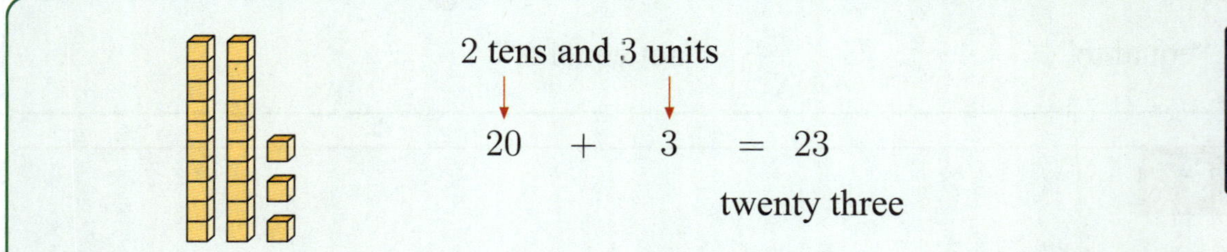

Exercise 7

What numbers are shown?

a 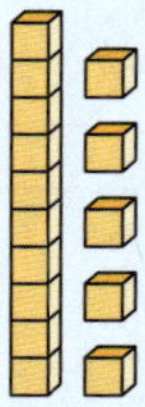

☐ ten and ☐ units

☐ + ☐ = ☐

____

b 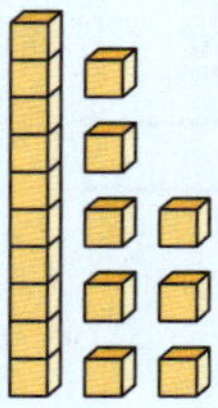

☐ ten and ☐ units

☐ + ☐ = ☐

____

c 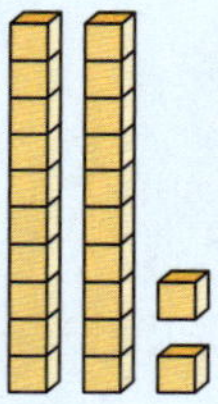

☐ tens and ☐ units

☐ + ☐ = ☐

____

d 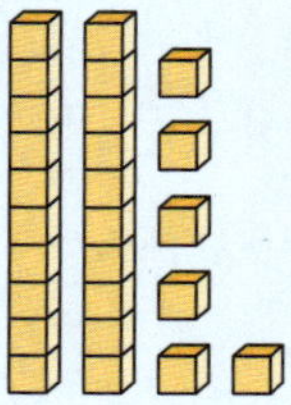

☐ tens and ☐ units

☐ + ☐ = ☐

____

e 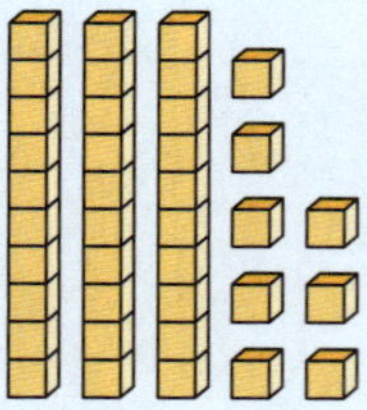

☐ tens and ☐ units

☐ + ☐ = ☐

____

f 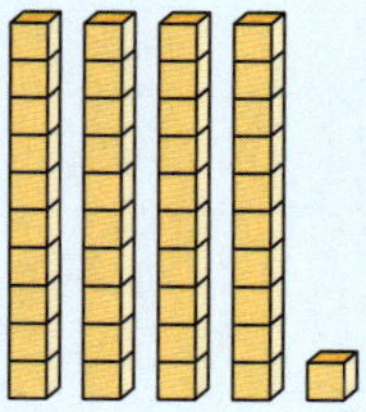

☐ tens and ☐ unit

☐ + ☐ = ☐

____

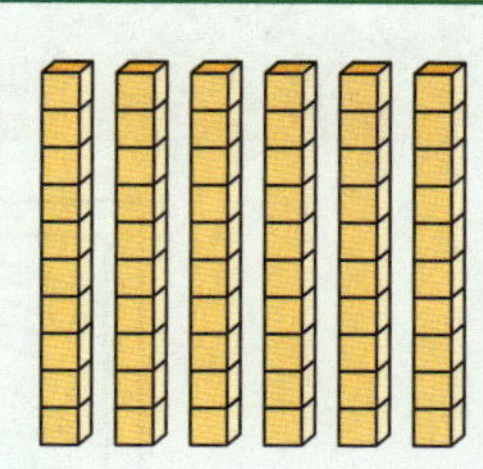   6 tens + 0 units = 60

"sixty"

There are zero units.

Exercise 8

a 

☐ tens + ☐ units = ☐

There are ___________ units.

b 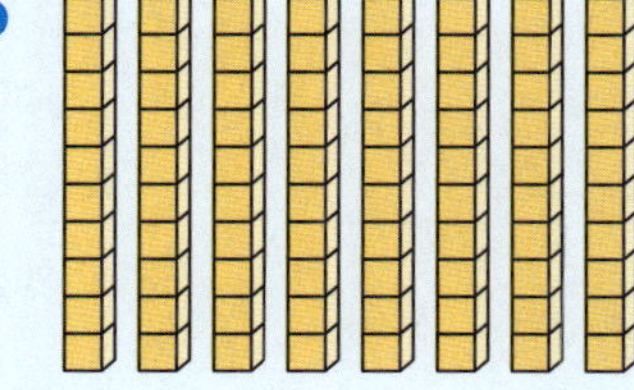

☐ tens + ☐ units = ☐

There are ___________ units.

c 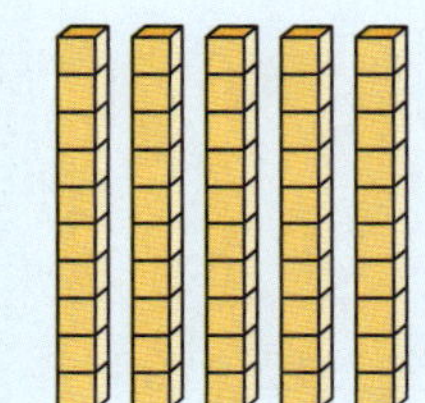

☐ tens + ☐ units = ☐

There are ___________ units.

Discussion

Where do we see zero in the world around us?

What other words do we use that mean "zero"?

Activity

This chart shows the numbers from 1 to 100.

1	2	3	4	5	6	7	8	9	10
11	12	13	14	15	16	17	18	19	20
21	22	23	24	25	26	27	28	29	30
31	32	33	34	35	36	37	38	39	40
41	42	43	44	45	46	47	48	49	50
51	52	53	54	55	56	57	58	59	60
61	62	63	64	65	66	67	68	69	70
71	72	73	74	75	76	77	78	79	80
81	82	83	84	85	86	87	88	89	90
91	92	93	94	95	96	97	98	99	100

1 What do you notice about the numbers directly under 5?

____

2 What do you notice about the numbers in the same row as 36?

____

____

3 Complete these *pieces* of the chart:

a

	18	
	28	

b

	62	

c

37		

d

		45

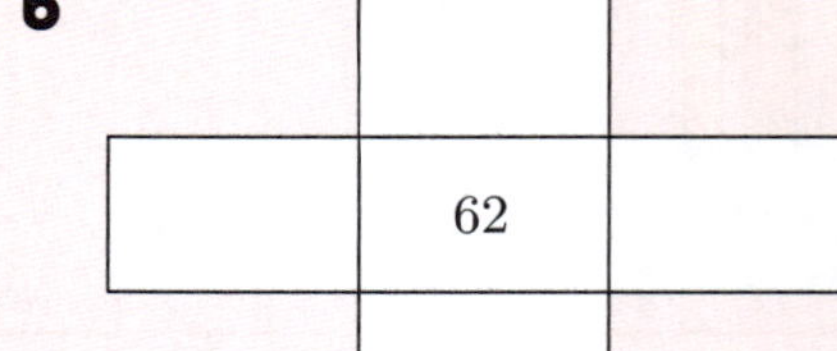

e

51		

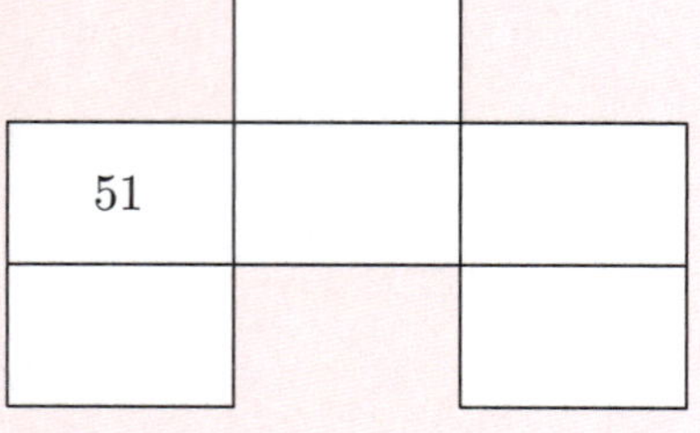

f

	98	

4 Look at the numbers which are shaded.

a Describe these numbers.

____

____

b Describe the pattern formed by these numbers.

____

____

Activity

Complete this chart.

	91				95		97	98	
80			83				87		89
	71	72		74		76			
		62		64			67	68	
	51		53			56			59
40				44		46		48	
	31		33		35				39
		22		24	25		27		
10			13			16			19
0		2			5			8	

What patterns can you see?

1 ____

____

2 ____

____

3 ____

Exercise 9

What numbers are shown?

a 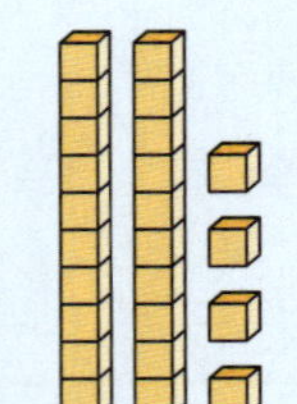 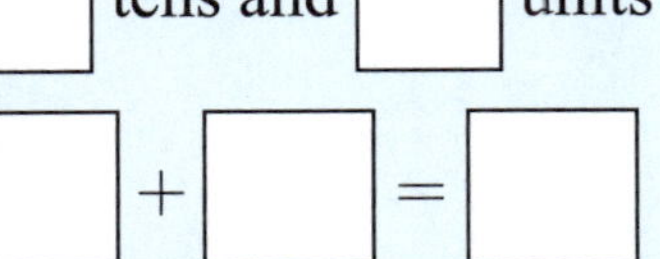 ☐ tens and ☐ units

☐ + ☐ = ☐

b  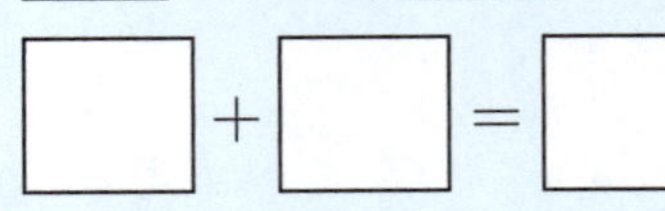 ☐ ten and ☐ units

☐ + ☐ = ☐

c 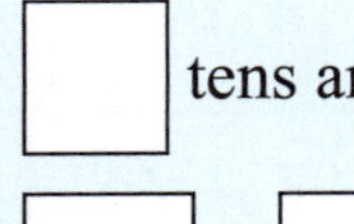 ☐ tens and ☐ units

☐ + ☐ = ☐

d 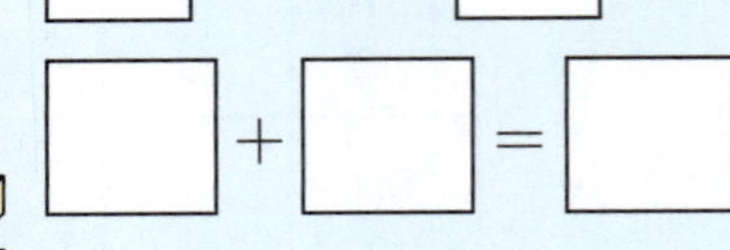 ☐ tens and ☐ units

☐ + ☐ = ☐

e 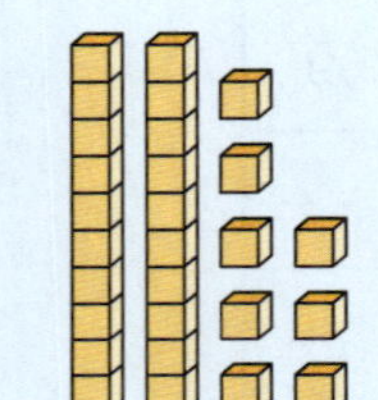 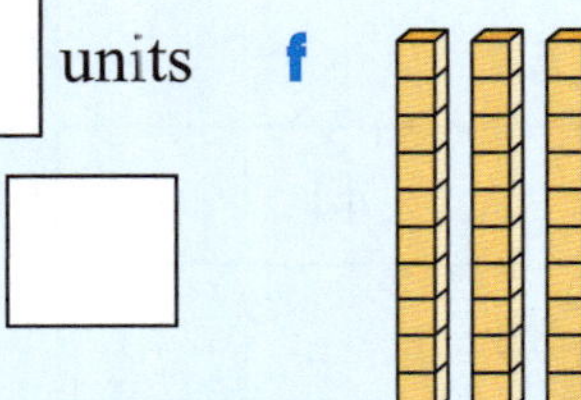 ☐ tens and ☐ units

☐ + ☐ = ☐

f 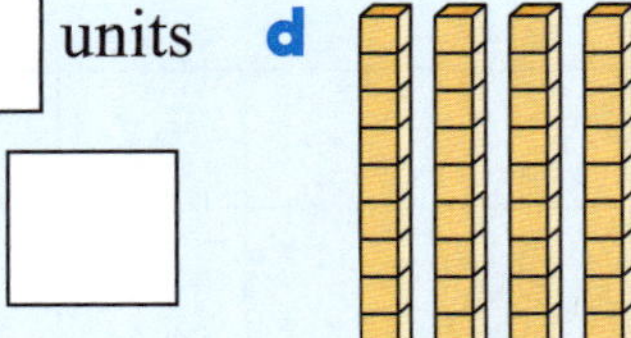 ☐ tens and ☐ unit

☐ + ☐ = ☐

g 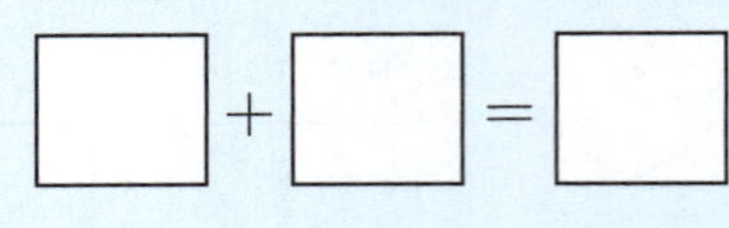 ☐ tens and ☐ units

☐ + ☐ = ☐

h 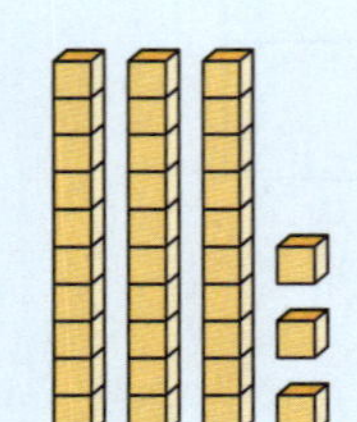 ☐ tens and ☐ units

☐ + ☐ = ☐

i 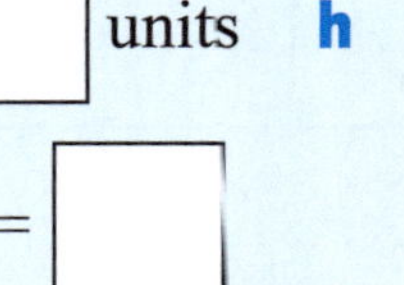 ☐ tens and ☐ units

☐ + ☐ = ☐

Exercise 10

Partition each number using place values:

4   6

4 tens   6 units

$40 + 6$

forty six

a   1   7

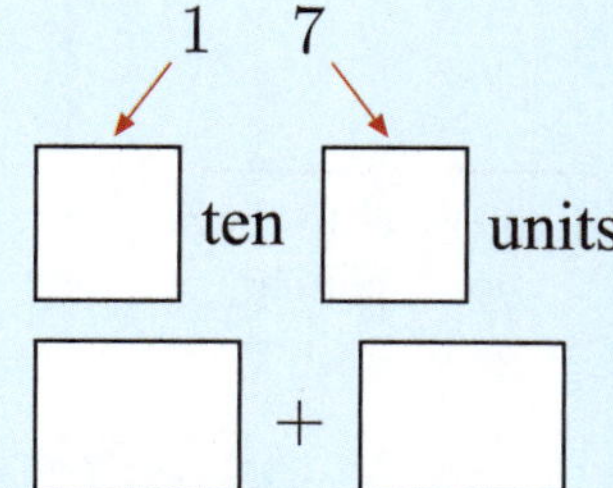

☐ ten ☐ units

☐ + ☐

b   2   8

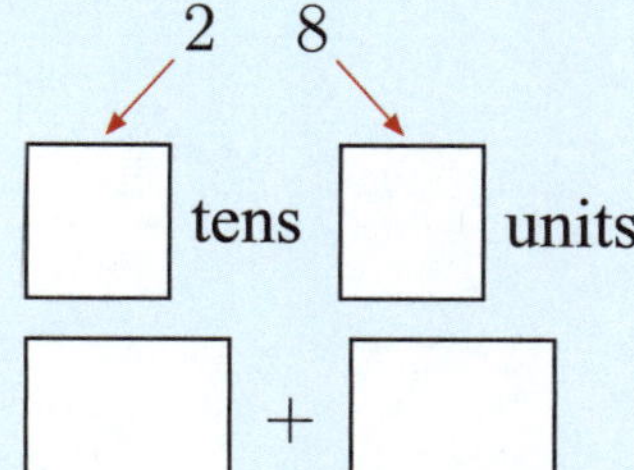

☐ tens ☐ units

☐ + ☐

c   4   3

☐ tens ☐ units

☐ + ☐

d   3   5

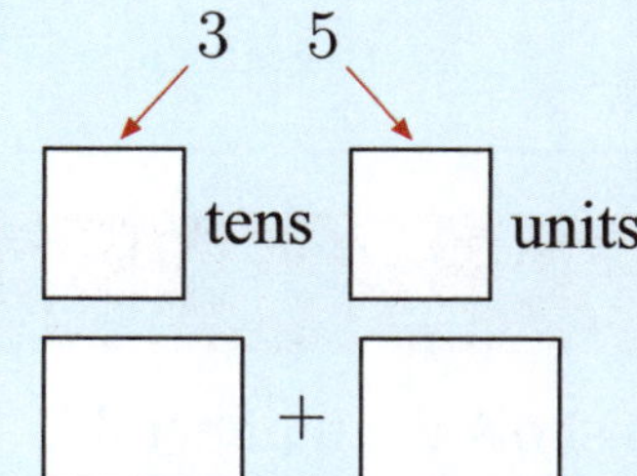

☐ tens ☐ units

☐ + ☐

e   7   1

☐ tens ☐ unit

☐ + ☐

f   9   0

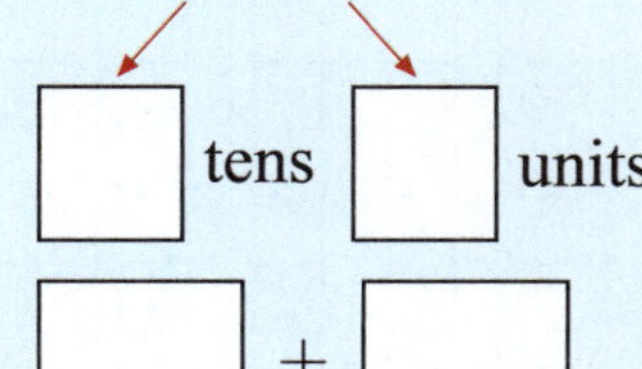

☐ tens ☐ units

☐ + ☐

g   6   5

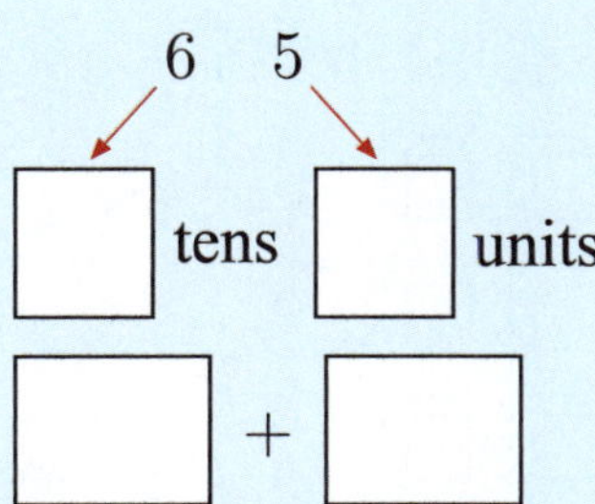

☐ tens ☐ units

☐ + ☐

h   8   2

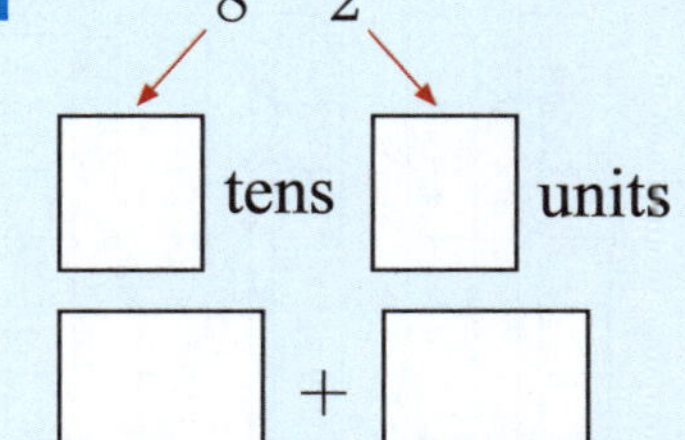

☐ tens ☐ units

☐ + ☐

Exercise 11

Write these numbers in words:

a 29 _______________________

b 37 _______________________

c 44 _______________________

d 58 _______________________

e 63 _______________________

f 79 _______________________

g 81 _______________________

h 92 _______________________

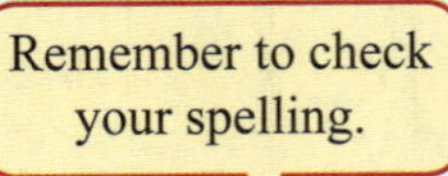

Activity

a Colour blocks to show the number 36.

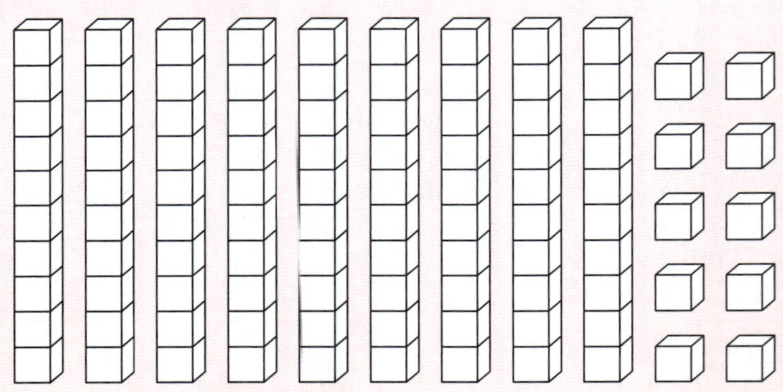

_______ blocks are *not* coloured.

b Colour blocks to show the number 83.

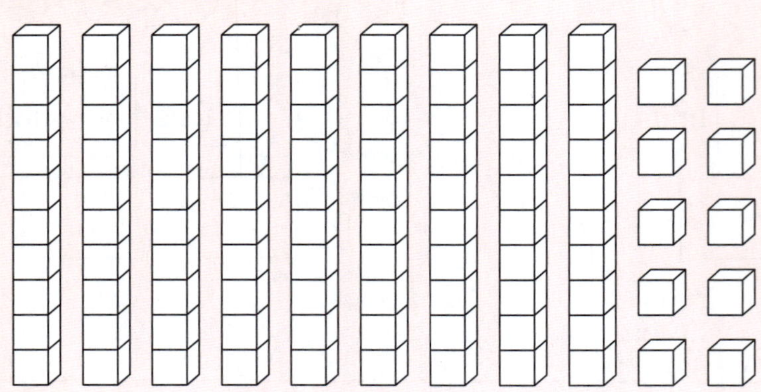

_______ blocks are *not* coloured.

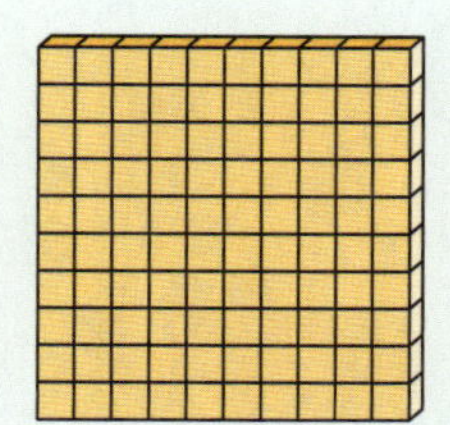

This block is a **hundred**.

It represents 10 tens

or   100 units.

Hundreds are also a **place value**.

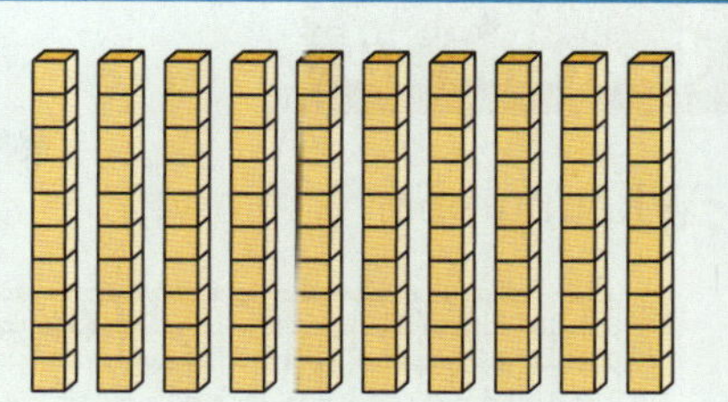

Exercise 12

What number is shown?

1 hundred   6 tens   2 units

The number is 162

or one hundred and sixty two.

a   ☐ hundred   ☐ tens   ☐ units

The number is ☐

or _______________________________________

_______________________________________ .

b   ☐ hundreds

☐ tens

☐ units

The number is ☐

or _______________________________________

_______________________________________ .

Exercise 13

What number is shown?

a

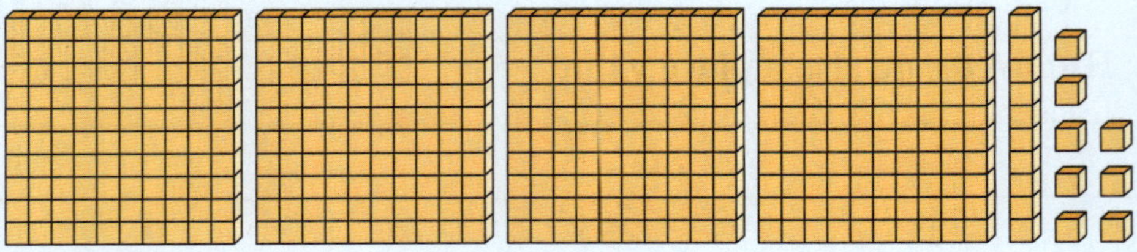

☐ hundreds   ☐ ten   ☐ units

The number is ☐ or

____

____ .

b

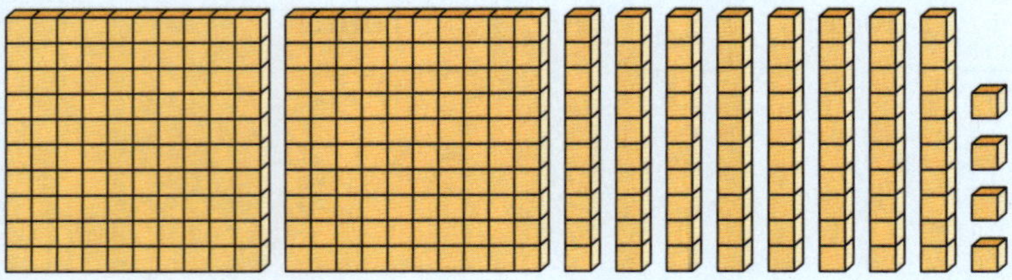

☐ hundreds   ☐ tens   ☐ units

The number is ☐ or

____

____ .

c

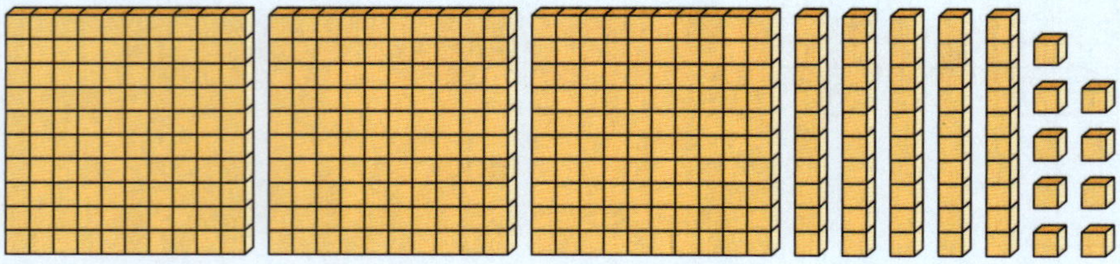

☐ hundreds   ☐ tens   ☐ units

The number is ☐ or

____

____ .

Exercise 14

Partition each number using place values:

a

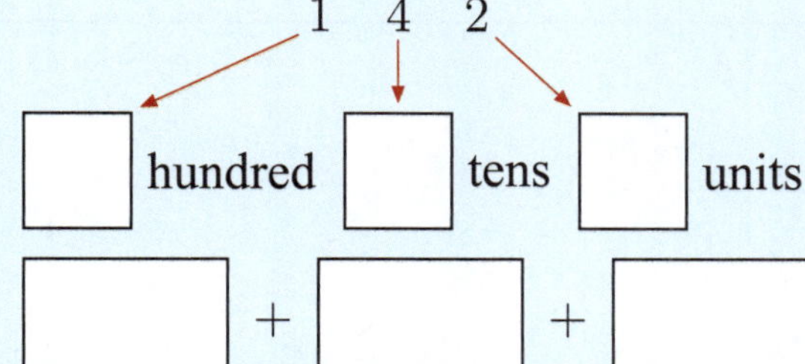

1 4 2

☐ hundred ☐ tens ☐ units

☐ + ☐ + ☐

b

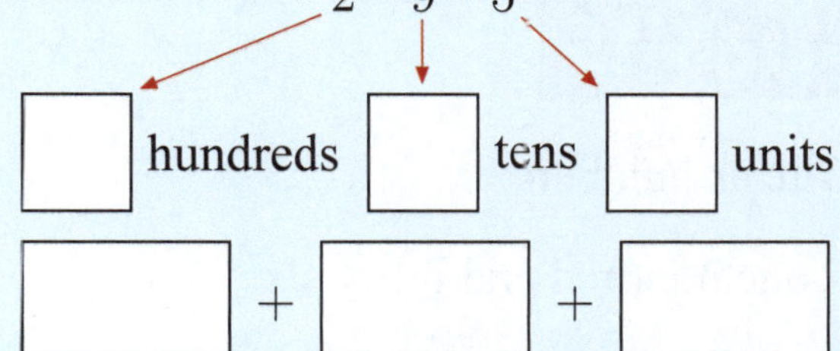

2 9 5

☐ hundreds ☐ tens ☐ units

☐ + ☐ + ☐

c

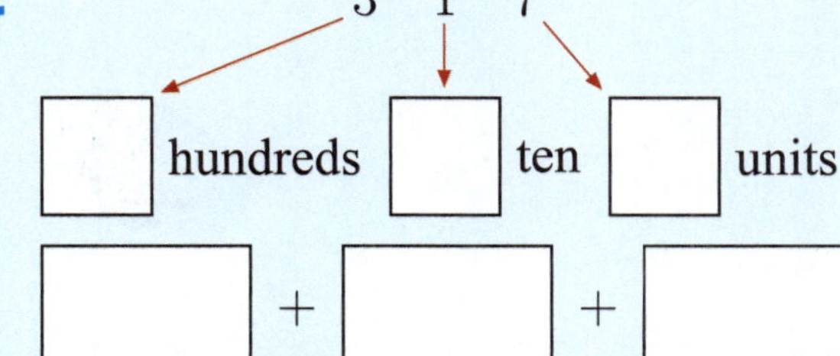

3 1 7

☐ hundreds ☐ ten ☐ units

☐ + ☐ + ☐

d

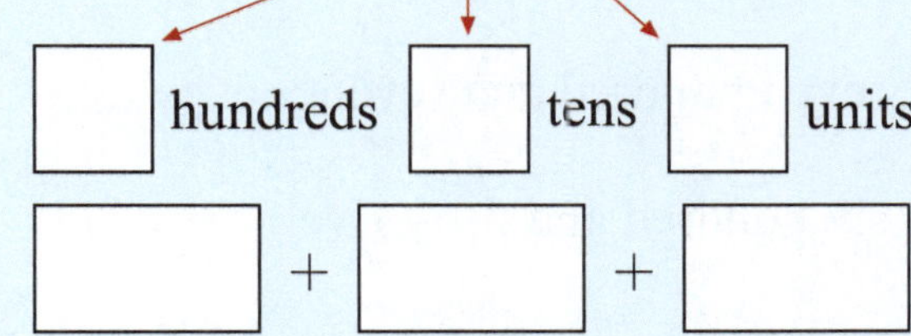

7 8 3

☐ hundreds ☐ tens ☐ units

☐ + ☐ + ☐

e

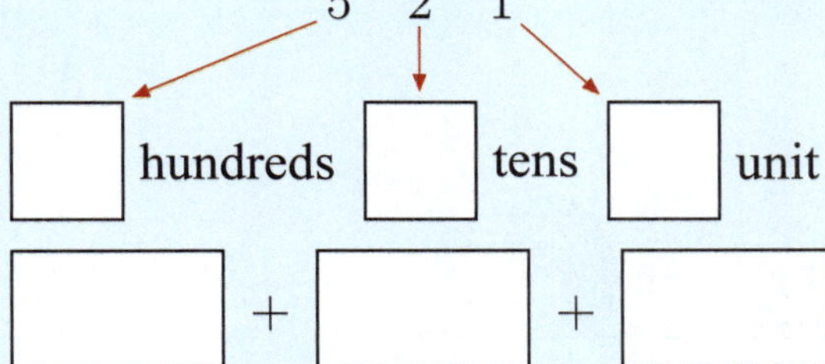

5 2 1

☐ hundreds ☐ tens ☐ unit

☐ + ☐ + ☐

f

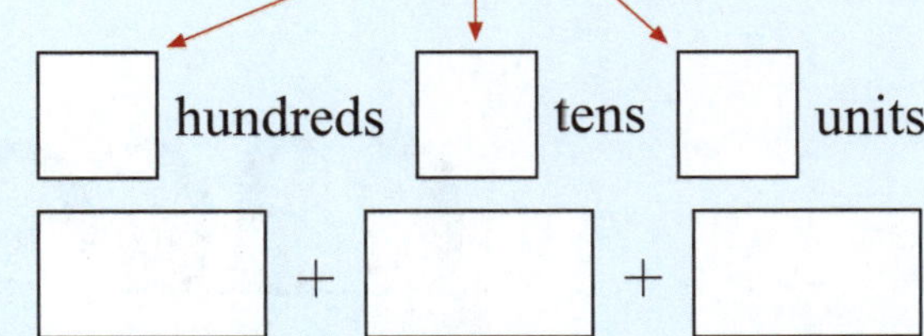

9 6 6

☐ hundreds ☐ tens ☐ units

☐ + ☐ + ☐

g

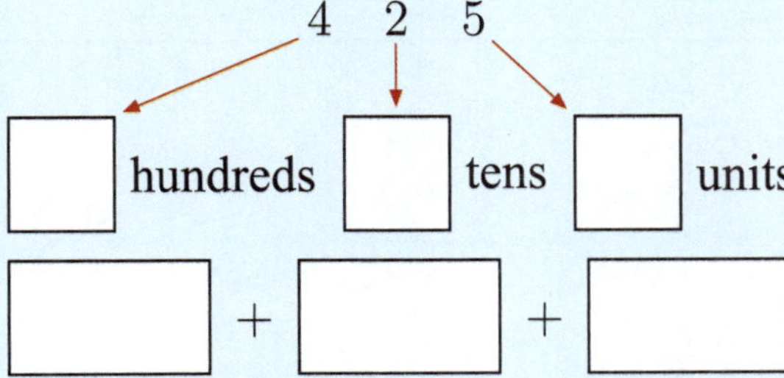

4 2 5

☐ hundreds ☐ tens ☐ units

☐ + ☐ + ☐

h

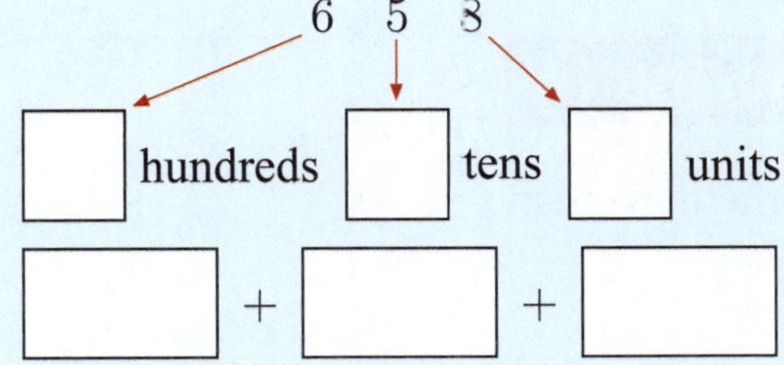

6 5 3

☐ hundreds ☐ tens ☐ units

☐ + ☐ + ☐

i

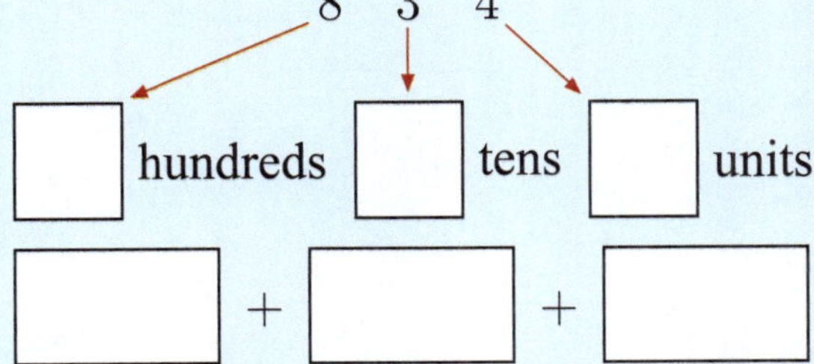

8 3 4

☐ hundreds ☐ tens ☐ units

☐ + ☐ + ☐

j

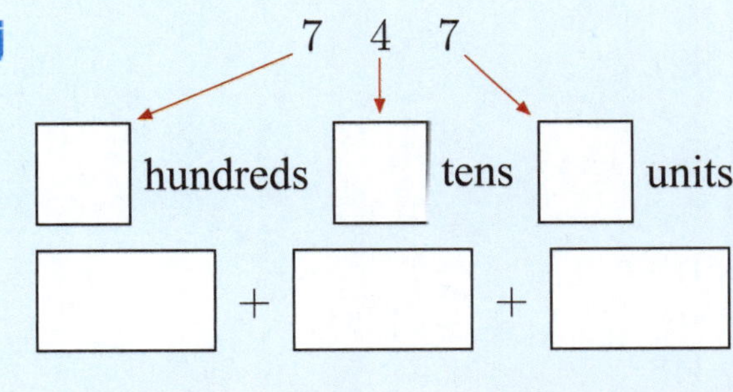

7 4 7

☐ hundreds ☐ tens ☐ units

☐ + ☐ + ☐

Activity

Click to practise listening to and reading numbers.

Exercise 15

Write as numbers:

a one hundred and thirty six _______________

b two hundred and sixty eight _______________

c eight hundred and fifty seven _______________

d seven hundred and seventy one _______________

e six hundred and forty five _______________

f nine hundred and sixteen _______________

g five hundred and twenty two _______________

Exercise 16

Write in words:

a 125 ____

b 493 ____

c 821 ____

d 682 ____

e 913 ____

f 348 ____

Exercise 17

What number is shown?

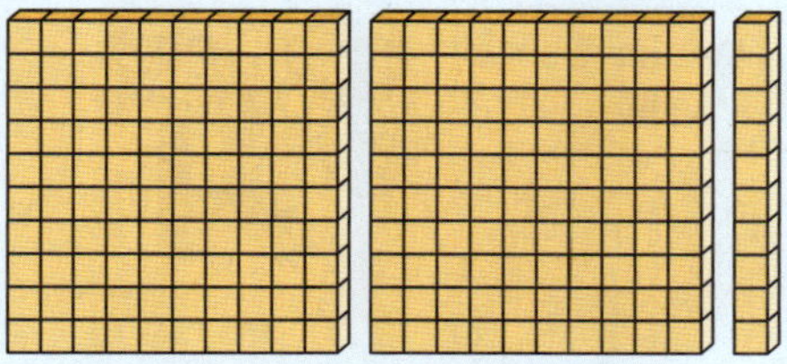

2 hundreds   **1** ten   **0** units

The number is 210 or two hundred and ten.

a ☐ hundred   ☐ tens   ☐ units

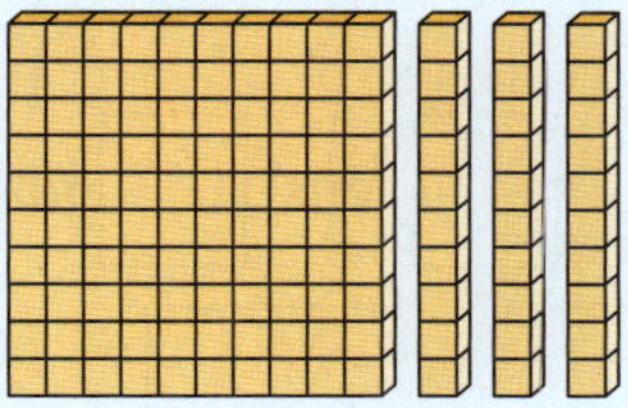

The number is ☐ or ________________ .

b ☐ hundreds   ☐ tens   ☐ units

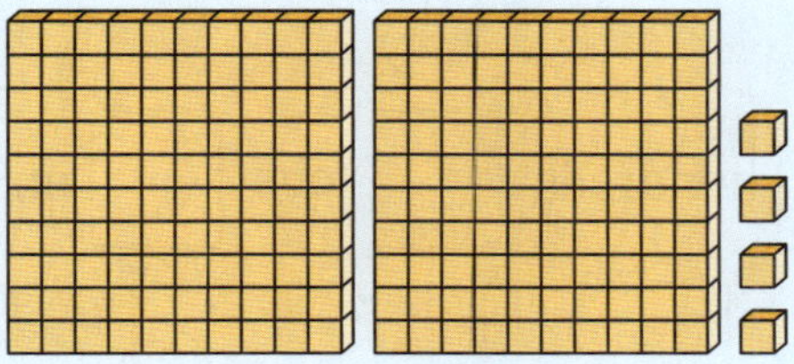

The number is ☐ or ________________ .

c ☐ hundreds

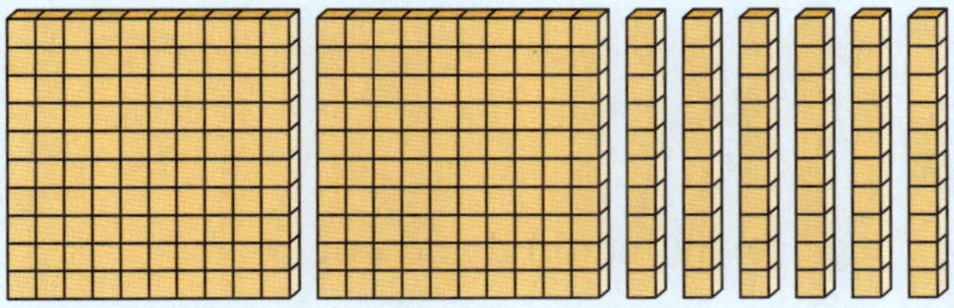

☐ tens

☐ units

The number is ☐ or ________________ .

d ☐ hundreds

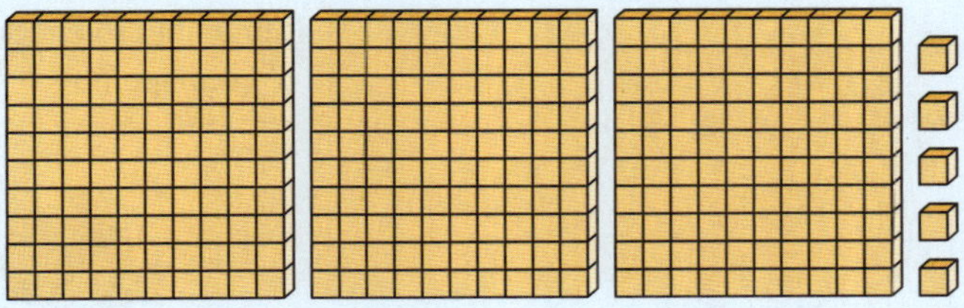

☐ tens

☐ units

The number is ☐ or ________________ .

Exercise 18

3   0   4

3 hundreds   0 tens   4 units

$300 + 4$

a

1   6   0

☐ hundred   ☐ tens   ☐ units

☐ + ☐

b

2   0   8

☐ hundreds   ☐ tens   ☐ units

☐ + ☐

c

4   0   7

☐ hundreds   ☐ tens   ☐ units

☐ + ☐

d

3   5   0

☐ hundreds   ☐ tens   ☐ units

☐ + ☐

Exercise 19

Write as numbers:

a nine hundred and seventy   __________

b seven hundred and one   __________

Exercise 20

Write these numbers in words.

a 107   __________

b 620   __________

c 309   __________

d 205   __________

e 340   __________

f 804   __________

Listening Activity

Click and follow the instructions.

1 __________ 2 __________ 3 __________ 4 __________ 5 __________

Discussion

How are 470 and 407 different?

____

____

Exercise 21

What is the value of the red digit?

243

The 4 means 4 tens or 40.

a 281 __________________________ b 382 __________________________

c 536 __________________________ d 476 __________________________

e 820 __________________________ f 664 __________________________

g 189 __________________________ h 307 __________________________

i 945 __________________________ j 218 __________________________

Game

Click to practise working with place values.

Group Activity

You will need: 4 people in your group

What to do:

1 Decide who will:

- represent hundreds (H)
- represent tens (T)
- represent units (U)
- guess the number (G).

2 Sit together with your group like this:

3 If you are H, T, or U, choose a number to show with your fingers.

Make sure your palms face away from you:

You can use two fists to show zero.

4 If you are G, try to say the number your group shows you.

5 Take turns sitting in each spot.

Puzzle

Colour the matching tiles the same:

318	$300 + 90 + 7$	seven hundred and sixty three
123	$400 + 8$	three hundred and ninety seven
210	$100 + 20 + 3$	three hundred and eighteen
397	$700 + 60 + 3$	four hundred and eight
830	$300 + 10 + 8$	eight hundred and thirty
408	$800 + 30$	two hundred and ten
763	$200 + 10$	one hundred and twenty three

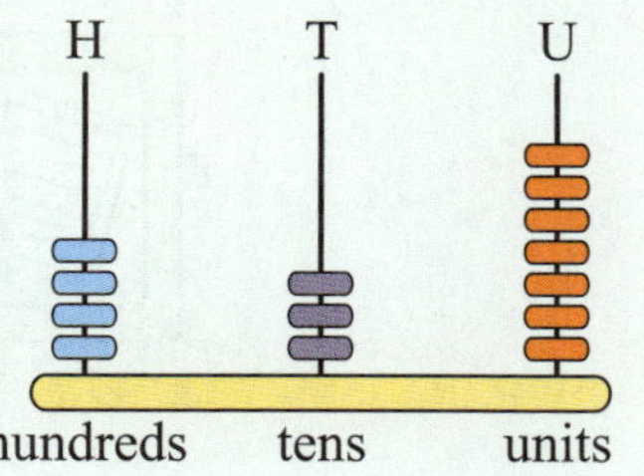

This is an **abacus**. It helps us to see place values.

Each column represents a different place value.

4 hundreds + 3 tens + 7 units = 437

Exercise 22

What number does each abacus show?

a 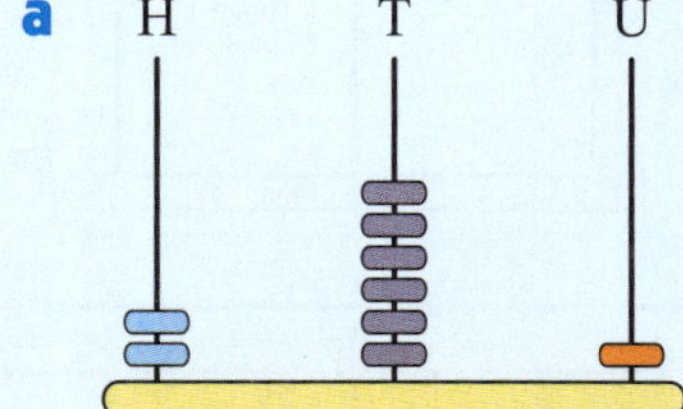

b 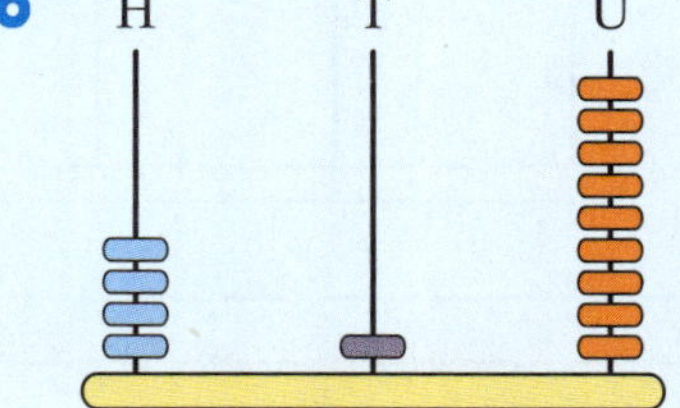

c 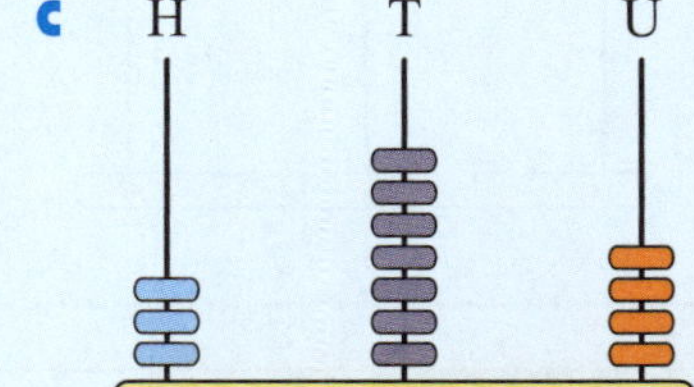

d 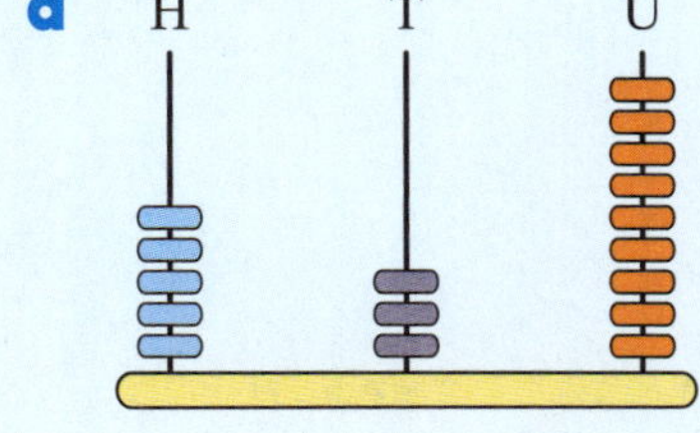

e 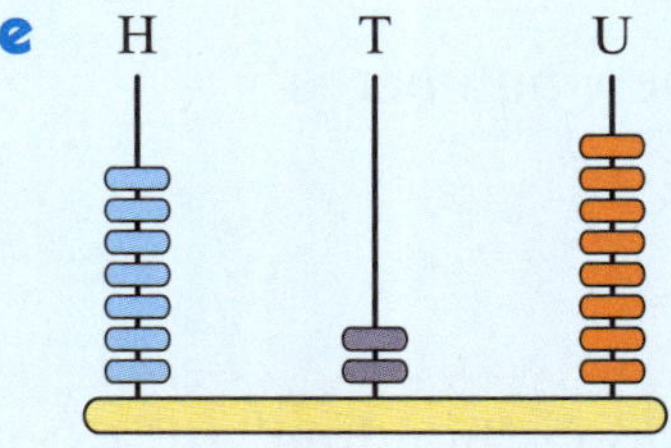

f 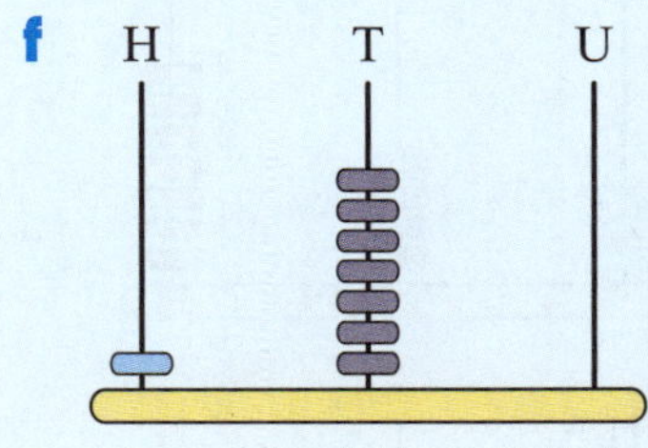

g 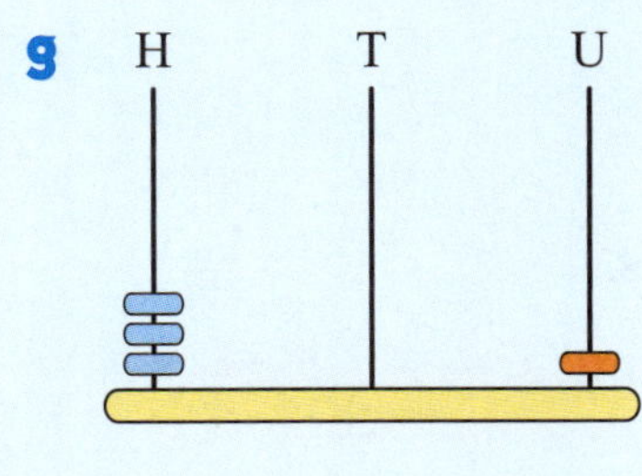

h 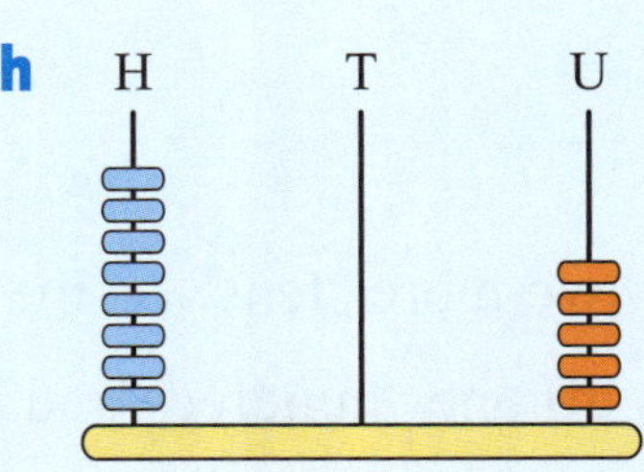

i 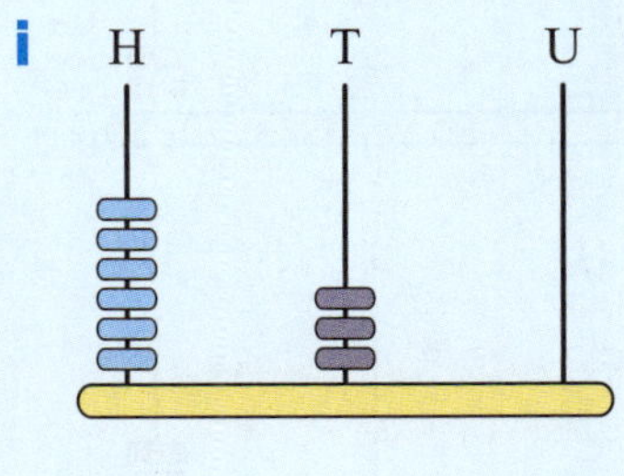

Discussion

Who invented the abacus?

Exercise 23

Show the number on the abacus:

a 284

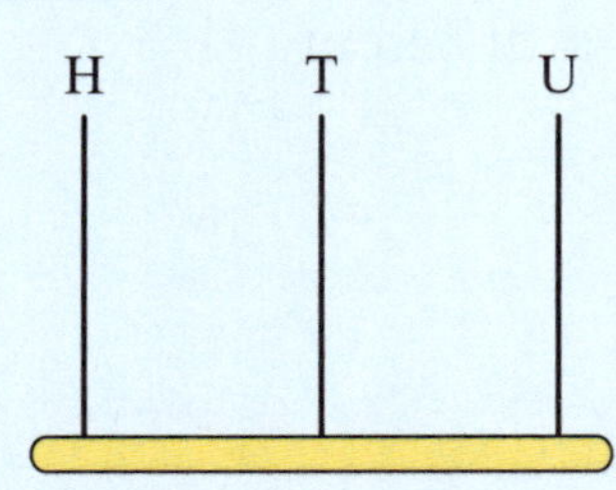

b 327

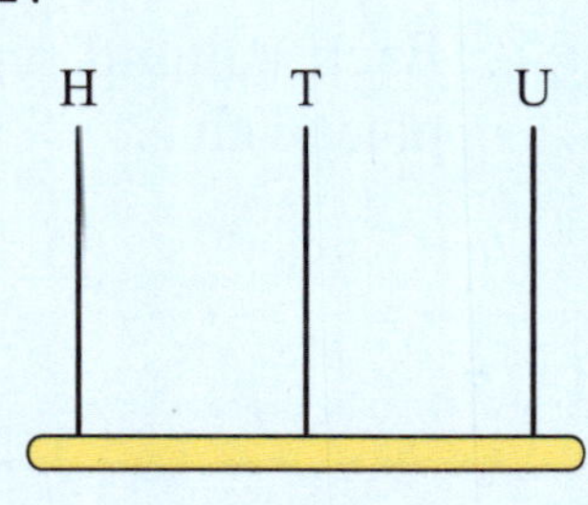

c 932

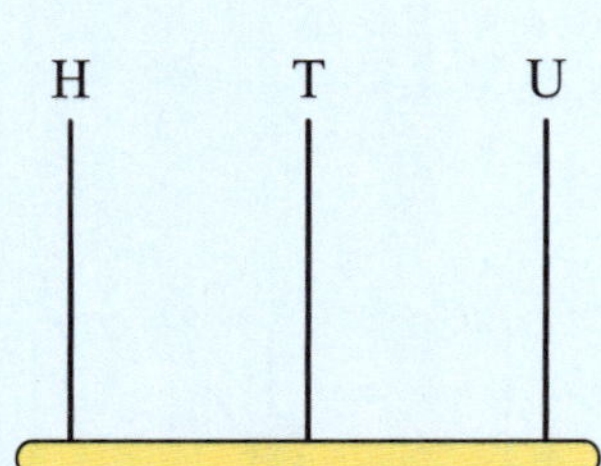

d 640

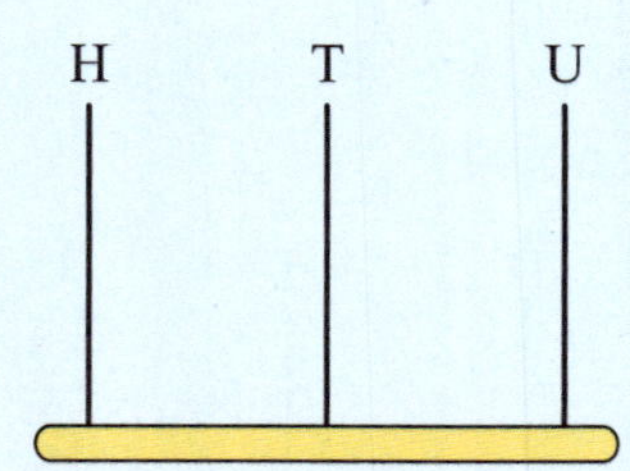

e 702

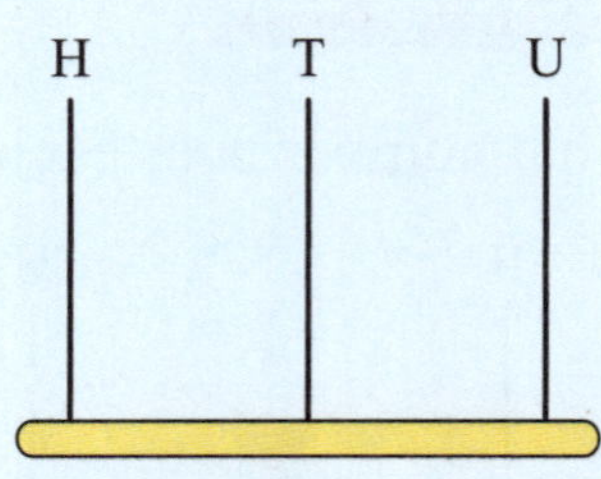

Exercise 24

a

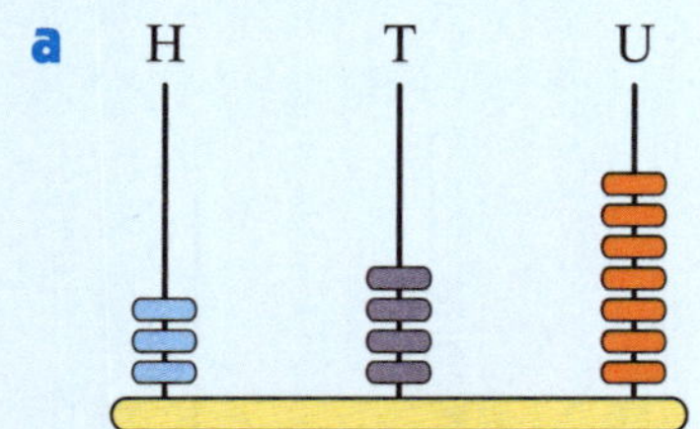

Draw one more "ten" on the abacus.

The new number is _______________ .

b

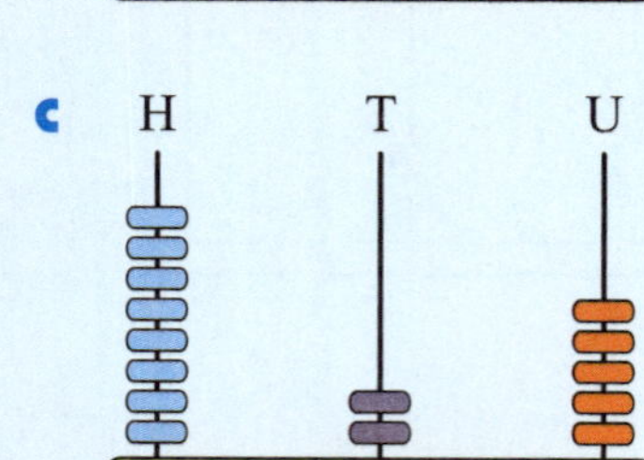

Cross one "hundred" off the abacus.

The new number is _______________ .

c

Draw two more "tens" on the abacus.

Cross off one "hundred" and one "unit".

The new number is _______________ .

Group Activity

List some places where you see numbers that are:

- from 10 to 99

- from 100 to 999

5 is less than 8.

$5 < 8$

There are fewer deer than goats.

8 is greater than 5.

$8 > 5$

There are more goats than deer.

Exercise 25

Practise your writing.

| > | greater than | | more than |

| < | less than | | fewer than |

Exercise 26

Use *greater than* or *less than*.

a

3 is _______________ 4.

b

8 is _______________ 6.

c

6 is _______________ 3.

d

7 is _______________ 9.

Exercise 27

Use > or <.

a

5 _____ 3

b

4 _____ 5

c

9 _____ 8

d

6 _____ 7

Discussion

Think about the numbers   5, 8, and 3.

Which of these numbers is the least?   _______

Which of these numbers is the greatest?   _______

A **number line** lists whole numbers in order from least to greatest.

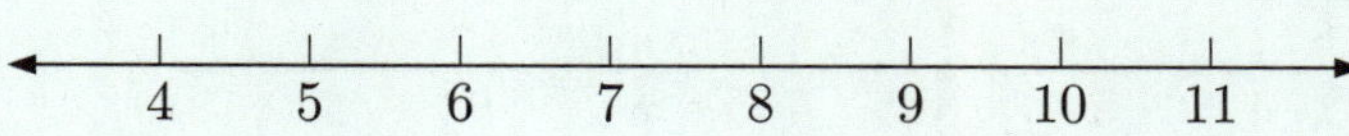

Exercice 28

Complete each number line:

a

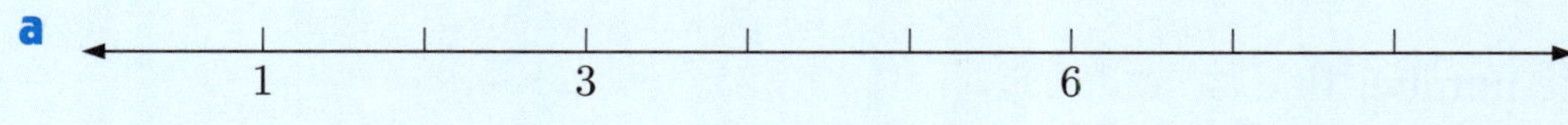

b

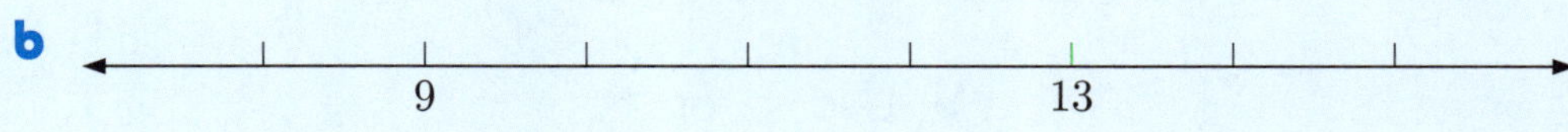

c

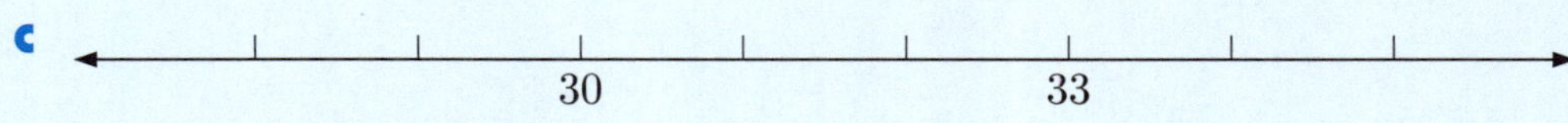

d

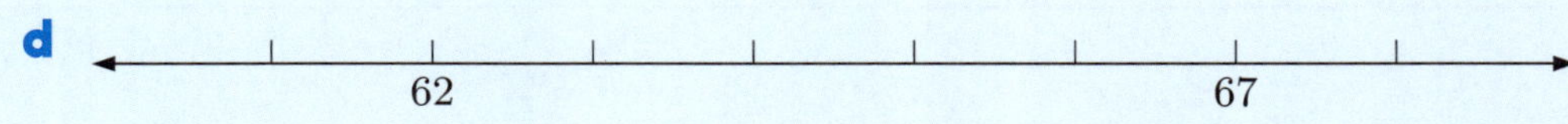

e

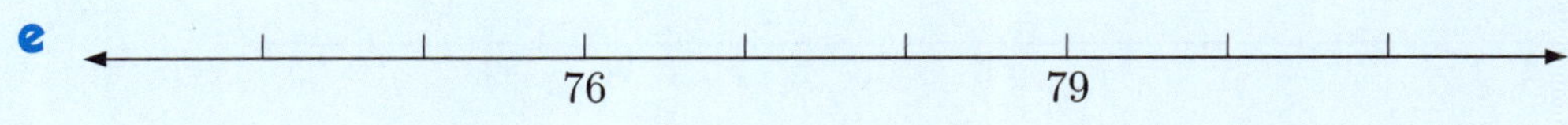

f

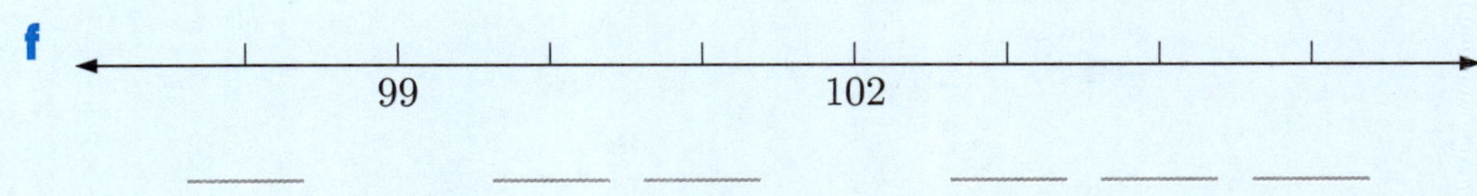

Exercice 29

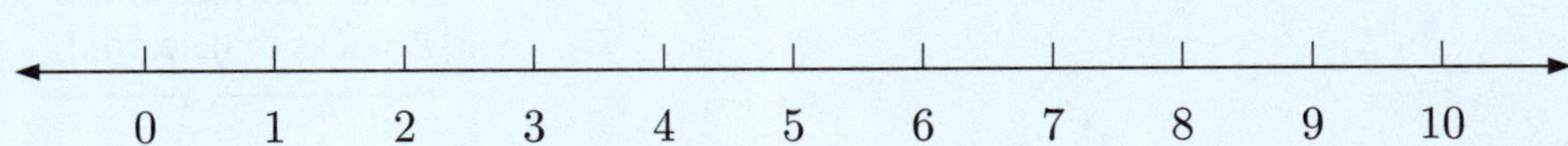

Circle the numbers *greater than* 4 in *blue*.

Circle the numbers *less than* 4 in *red*.

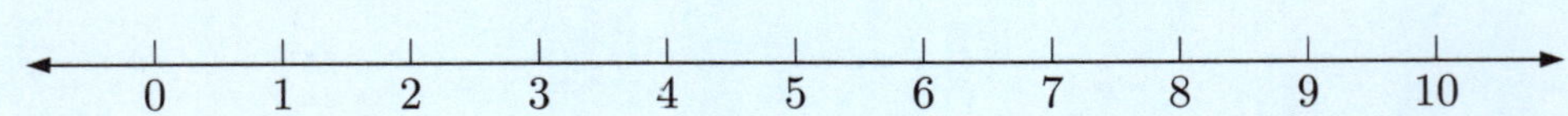

Numbers to the left of 6 are less than 6.

Numbers to the right of 6 are greater than 6.

Exercise 30

Mark 2 and 5 on the number line.

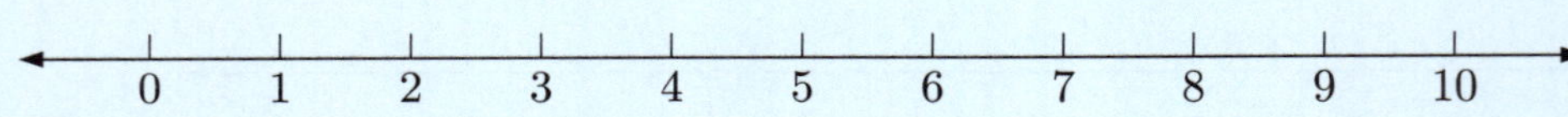

a Use *left* or *right*.

2 is to the __________ of 5.

5 is to the __________ of 2.

b Use $<$ or $>$.

2 ____ 5

5 ____ 2

Exercise 31

Mark 3 and 8 on the number line.

a Use *left* or *right*.

3 is to the __________ of 8.

8 is to the __________ of 3.

b Use $<$ or $>$.

3 ____ 8

8 ____ 3

Exercise 32

Use $<$ or $>$.

a 7 ____ 9

b 4 ____ 2

c 3 ____ 4

d 8 ____ 6

e 5 ____ 1

f 6 ____ 10

When we run a race, we finish in **order**.

 first    second    third    fourth

The **final** person to finish is **last**.

Listening Activity

This Listening Activity will help you learn and say the words for numbers that describe order.

Exercise 33

Copy each place number and learn its spelling.

1st	first	____________________
2nd	second	____________________
3rd	third	____________________
4th	fourth	____________________
5th	fifth	____________________
6th	sixth	____________________
7th	seventh	____________________
8th	eighth	____________________
9th	ninth	____________________
10th	tenth	____________________
11th	eleventh	____________________
12th	twelfth	____________________

13th	thirteenth	_____________________
14th	fourteenth	_____________________
15th	fifteenth	_____________________
16th	sixteenth	_____________________
17th	seventeenth	_____________________
18th	eighteenth	_____________________
19th	nineteenth	_____________________
20th	twentieth	_____________________
21st	twenty first	_____________________
22nd	twenty second	_____________________
23rd	twenty third	_____________________
24th	twenty fourth	_____________________
25th	twenty fifth	_____________________
26th	twenty sixth	_____________________
27th	twenty seventh	_____________________
28th	twenty eighth	_____________________
29th	twenty ninth	_____________________
30th	thirtieth	_____________________
31st	thirty first	_____________________

Activity

Claude is 1st to walk off the stage.

a ___________________________ is 2nd to walk off the stage.

b ___________________________ is 5th to walk off the stage.

c Chris is ______ rd, and Ella is ______ th.

d Who will be 6th? ___________________________

e In which position is Sarah? ___________________

f Who will be 8th? ___________________________

g Betty, in ________ th place, is the ___________________ to leave.

Discussion

The *winner* of a race finishes ___________________________.

The *runner-up* finishes ___________________________.

What do we mean by a *dead heat*?

____

____

Activity

The African animals ran a race.

1 a Which animal came second?

   b Which animal came fifth?

   c In which place did the zebra come?

   d In which place did the crocodile come?

   e Which animal finished immediately after the animal in third place?

2 How did the cheetah win?

    f   **c**   **d** **f** **c** **e** **a** **c** **b**

   a the third letter of antelope   **b** the fourth letter of wildebeest

   c the seventh letter of giraffe   **d** the sixth letter of rhinoceros

   e the fifth letter of zebra   **f** the first letter of hippopotamus

Discussion

Here are some words we can use to describe the **position** of an object.

With your class, think of a sentence which uses each word or words.

Copy the words to practise your handwriting.

to the left of		to the right of	
above		over	
below		under	
behind		in front of	
alongside		next to	
near to		far from	

Exercise 1

Complete each sentence:

a The cage is below the _______________ .

b The bird is above the _______________ .

c The tree is next to the _______________ .

d The flower is alongside the _______________ .

Exercise 2

Complete each statement using *near*, *far from*, *over*, and *under*.

The swimmer is swimming:

- _______________ the fish
- _______________ the boat
- _______________ the mountain
- _______________ the bird.

Discussion

What words could we use to complete the sentence
"The bin is _______________ the desk"?

____

____

____

____

Exercise 3

Look around your room.

Fill in each space with a classmate's name:

a I am sitting to the left of _______________ .

b I am sitting to the right of _______________ .

c _______________ is sitting in front of _______________ .

d _______________ is sitting behind _______________ .

Puzzle

Print the animals and cut them out.

Locate what position each animal must be in using the rules:

- The alpaca is above the horse.
- The sheep is to the left of the alpaca.
- The goat is to the right of the horse.
- The goat is below the dog.

When you have them all positioned, stick them in.

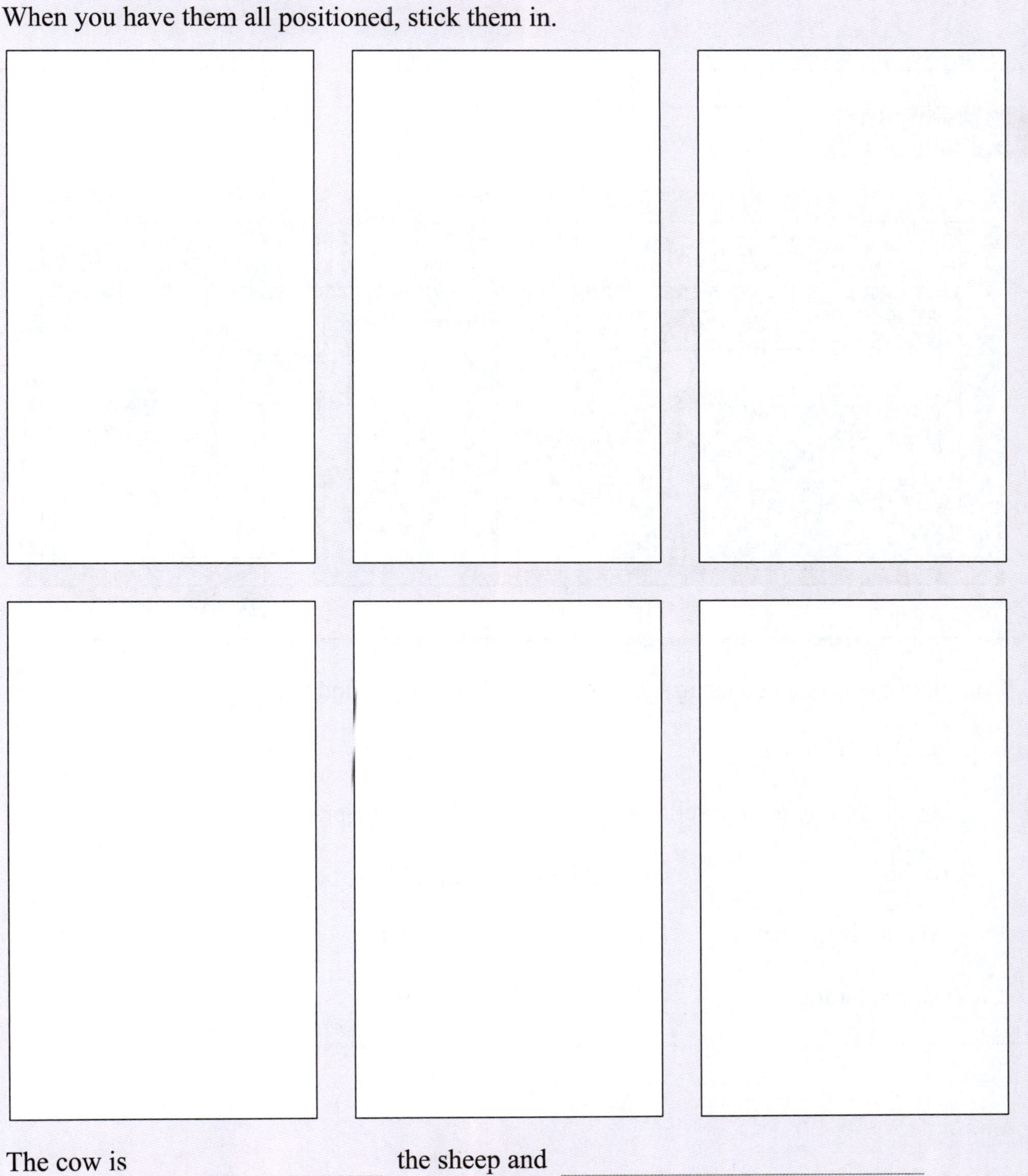

The cow is _________________ the sheep and _________________________________

of the horse.

Exercise 4

The treasure is *inside* the chest.

The mouse is *outside* the chest.

Practise your spelling.

inside   outside

___________________   ___________________

___________________   ___________________

Exercise 5

Complete each sentence using *left*, *inside*, *outside*, *next to*, and *over*:

Laura is standing ___________________ the pharmacy.

She is holding an umbrella ___________________ the pram.

To the ___________________ of the pharmacy is the grocer.

Conrad the grocer is ___________________ his store.

He is standing ___________________ the pineapples.

Discussion

What do we mean by an *opposite*?

Exercise 6

Match the words with *opposite* meanings.

over

far

left

right

outside

near

in front of

under

behind

inside

Exercise 7

The bullseye is in the *middle* or *centre* of the target.
Practise your spelling.

middle   centre

_______________   _______________

_______________   _______________

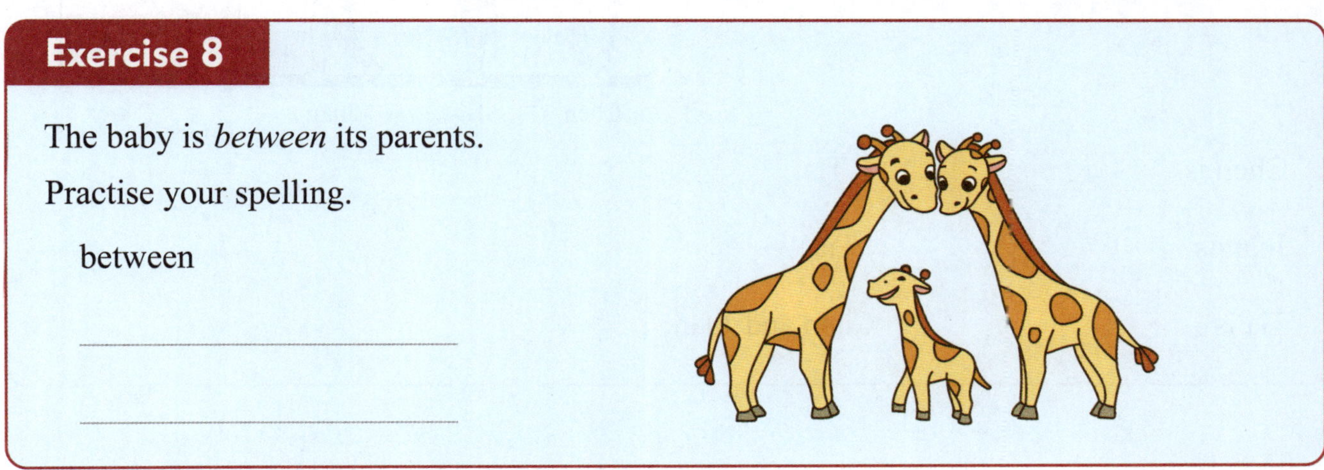

Exercise 8

The baby is *between* its parents.
Practise your spelling.

between

____

____

Exercise 9

Complete each sentence describing a position:

a The ________________ is between the fire and the painting.

b The open door is to the ________________ of the clock.

c The trophy is ________________ to the red books.

d The closed cupboards are ________________ the shelves.

e The photo is on the ________________ of the ________________ shelf.

Draw a vase of flowers on the right side of the top shelf.

Exercise 10

Complete each sentence about the queue.
Use *in front of*, *between*, and *behind*.

Chen is ________________ Tia.

Juan is ________________ Tia.

Tia is ________________ Chen and Juan.

Activity

A **plan** is a "bird's eye view" looking down from above.

Here is a plan of Natsuki's bedroom.

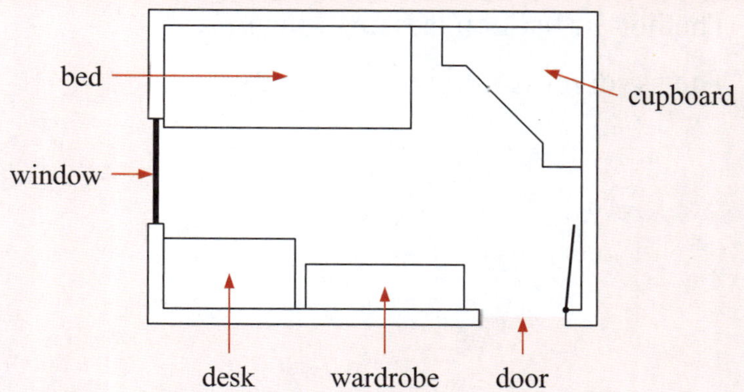

In this Activity you will draw a plan of *your* bedroom.

You will need: pencil, ruler

What to do:

1   Imagine you are standing at your bedroom door. Draw the *shape* of your room from the doorway.

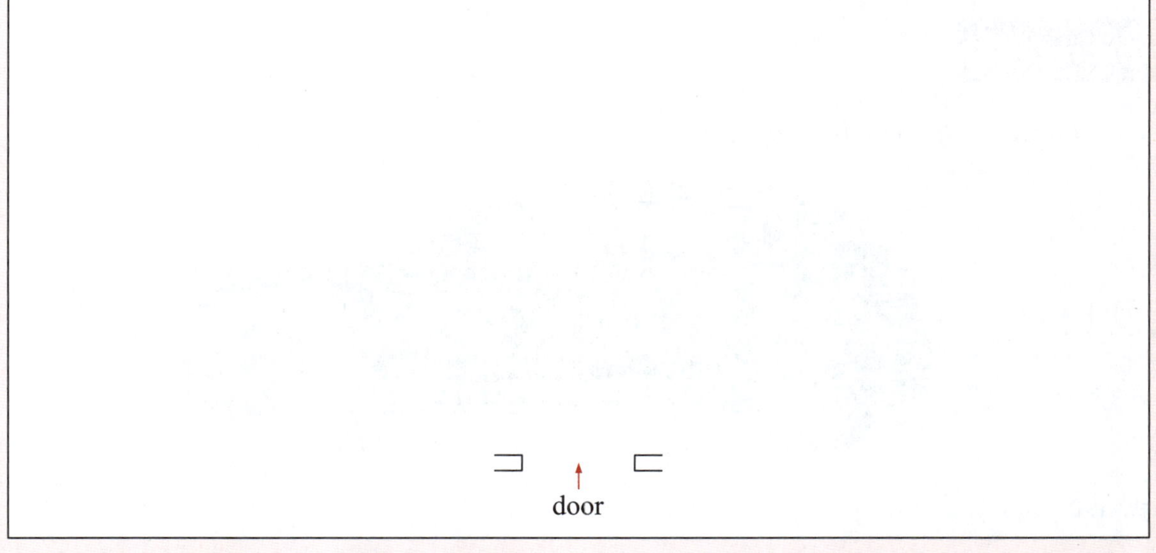

2   Mark any windows with a single line ——

3   Draw any furniture you have, for example:

- bed
- wardrobe
- desk
- bookshelf
- chair
- cupboard

Game

Kittens *love* boxes.

Click on the icon to position your kitten!

Exercise 11

The hot-air balloon is rising *upwards*.

The skydiver is falling *downwards*.

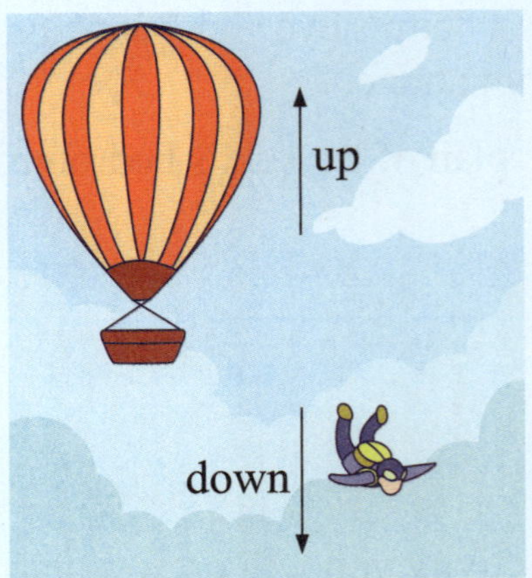

Practise your spelling.

upwards ________________   downwards ________________

________________   ________________

Exercise 12

A car can move *forwards* or *backwards*.

Practise your spelling.

forwards ________________   backwards ________________

________________   ________________

Exercise 13

A crab can move *sideways* to the left or the right.

Practise your spelling.

sideways ________________

____

Discussion

A **turn** is a change of direction.

What do we mean by:

- a turn to the left
- a turn to the right?

Exercise 14

Stu is at the entrance to the zoo.

To visit the Asian Animals, he needs to take the *second* turn to the *right*.

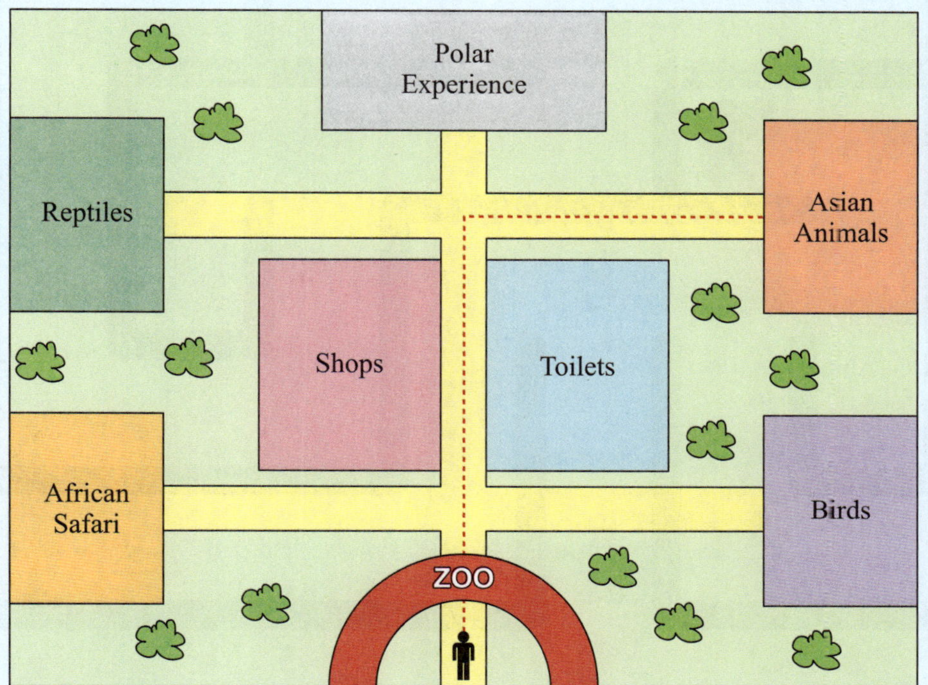

a To visit the African Safari, Stu needs to take the first turn to the _______________ .

b To visit the Birds, Stu needs to take the _______________ turn to the
_______________ .

c To visit the Reptiles, Stu needs to take the _______________ turn to the
_______________ .

Discussion

Look at the map of the zoo in **Exercise 14**.

How would you direct Stu to the Polar Experience?

____

Puzzle

Sir Gawain needs directions to the treasure.

At the _______________________ turn ________________ and then immediately

__________________ again.

Follow the path until you see an opening on your _________________ .

Enter this opening.

At the ________________________ , turn __________________ .

Follow the path until you reach a junction.

Turn ________________ , and the treasure is ____________________ of you.

Discussion

What words tell us that we need to add?

______________________   ______________________   ______________________

______________________   ______________________

What is $4 + 2$?

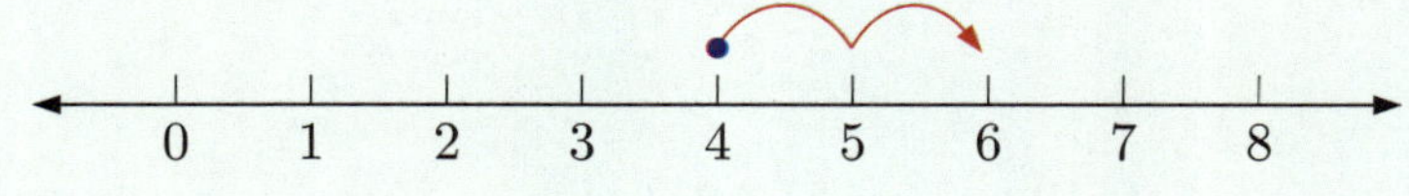

$4 + 2 = 6$

Exercise 1

Use the number line to help you add:

a

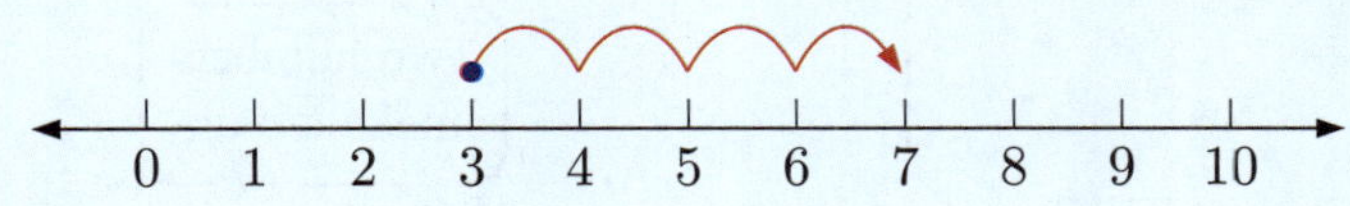

$3 + 4 = \underline{\hspace{1.5cm}}$

b

$6 + 2 = \underline{\hspace{1.5cm}}$

c

$2 + 7 = \underline{\hspace{1.5cm}}$

d

$5 + 5 = \underline{\hspace{1.5cm}}$

e

$1 + 6 = \underline{\hspace{1.5cm}}$

Exercise 2

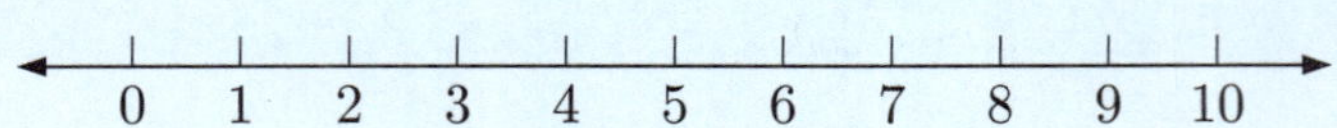

Write down the result of each addition:

a $2 + 3 =$ ________

b $2 + 2 =$ ________

c $1 + 4 =$ ________

d 3 add 1 $=$ ________

e 2 plus 4 $=$ ________

f 3 add 3 $=$ ________

g $4 + 5 =$ ________

h 3 plus 6 $=$ ________

i $5 + 3 =$ ________

Exercise 3

a The sum of 8 and 1 is ________ .

b The total of 2 and 5 is ________ .

c The sum of 7 and 2 is ________ .

d The total of 6 and 4 is ________ .

e The number 3 more than 4 is ________ .

f The number 6 more than 2 is ________ .

Exercise 4

Write pairs of numbers that add to 5.

$5 = \square + \square$ $5 = \square + \square$

$5 = \square + \square$ $5 = \square + \square$

Exercise 5

Write pairs of numbers that total 8.

$8 = \square + \square$ $8 = \square + \square$ $8 = \square + \square$

$8 = \square + \square$ $8 = \square + \square$ $8 = \square + \square$

$8 = \square + \square$

Exercise 6

Write pairs of numbers that total 10.

$10 = \boxed{} + \boxed{}$ $10 = \boxed{} + \boxed{}$ $10 = \boxed{} + \boxed{}$

$10 = \boxed{} + \boxed{}$ $10 = \boxed{} + \boxed{}$ $10 = \boxed{} + \boxed{}$

$10 = \boxed{} + \boxed{}$ $10 = \boxed{} + \boxed{}$ $10 = \boxed{} + \boxed{}$

Exercise 7

Use the number line to help you add:

a

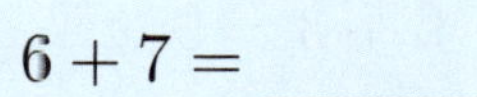

$6 + 7 =$ _______

b

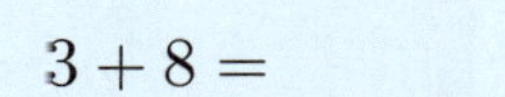

$8 + 6 =$ _______

c

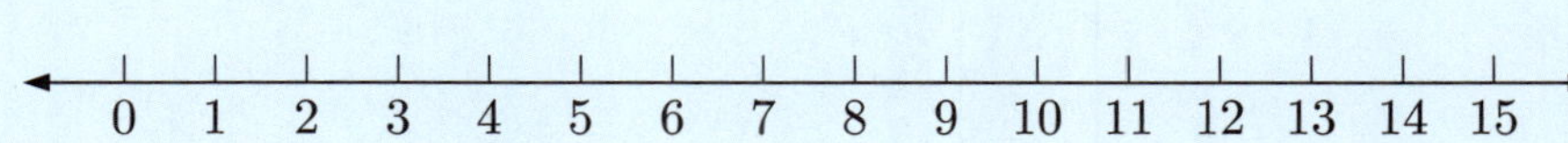

$3 + 8 =$ _______

d

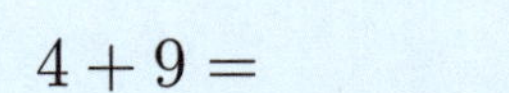

$4 + 9 =$ _______

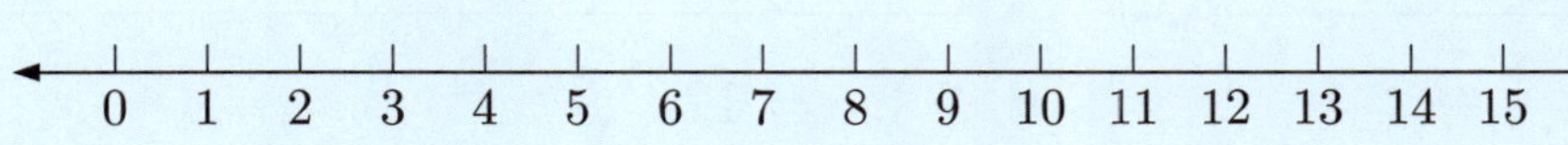

Exercise 8

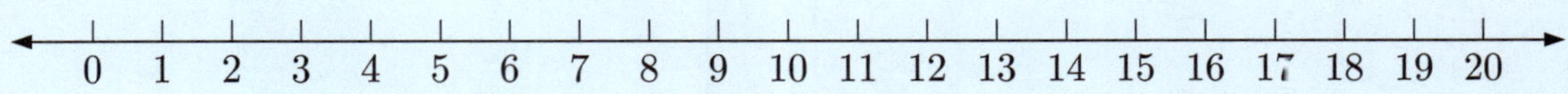

Write down the result of each addition:

a 7 add $4 =$ _______   **b** $5 + 9 =$ _______   **c** 8 plus $3 =$ _______

d $7 + 9 =$ _______   **e** 6 plus $6 =$ _______   **f** 9 add $8 =$ _______

Exercise 9

Colour the objects to match the sum.

a $3 + 5 = 8$

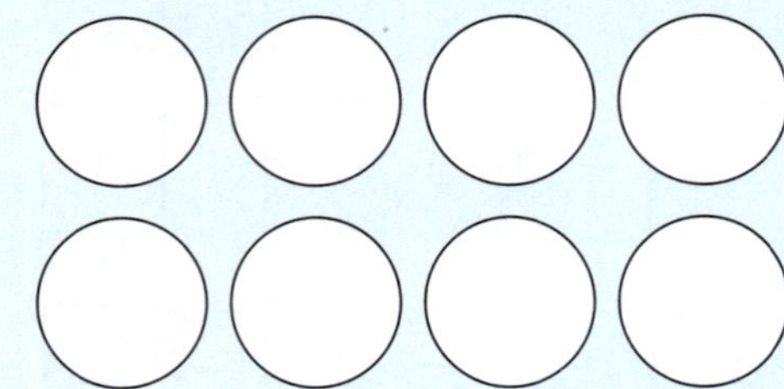

b $4 + 7 = 11$

Exercise 10

Draw and colour objects to match the sum.

a $6 + 3 = 9$

b $5 + 8 = 13$

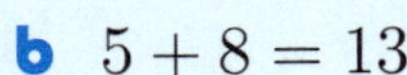

c $9 + 6 = 15$

d $7 + 5 = 12$

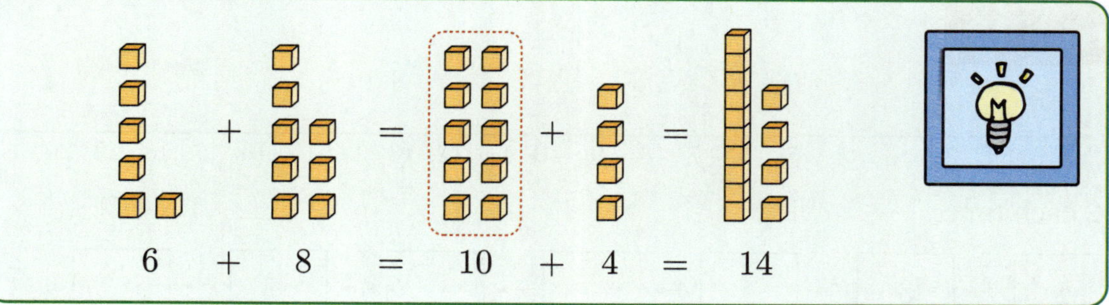

Exercise 11

Use blocks to help you add:

a $5 + 6 =$ _____

b $8 + 3 =$ _____

c $9 + 9 =$ _____

d $7 + 6 =$ _____

e $3 + 7 =$ _____

f $8 + 7 =$ _____

g $8 + 8 =$ _____

h $9 + 5 =$ _____

i $2 + 9 =$ _____

Game

Click the icon to practise your addition.

Exercise 12

Write pairs of numbers that total 12. How many can you find?

Exercise 13

0 1 2 3 4 5 6 7 8 9 10 11 12 13 14 15 16 17 18 19 20

Add 2 each time:

a | 1 | → | 3 | → | | → | | → | | → | | → | |

b | 4 | → | | → | | → | | → | | → | | → | |

Add 3 each time:

c | 2 | → | | → | | → | | → | | → | |

d | 6 | → | | → | | → | | → | |

Add 4 each time:

e | 1 | → | | → | | → | | → | |

f | 3 | → | | → | | → | | → | |

Exercise 14

Describe what is happening in each sequence:

a 3 → 6 → 9 → 12 → 15

I start with _______ , and add _______ each time.

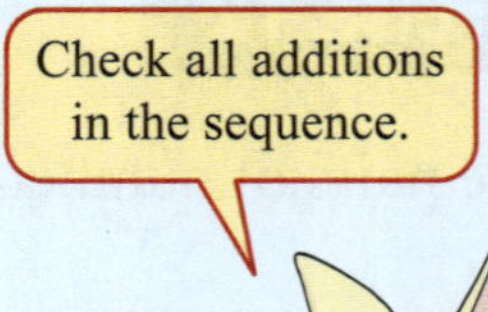

b 2 → 4 → 6 → 8 → 10

c 1 → 5 → 9 → 13 → 17

d 5 → 8 → 11 → 14 → 17

Exercise 15

These dice show a "double 3".

"Double 3" is $3 + 3 = 6$.

Complete this table of doubles.

	double 1	$1 + 1 =$ _______

Exercise 16

Complete these additions.

a $4 + 4 =$ _______   **b** $7 + 7 =$ _______   **c** $5 + 5 =$ _______

d $9 + 9 =$ _______   **e** $6 + 6 =$ _______   **f** $8 + 8 =$ _______

Discussion

If you know what $7 + 7$ is, how can you find:

- $7 + 8$
- $7 + 9$?

Exercise 17

Complete these pairs of additions.

a

$4 + 2 =$ _______

$2 + 4 =$ _______

b

$3 + 7 =$ _______

$7 + 3 =$ _______

c

$8 + 4 =$ _______

$4 + 8 =$ _______

d

$6 + 9 =$ _______

$9 + 6 =$ _______

Discussion

Can you see patterns in the pairs of additions above?

Can you explain *why* this happens?

Exercise 18

Write *two* additions for each group of objects:

a

____

____

b
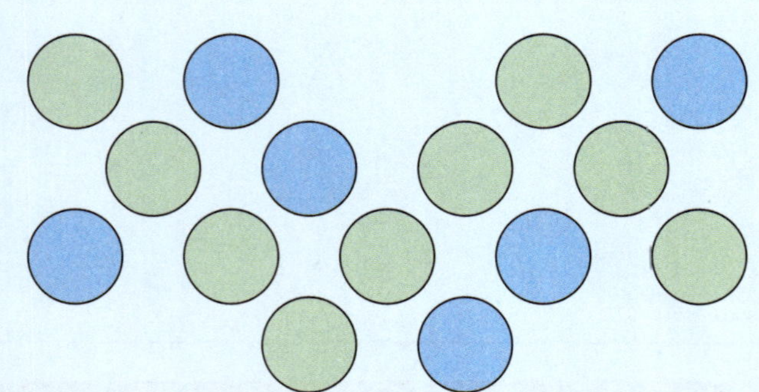

____

____

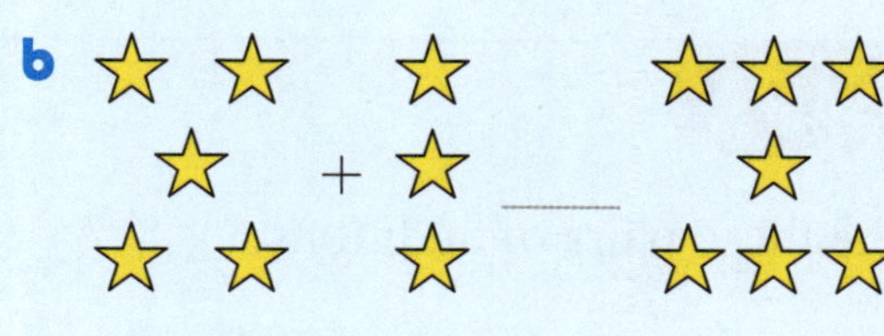

$3 + 3 = 6$

3 plus 3 is equal to 6.

3 plus 3 equals 6.

$3 + 3 \neq 5$

3 plus 3 is not equal to 5.

3 plus 3 does not equal 5.

Exercise 19

Complete:

a

$$4 \quad + \quad 2 \quad \underline{} \quad 6$$

4 plus 2 ______________________ 6

b

$$5 \quad + \quad 3 \quad \underline{} \quad 7$$

5 plus 3 ______________________ 7

c

$$3 \quad + \quad 2 \quad \underline{} \quad 4$$

3 plus 2 ______________________ 4

d

$$7 \quad + \quad 2 \quad \underline{} \quad 9$$

7 plus 2 ______________________ 9

Exercise 20

Use $=$ or $\neq$.

a $3 + 1 \ \underline{} \ 4$

b $6 + 3 \ \underline{} \ 8$

c $5 + 2 \ \underline{} \ 8$

d $4 + 3 \ \underline{} \ 7$

e $2 + 2 \ \underline{} \ 4$

f $5 + 4 \ \underline{} \ 9$

g $6 + 5 \ \underline{} \ 11$

h $8 + 4 \ \underline{} \ 14$

i $7 + 8 \ \underline{} \ 15$

j $9 + 3 \ \underline{} \ 13$

k $6 + 9 \ \underline{} \ 17$

l $9 + 5 \ \underline{} \ 14$

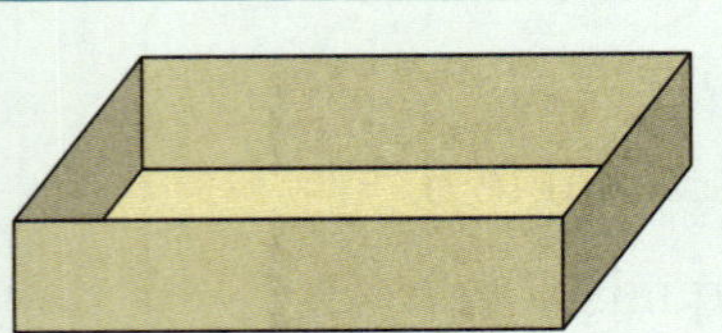

There are no balloons in the box.

We say there are **zero** balloons or 0 balloons.

$0 + 8 = 8$ 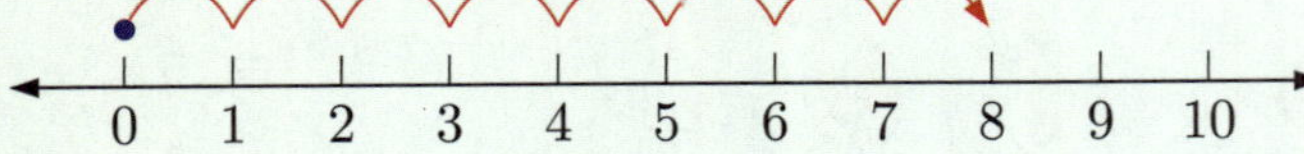

I start with nothing and add 8.

$8 + 0 = 8$ 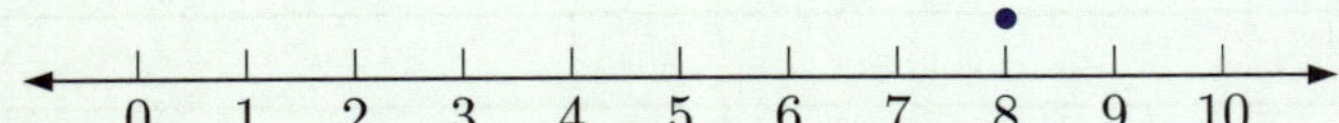

I start with 8 and add nothing.

Exercise 21

Complete these pairs of additions.

a $5 + 0 =$ _______

   $0 + 5 =$ _______

b $0 + 7 =$ _______

   $7 + 0 =$ _______

c $12 + 0 =$ _______

   $0 + 12 =$ _______

Exercise 22

Complete these additions.

a $2 +$ _______ $= 10$

b $6 +$ _______ $= 10$

c $4 +$ _______ $= 10$

d $0 +$ _______ $= 10$

e _______ $+ 7 = 10$

f _______ $+ 5 = 10$

g _______ $+ 3 = 10$

h $8 +$ _______ $= 10$

i _______ $+ 9 = 10$

Exercise 23

Complete these additions.

a $3 +$ _______ $= 8$

b _______ $+ 2 = 9$

c $4 +$ _______ $= 7$

d $1 +$ _______ $= 6$

e _______ $+ 5 = 11$

f $8 +$ _______ $= 13$

g $4 +$ _______ $= 12$

h _______ $+ 6 = 15$

i $9 +$ _______ $= 17$

Exercise 24

Complete these additions.

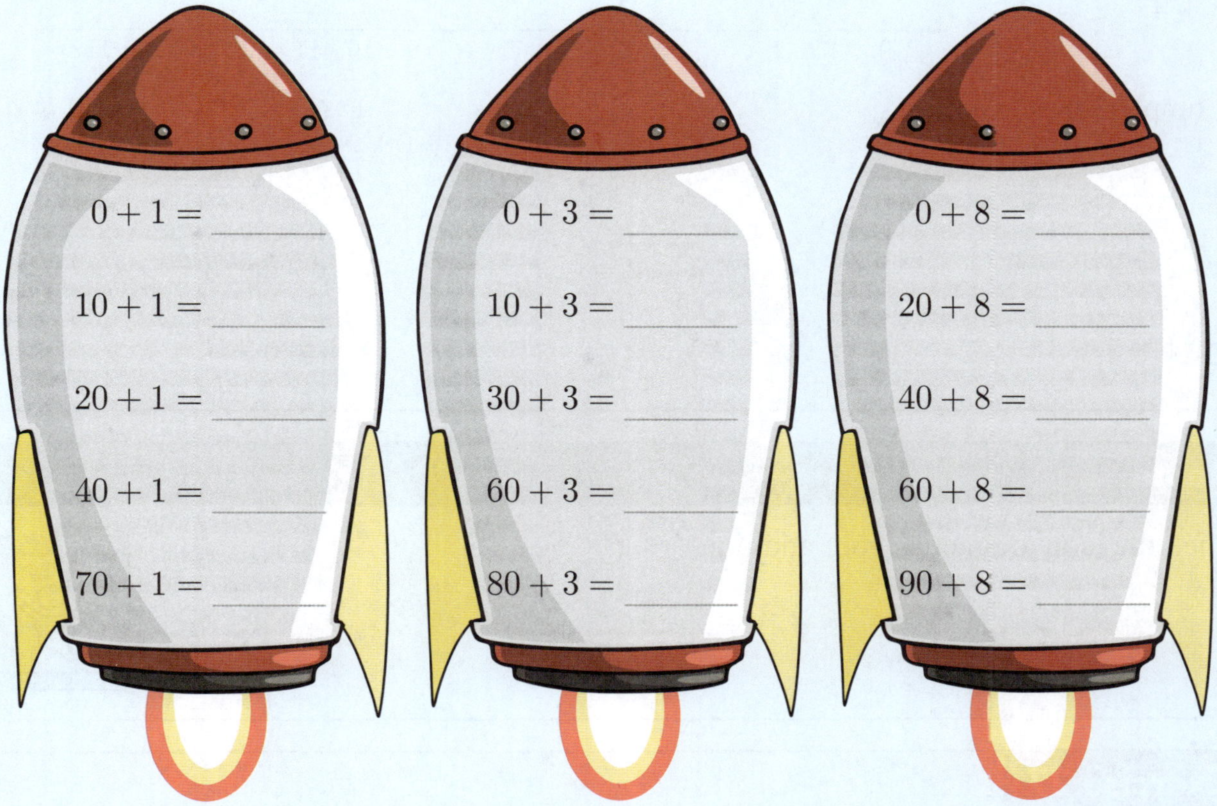

Exercise 25

Complete these additions.

Exercise 26

$5 + 4 + 2 = 11$

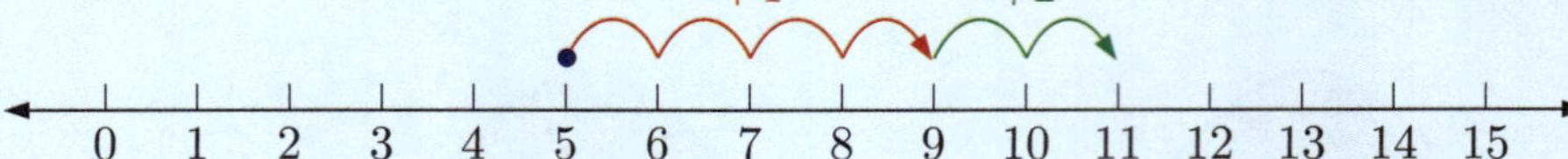

Complete these additions.

a $1 + 2 + 1 = $ _______

b $2 + 3 + 1 = $ _______

c $3 + 1 + 3 = $ _______

d $2 + 3 + 3 = $ _______

e $4 + 1 + 2 = $ _______

f $5 + 2 + 3 = $ _______

g $4 + 5 + 3 = $ _______

h $6 + 5 + 2 = $ _______

i $5 + 6 + 4 = $ _______

Game

Click the icon to practise your addition.

Exercise 27

Jill has 7 red apples and 2 green apples.

In total, Jill has $7 + 2 = 9$ apples.

Write an addition to solve these problems.

a Lucy has 3 blue cars and 4 red cars.

Altogether, Lucy has _______________ = _______ cars.

b Marius has 7 glasses and 8 mugs.

In total, Marius has _______________ = _______ items.

c Ali has 6 pens, 4 pencils, and 2 crayons.

In total, Ali has _______________ = _______ items.

d Mo has 5 daisies, 6 roses, and 6 lillies.

Altogether, Mo has _______________ = _______ flowers.

Exercise 28

Write an addition to find the total number of:

a red items _______________________

b black items _______________________

c yellow items _______________________

d baseball caps _______________________

e T-shirts _______________________

Activity

You will need: a blank sheet of paper

What to do:

1 Write your own worded problem which involves addition.

2 Swap questions with someone else in your class. Check that the answer they find for your question is the answer you expect.

Puzzle

You have 12 counters.

Place the counters in *three* groups with a *different* number of counters in each group.

Write a sum for each different answer you find.

Puzzle

You have 10 counters.

Place the counters in the circles so that the totals along the coloured lines are all *different*.

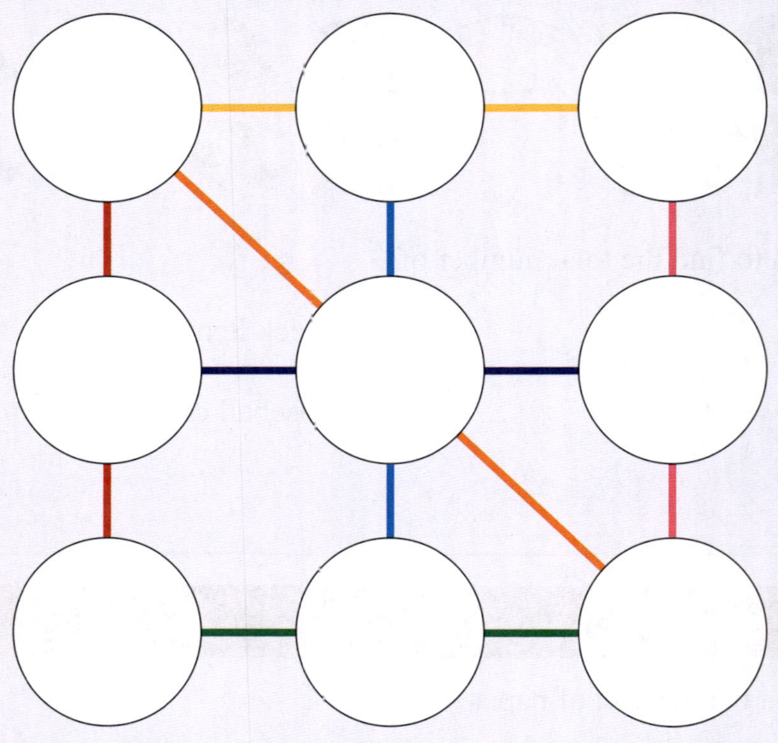

Exercise 29

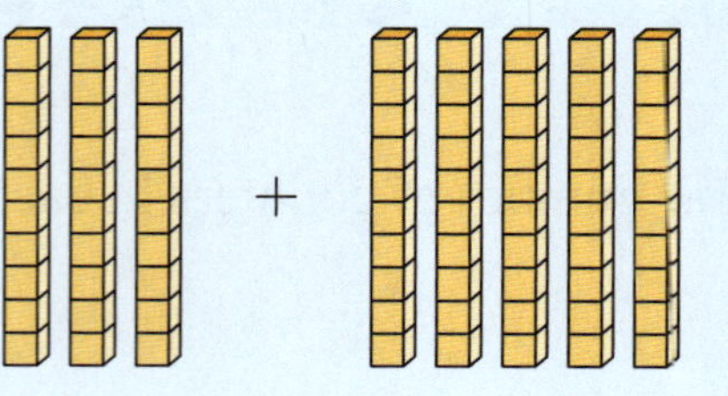   +   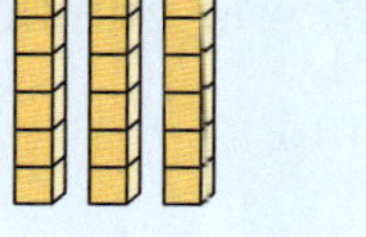   =   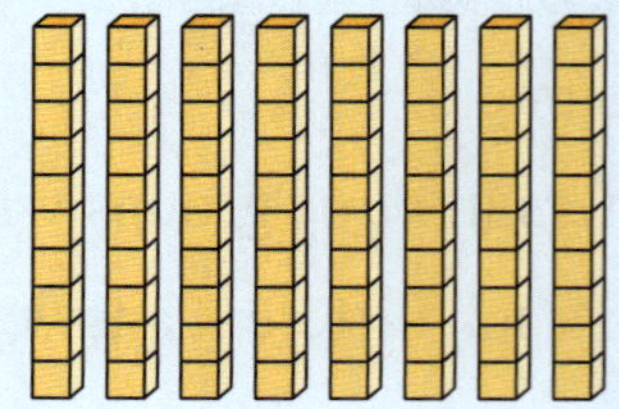

$$3 + 5 = 8$$

$$30 + 50 = 80$$

Complete these additions.

a $5 + 1 =$ _______   b $4 + 3 =$ _______   c $2 + 6 =$ _______

   $50 + 10 =$ _______ $40 + 30 =$ _______ $20 + 60 =$ _______

d $4 + 4 =$ _______   e $8 + 1 =$ _______   f $3 + 7 =$ _______

   $40 + 40 =$ _______ $80 + 10 =$ _______ $30 + 70 =$ _______

To add numbers with more than one digit, we can use **column addition**.

We write the numbers in columns so the place values line up.

Add each column, starting with the units.

$25 + 4$

1 Set out your addition.

2 Add the units:

$5 + 4 = 9$

3 Add the tens:

$2 + 0 = 2$

	T	U
	2	5
+		4
	2	9

$25 + 4 = 29$

Exercise 30

Complete each addition:

a

	T	U
	1	2
+		6

b

	T	U
	1	4
+		4

c

	T	U
	2	3
+		6

d

	T	U
	2	1
+		7

e

	T	U
	3	1
+		8

f

	T	U
	4	2
+		5

$23 + 14$

1 Set out your addition.

2 Add the units:

$3 + 4 = 7$

3 Add the tens:

$2 + 1 = 3$

	T	U
	2	3
+	1	4
	3	7

$23 + 14 = 37$

Exercise 31

Complete each addition:

a

	T	U
	2	1
+	1	3

b

	T	U
	2	7
+	1	2

c

	T	U
	3	3
+	1	0

d

	T	U
	4	1
+	2	2

e

	T	U
	5	0
+	2	6

f

	T	U
	4	2
+	3	7

Exercise 32

Set out each addition and find the answer:

a $24 + 15$

b $36 + 21$

c $43 + 26$

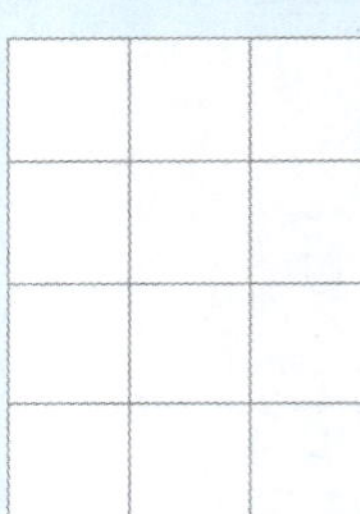

Exercise 33

a At a party there are 23 pink balloons and 16 blue balloons.

How many balloons are there in total?

There are [ ] balloons in total.

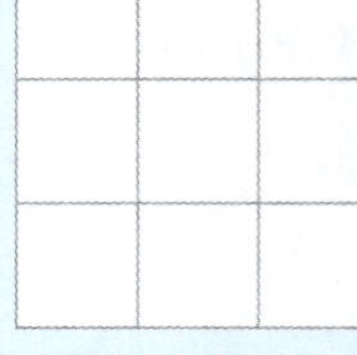

b In the garden there are 26 roses and 33 camellias.

How many flowers are there in total?

There are [ ] flowers in total.

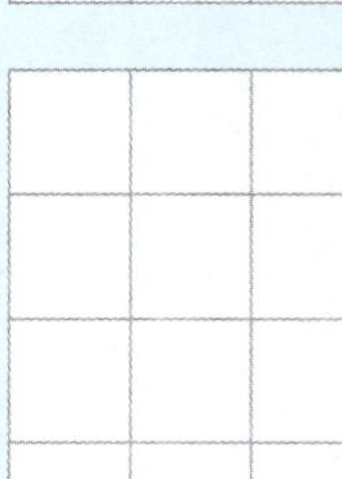

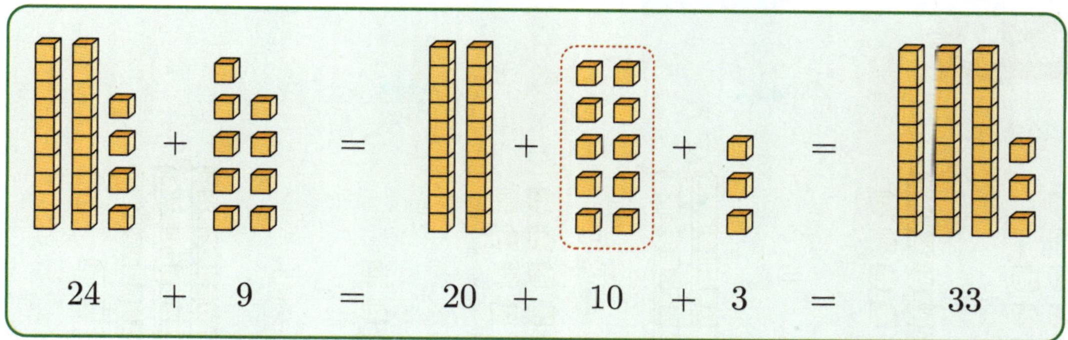

$$24 + 9 = 20 + 10 + 3 = 33$$

Exercise 34

Complete:

a

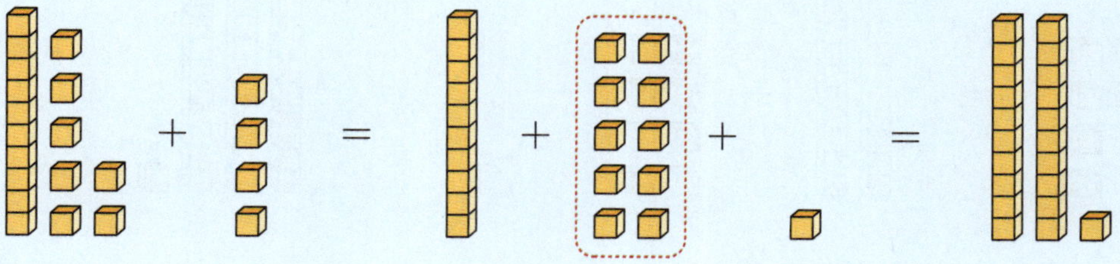

_____ + _____ = _____ + 10 + _____ = _____

b

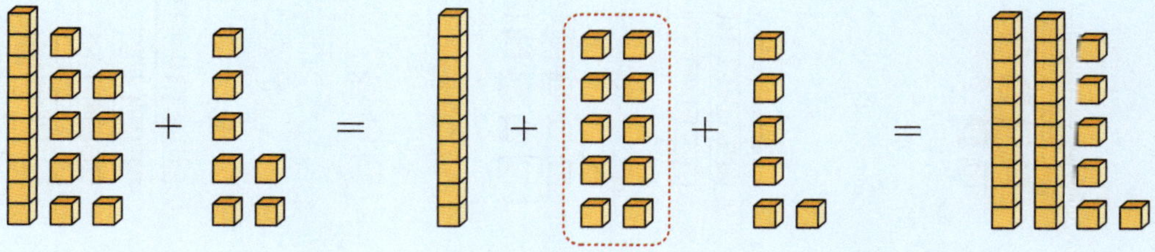

_____ + _____ = _____ + _____ + _____ = _____

c

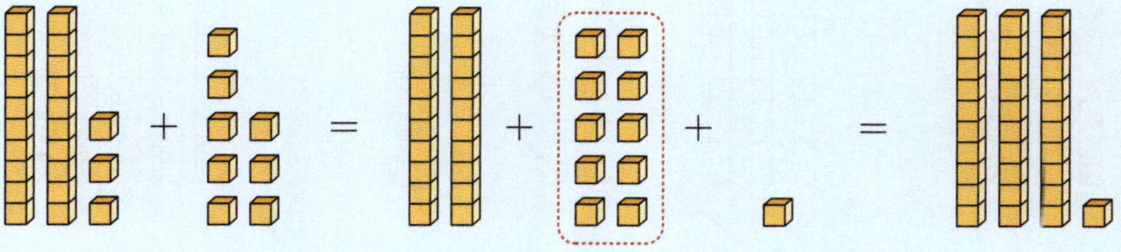

_____ + _____ = _____ + _____ + _____ = _____

Exercise 35

Use blocks to help you add:

a $16 + 6 =$ _____

b $18 + 3 =$ _____

c $21 + 9 =$ _____

d $14 + 8 =$ _____

e $25 + 7 =$ _____

f $29 + 4 =$ _____

Exercise 36

Complete:

a 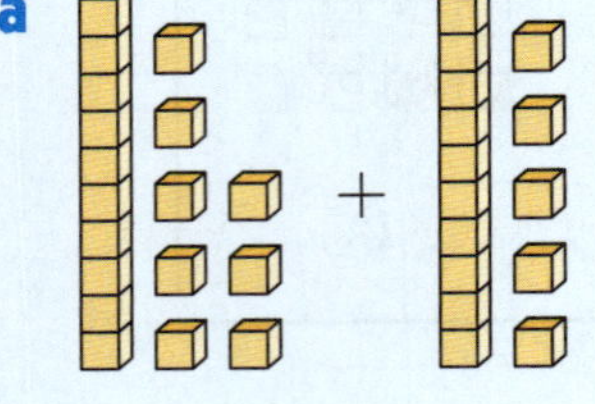 $+$ 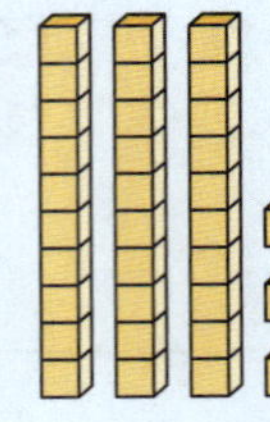 $+$ 

_____ $+$ _____ $=$ _____ $+$ 10 $+$ _____ $=$ _____

b 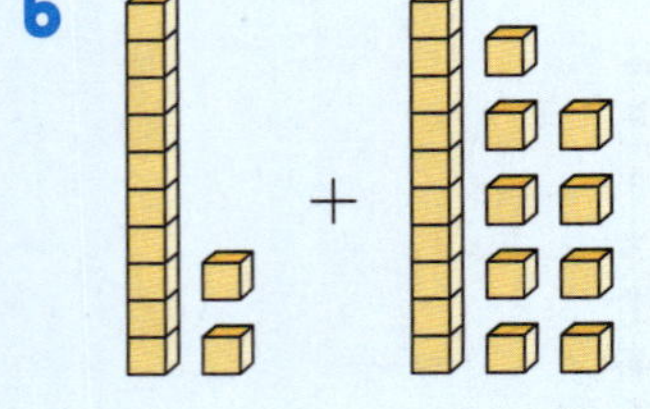 $+$ 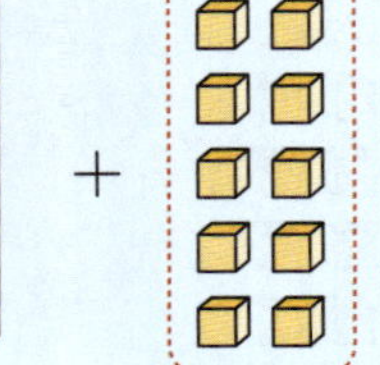 $+$ 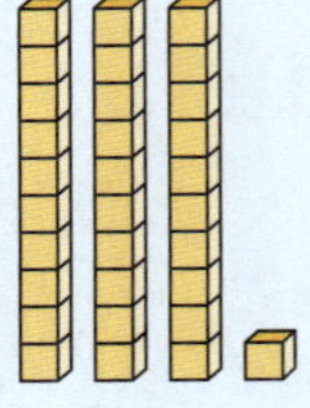

_____ $+$ _____ $=$ _____ $+$ _____ $+$ _____ $=$ _____

c 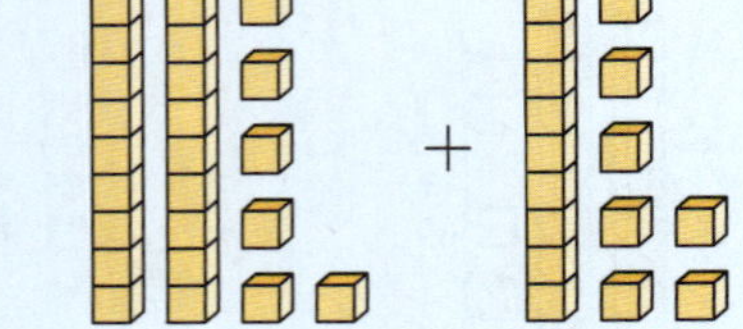 $+$ 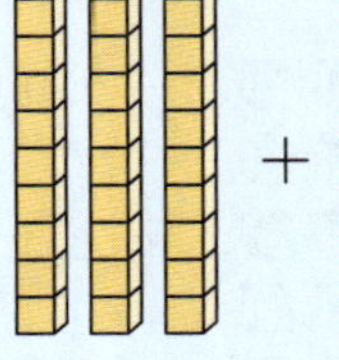

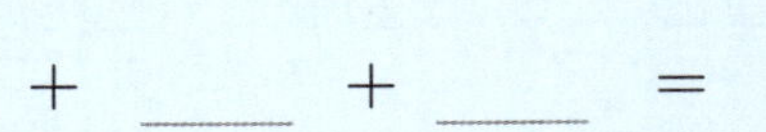

_____ $+$ _____ $=$ _____ $+$ _____ $+$ _____ $=$ _____

d 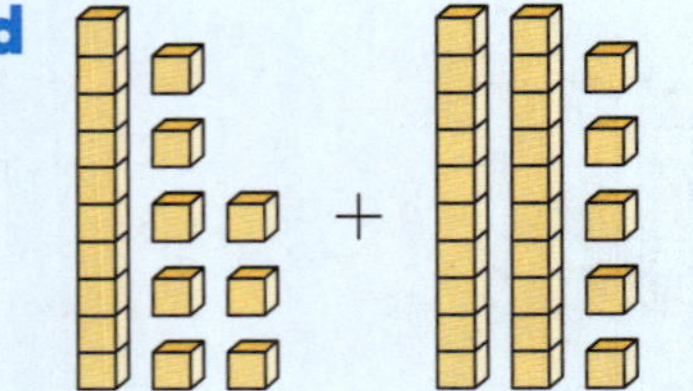 $+$ 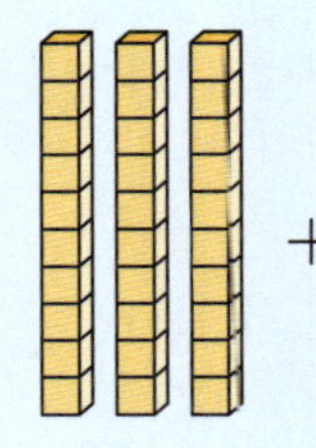

_____ $+$ _____ $=$ _____ $+$ _____ $+$ _____ $=$ _____

Exercise 37

Use blocks to help you add:

a $17 + 14 =$ _____

b $22 + 18 =$ _____

c $19 + 18 =$ _____

d $18 + 28 =$ _____

e $27 + 17 =$ _____

f $13 + 29 =$ _____

A **straight line** never changes direction.   A **curve** is not straight.

A shape is **2-dimensional** if it can be drawn on a **surface**.

Exercise 1

These 2-dimensional shapes have straight sides.

a Practise your spelling. Copy the name of each shape.

b Complete the number of sides.

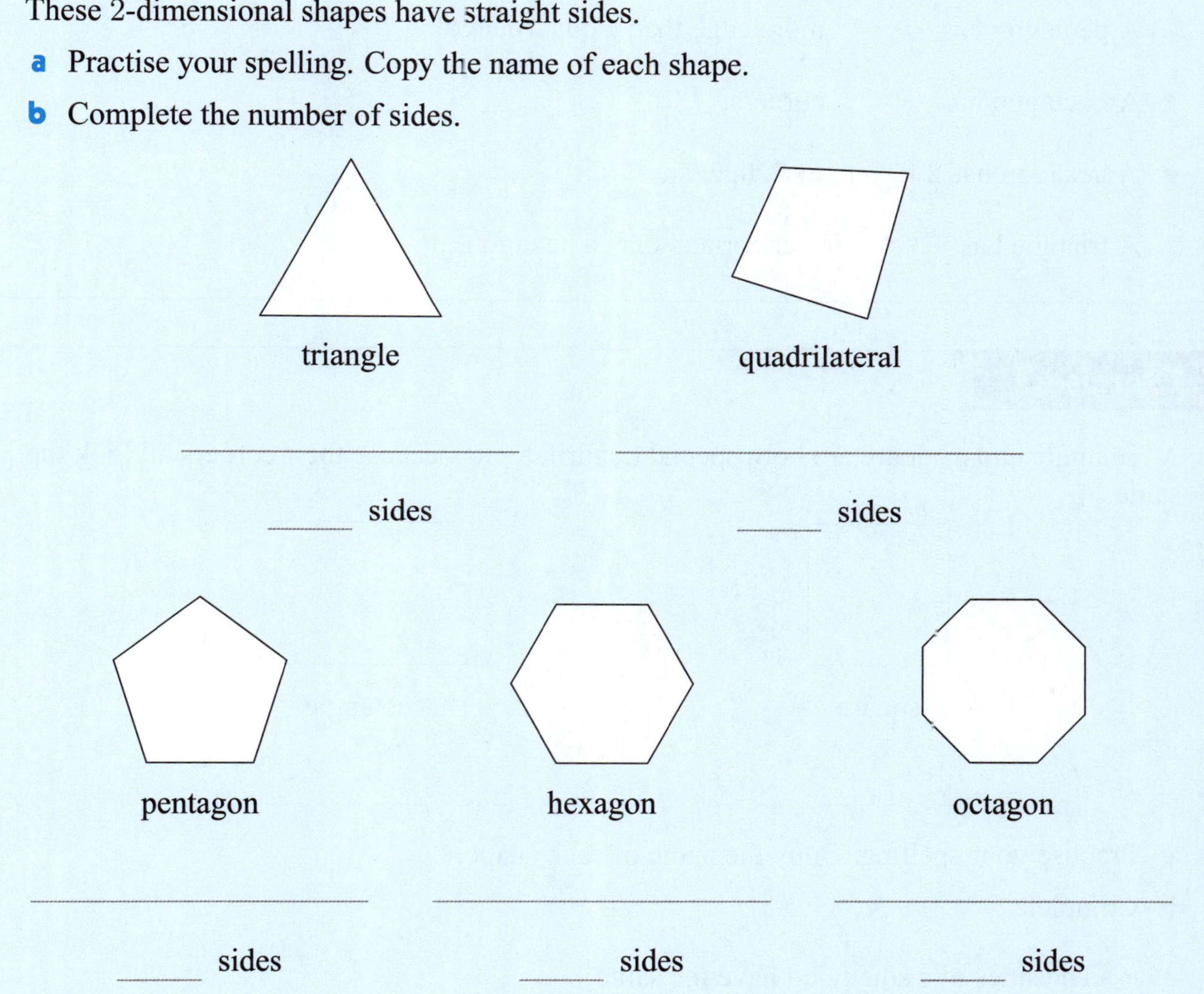

Each side of a shape can also be called an **edge**.

Two edges meet at a **corner**.

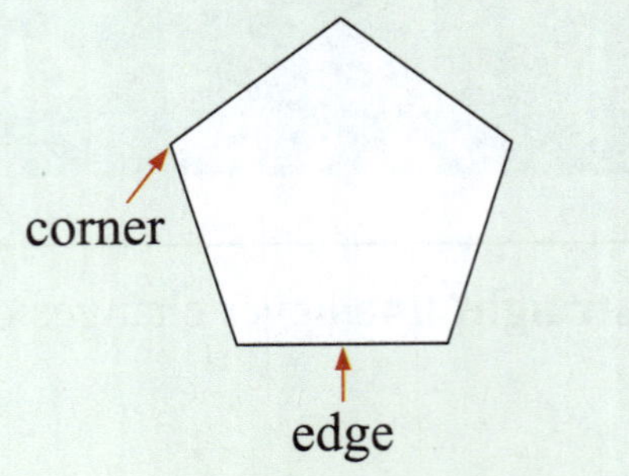

Exercise 2

Complete:

a A quadrilateral has _______ corners.

b A pentagon has _______ more edge than a quadrilateral.

c An octagon has _______ corners.

d A hexagon has 2 fewer edges than an _______________ .

e A triangle has _______ fewer corners than a hexagon.

Exercise 3

A rectangle and a square are both special quadrilaterals because their corners all look the same.

square

rectangle

_________________________   _________________________

a Practise your spelling. Copy the name of each shape.

b Complete:

- The sides of a square all have the same _______________ .

- The *opposite* sides of a rectangle have the same _______________ .

Exercise 4

These shapes are made using a curve.

Practise your spelling. Copy each name.

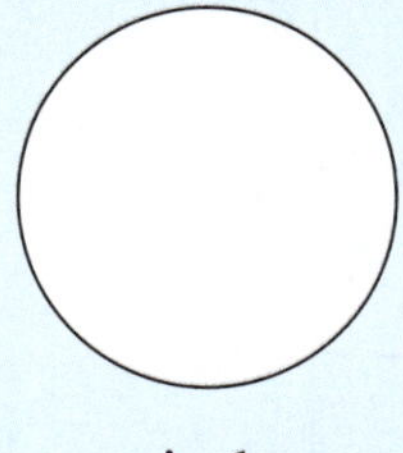

circle

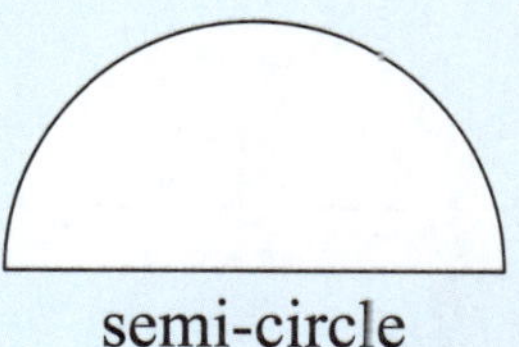

oval   or   ellipse

____

_______________________   or   _______________________

A semi-circle has a curve and a straight side.

semi-circle

____

Exercise 5

Match each food to its shape:

| rectangle |

| circle |

| triangle |

| square |

| ellipse |

| semi-circle |

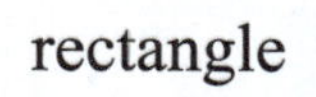

Exercise 6

Complete:

Shape	Edges	Corners
square		4
	3	
hexagon		
	5	
		8

Exercise 7

Draw each shape. Use your ruler if needed.

a a triangle

b a rectangle

c a circle

d a pentagon

e an ellipse

f an octagon

Activity

In the picture below, describe where you see:

a a triangle

b a square

c a pentagon

d a hexagon

e a circle

f an octagon

g a semi-circle

h a rectangle

Exercise 8

I can *describe* a pentagon as a shape with 5 sides and 5 corners.

How would you *describe* these shapes to someone?

a a triangle

b a square

c an octagon

d a circle

Puzzle

Without lifting your pencil, draw *five* lines which create *two* triangles.

Listening Activity

Listen to the audio. Follow the instructions.

Activity

Can you draw .... ?

a a car using a large rectangle, 2 circles, and 2 squares

b an animal using triangles

Puzzle

In each figure, count how many triangles and squares you can find:

a 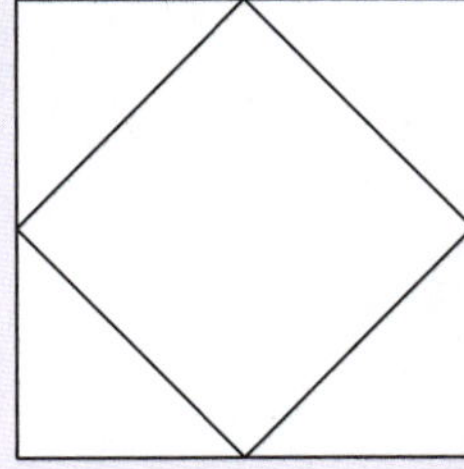

_________ squares

_________ triangles ?

b 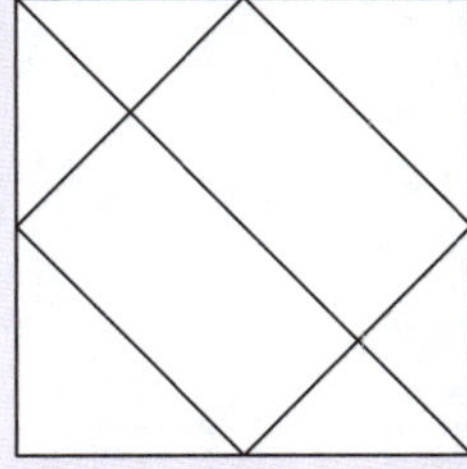

_________ squares

_________ triangles

c 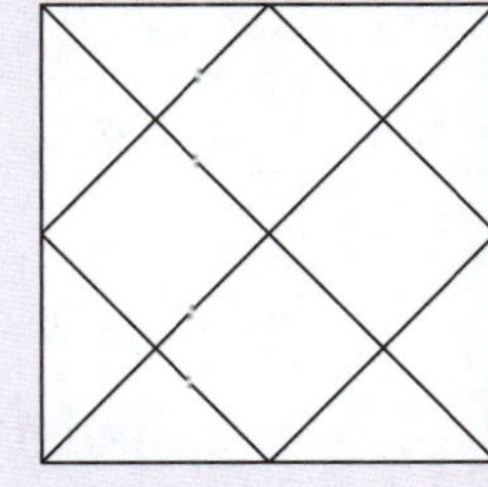

_________ squares

_________ triangles

A **solid** is a shape which takes up **space**.

We say that solids are **3-dimensional**.

We can *draw* solids as 2-dimensional shapes on a page.

Exercise 9

Here are some 3-dimensional solids. Practise your writing.

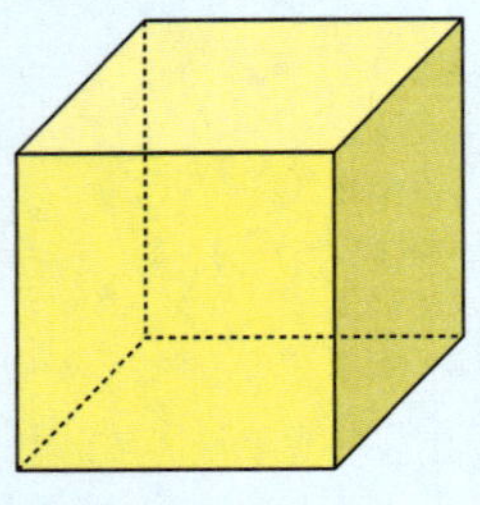

cube

____

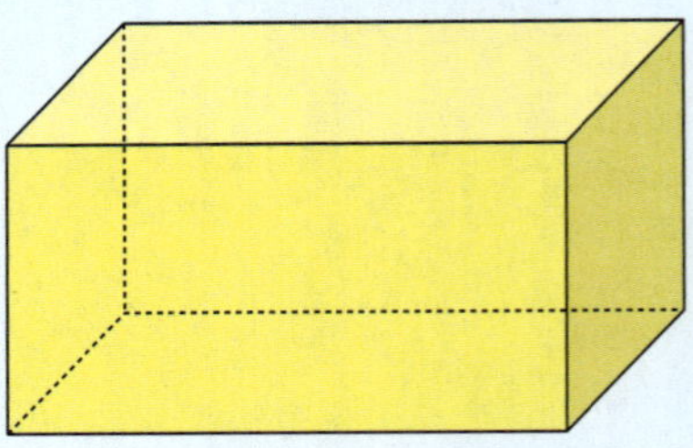

rectangular prism

____

____

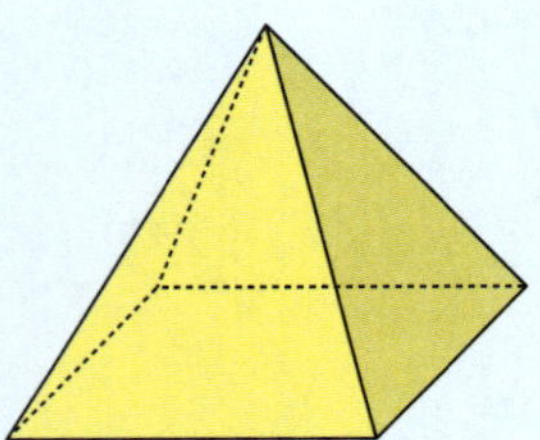

square-based pyramid

____

____

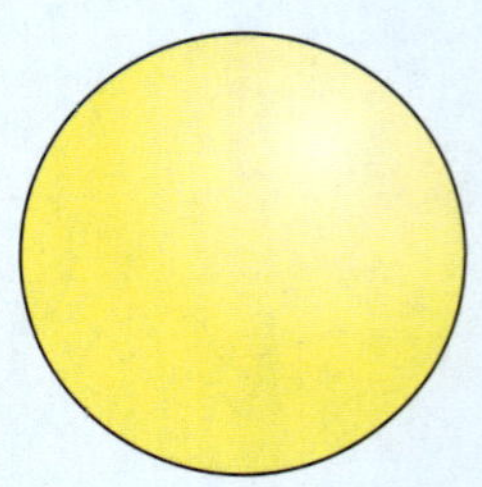

sphere

____

cylinder

____

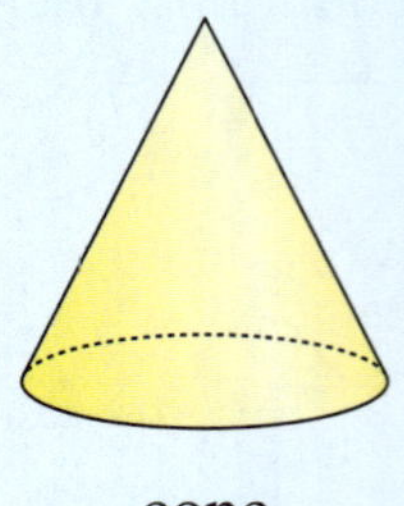

cone

____

Click on each solid to view it from other angles.

Any flat surface of a solid is called a **face**.

Any line where two faces meet is called an **edge**.

Any point where three faces meet is called a **corner**.

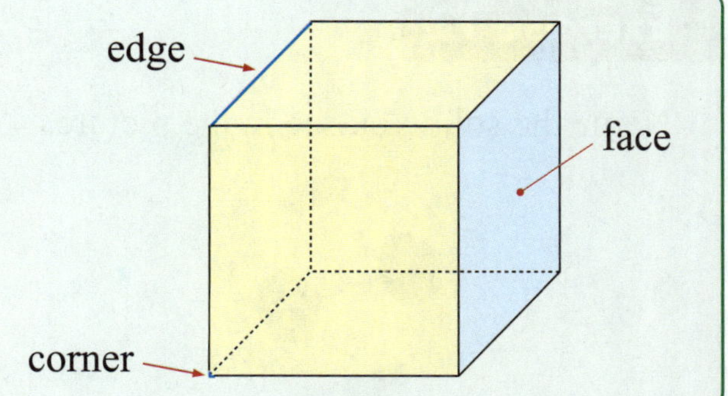

Exercise 10

Count the faces, edges, and corners of each solid:

Solid	Faces	Edges	Corners
cube			
rectangular prism			
square-based pyramid			
sphere	1 curved surface		
cylinder	______ faces and 1 curved surface		

Exercise 11

A cube is a special kind of rectangular prism.

a What is special about its faces?

____

____

b What is special about its edges?

____

____

c Sketch a cube in the space alongside.

Exercise 12

Name the solid you see in the picture.

a

____

b

____

____

c
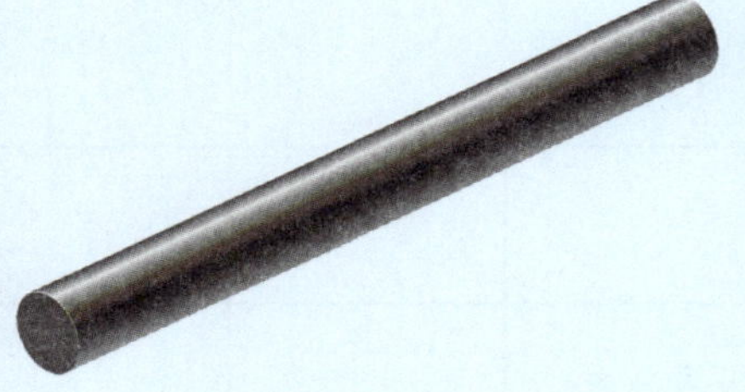

____

d

____

e

____

____

Activity

1 Use popsicle sticks and plasticine to make your own:

   a cube   **b** rectangular prism   **c** square-based pyramid.

2 Make a sphere using plasticine.

3 Can you make a cone using plasticine?

CHAPTER 5: SUBTRACTION

What words tell us that we need to subtract?

______________________   ______________________   ______________________

______________________   ______________________   ______________________

I had 5 apples.

My brothers ate 2 apples.

I have 3 apples left.

$5 - 2 = 3$

"Five minus two equals three."

Exercise 1

Use the pictures to write a subtraction:

a 

______ − ______ = ______

b 

______ − ______ = ______

c 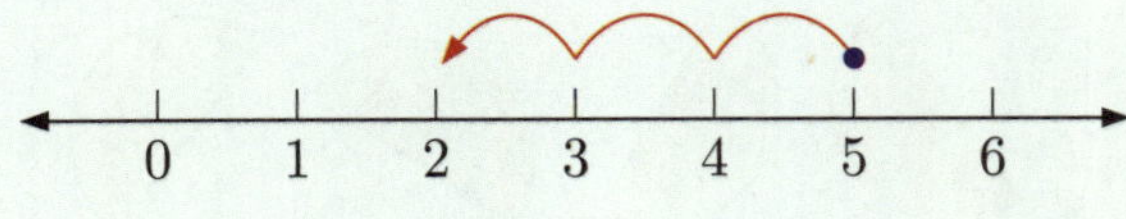

______ − ______ = ______

What is $5 - 3$?

$5 - 3 = 2$

Exercise 2

Use the number line to help you subtract:

a
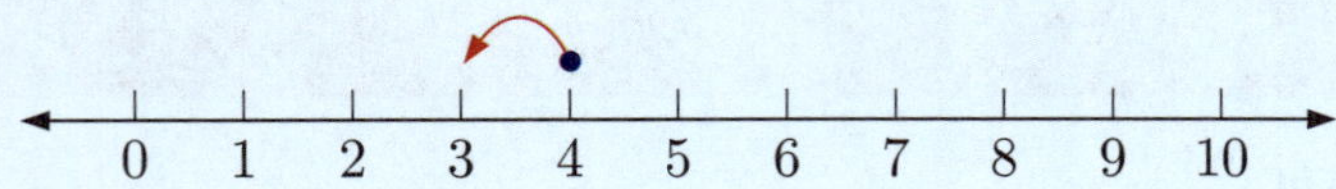

$4 - 1 =$ ________

b
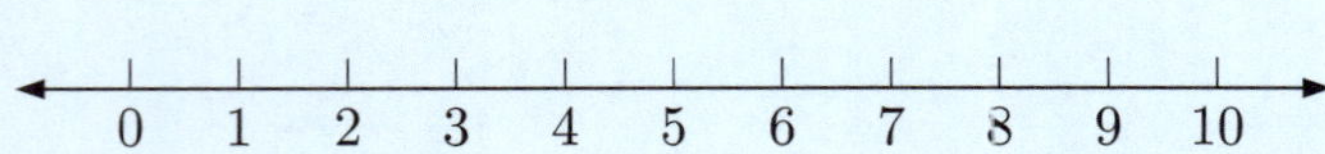

$6 - 4 =$ ________

c
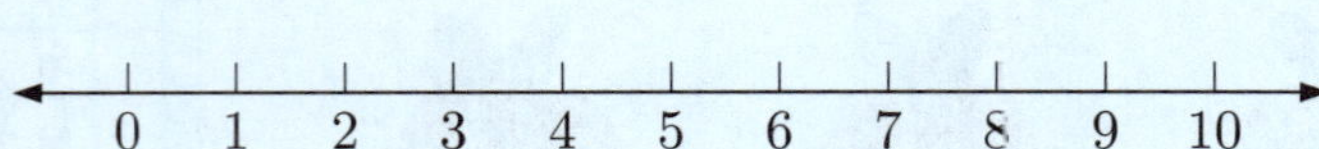

$9 - 5 =$ ________

d
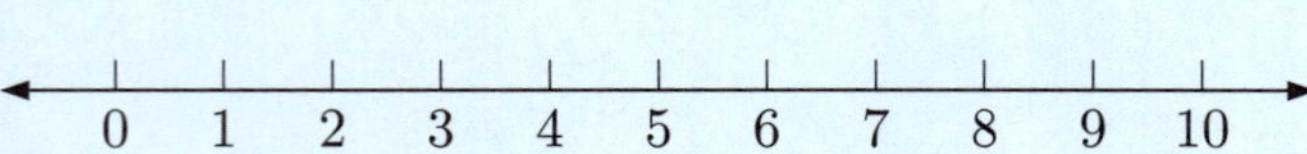

$8 - 6 =$ ________

e

$7 - 3 =$ ________

Exercise 3

Cross out pictures to help you subtract.

a
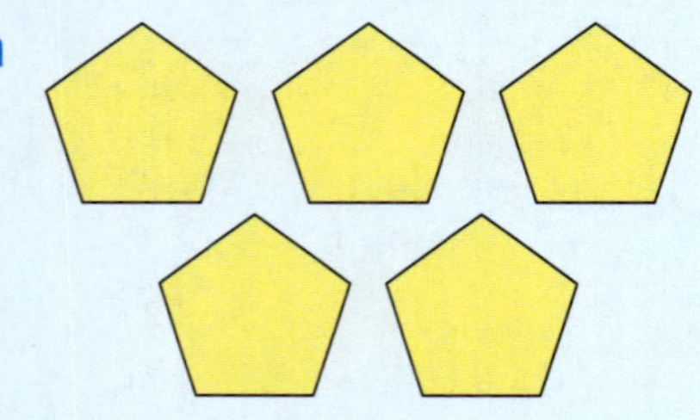

$5 - 4 =$ ________

b
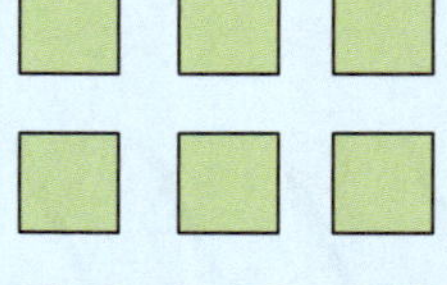

$9 - 4 =$ ________

c

$7 - 5 =$ ________

d
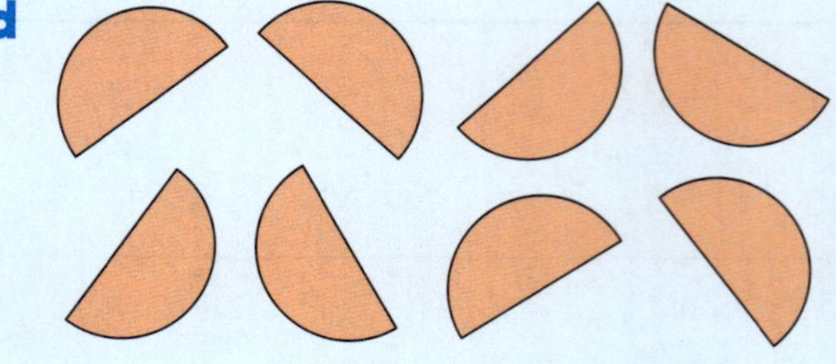

$8 - 1 =$ ________

Exercise 4

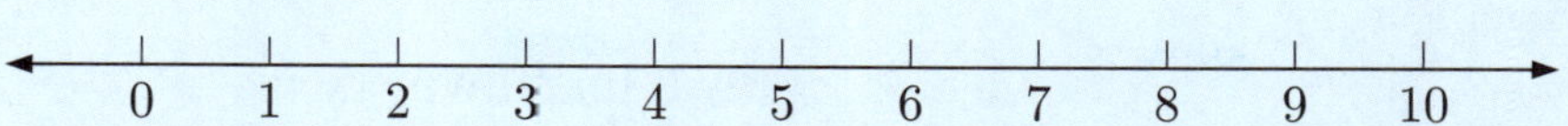

Write down the result of each subtraction:

a $5 - 2 =$ _______

b $6 - 1 =$ _______

c $4 - 3 =$ _______

d $6 - 3 =$ _______

e $8 - 7 =$ _______

f $9 - 2 =$ _______

g $4 - 2 =$ _______

h $7 - 6 =$ _______

i $9 - 3 =$ _______

j $8 - 4 =$ _______

k $9 - 6 =$ _______

l $8 - 2 =$ _______

To get 7 from 3, I need to add 4.

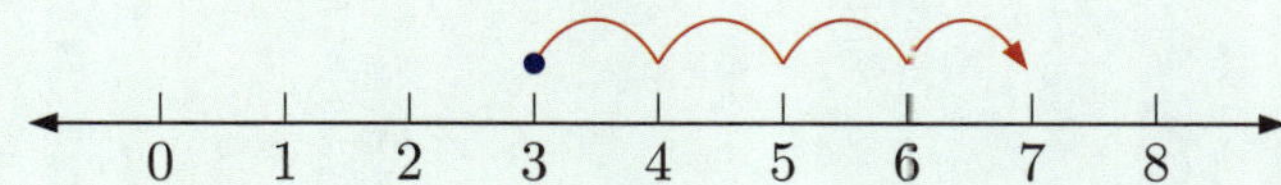

$7 = 3 + 4$

If I start with 7 and take away 4, I am left with 3.

$7 - 4 = 3$

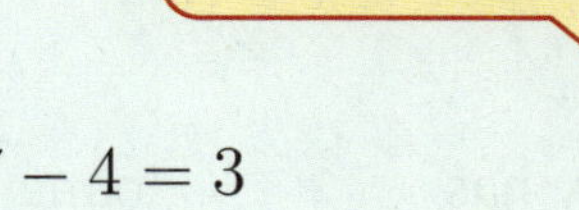

Exercise 5

Find the missing numbers.

a $5 = \boxed{} + 1$

so $5 - 1 = \boxed{}$.

b $7 = \boxed{} + 2$

so $7 - 2 = \boxed{}$.

c $8 = \boxed{} + 5$

so $8 - 5 = \boxed{}$.

d $9 = \boxed{} + 7$

so $9 - 7 = \boxed{}$.

Exercise 6

a 9 minus $1 =$ _______

b 6 take away $5 =$ _______

c 7 subtract $3 =$ _______

d Find the difference: $8 - 6 =$ _______

e Subtract 6 from 9: $9 - 6 =$ _______

f Take 3 away from 8: $8 - 3 =$ _______

g Find the difference: $7 - 1 =$ _______

h Subtract 2 from 6:   _______ $-$ _______ $=$ _______

i Take 4 away from 9:   _______ $-$ _______ $=$ _______

Exercise 7

Write a subtraction to solve each problem:

a A monkey has 6 bananas. She eats 2 bananas.

_______ $-$ _______ $=$ _______

The monkey has _______ bananas left.

b Bernard has 8 birds in a cage. 5 birds escape and fly away.

_______ $-$ _______ $=$ _______

Bernard has _______ birds left.

c A team has 9 players. 3 players are off the court.

_______ $-$ _______ $=$ _______

_______ players are on the court.

d Sally won 7 ribbons on Sports Day. Gemma won 2 fewer ribbons than Sally.

_______ $-$ _______ $=$ _______

Gemma won _______ ribbons on Sports Day.

Exercise 8

$$2 + 5 = 7$$
$$\text{so} \quad 7 - 2 = 5$$
$$\text{and} \quad 7 - 5 = 2$$

Use the addition to write *two* subtractions.

a $3 + 4 = 7$

so $7 - \underline{\qquad} = \underline{\qquad}$

and $7 - \underline{\qquad} = \underline{\qquad}$

b $2 + 7 = 9$

so $9 - \underline{\qquad} = \underline{\qquad}$

and $9 - \underline{\qquad} = \underline{\qquad}$

c $5 + 3 = \underline{\qquad}$

so $\underline{\qquad} - 5 = \underline{\qquad}$

and $\underline{\qquad} - 3 = \underline{\qquad}$

d $6 + 2 = \underline{\qquad}$

so $\underline{\qquad} - 6 = \underline{\qquad}$

and $\underline{\qquad} - 2 = \underline{\qquad}$

I had 6 honey pots and lost none.
I still have 6 honey pots left!

$$6 - 0 = 6$$

I had 6 carrots but someone stole them.
I have no carrots left!

$$6 - 6 = 0$$

Exercise 9

Complete these subtractions.

a $4 - 0 = \underline{\qquad}$

b $3 - 3 = \underline{\qquad}$

c $5 - 0 = \underline{\qquad}$

d $7 - 7 = \underline{\qquad}$

e $8 - 0 = \underline{\qquad}$

f $9 - 9 = \underline{\qquad}$

Exercise 10

Use the number lines to help with these subtractions.

Exercise 11

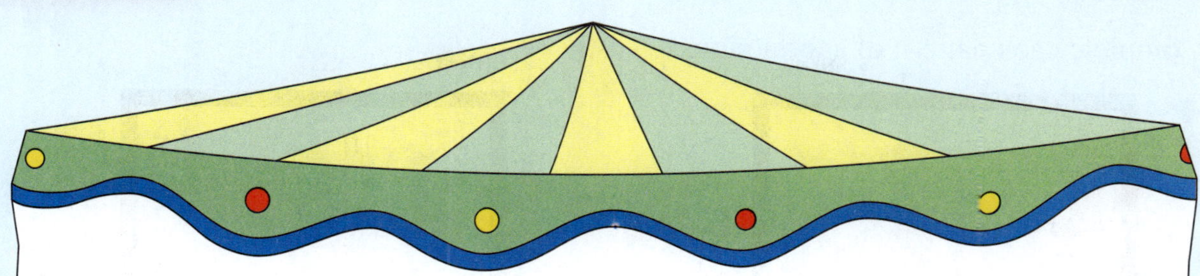

Complete these subtractions.

a $8 - 6 =$ _____

$9 - 6 =$ _____

$10 - 6 =$ _____

$11 - 6 =$ _____

$12 - 6 =$ _____

$13 - 6 =$ _____

b $8 - 7 =$ _____

$9 - 7 =$ _____

$10 - 7 =$ _____

$11 - 7 =$ _____

$12 - 7 =$ _____

$13 - 7 =$ _____

c $9 - 8 =$ _____

$10 - 8 =$ _____

$11 - 8 =$ _____

$12 - 8 =$ _____

$13 - 8 =$ _____

$14 - 8 =$ _____

Exercise 12

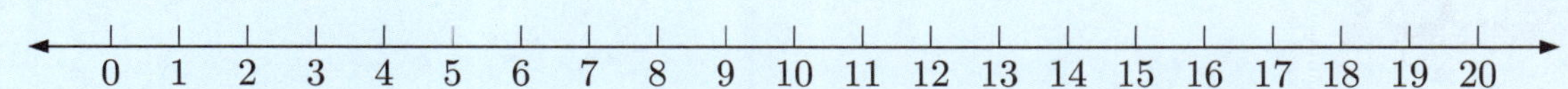

Use the number line to help with these subtractions.

a $17 - 3 =$ _____

d $16 - 6 =$ _____

g $11 - 2 =$ _____

j $15 - 7 =$ _____

b $14 - 2 =$ _____

e $13 - 5 =$ _____

h $19 - 8 =$ _____

k $17 - 9 =$ _____

c $19 - 4 =$ _____

f $18 - 7 =$ _____

i $12 - 5 =$ _____

l $14 - 6 =$ _____

Exercise 13

Use the addition to write *two* subtractions.

a $6 + 9 = 15$

so $15 -$ _____ $=$ _____

and $15 -$ _____ $=$ _____

b $8 + 5 =$ _____

so   _____ $- 8 =$ _____

and   _____ $- 5 =$ _____

Exercise 14

Complete each pattern of subtractions:

a

$10 - 10 =$ _____

$20 - 10 =$ _____

$30 - 10 =$ _____

$40 - 10 =$ _____

$50 - 10 =$ _____

$60 - 10 =$ _____

$70 - 10 =$ _____

$80 - 10 =$ _____

$90 - 10 =$ _____

b

$15 - 10 =$ _____

$25 - 10 =$ _____

$35 - 10 =$ _____

$45 - 10 =$ _____

$55 - 10 =$ _____

$65 - 10 =$ _____

$75 - 10 =$ _____

$85 - 10 =$ _____

$95 - 10 =$ _____

Exercise 15

Complete these subtractions.

a $12 - 10 =$ _____

b $16 - 10 =$ _____

c $14 - 10 =$ _____

d $23 - 10 =$ _____

e $27 - 10 =$ _____

f $31 - 10 =$ _____

Exercise 16

Complete these subtractions.

a $9 - 4 =$ _____

so $9 -$ _____ $= 4$

b $10 - 3 =$ _____

so $10 -$ _____ $= 3$

c $14 - 4 =$ _____

so $14 -$ _____ $= 4$

d $17 - 2 =$ _____

so $17 -$ _____ $= 2$

e $18 - 3 =$ _____

so $18 -$ _____ $= 3$

f $16 - 5 =$ _____

so $16 -$ _____ $= 5$

Puzzle

2	3	4	5	6	7	8	9	10

Use the cards *once each* to complete *three* subtractions.

☐ − ☐ = ☐

☐ − ☐ = ☐

☐ − ☐ = ☐

Exercise 17

Complete these subtractions.

a $11 - 4 =$ ________

b $13 - 5 =$ ________

c $12 - 7 =$ ________

d $15 - 5 =$ ________

e $14 - 8 =$ ________

f $11 - 9 =$ ________

g $12 - 4 =$ ________

h $16 - 7 =$ ________

i $15 - 8 =$ ________

j $18 - 9 =$ ________

k $19 - 9 =$ ________

l $17 - 8 =$ ________

Exercise 18

Complete these subtractions.

a $8 - 5 =$ ________

b $7 - 7 =$ ________

c $15 - 9 =$ ________

$18 - 5 =$ ________

$17 - 7 =$ ________

$25 - 9 =$ ________

$28 - 5 =$ ________

$27 - 7 =$ ________

$35 - 9 =$ ________

$38 - 5 =$ ________

$37 - 7 =$ ________

$45 - 9 =$ ________

$48 - 5 =$ ________

$47 - 7 =$ ________

$55 - 9 =$ ________

Exercise 19

Complete these subtractions.

a $4 - 1 =$ _____

   $40 - 10 =$ _____

b $7 - 2 =$ _____

   $70 - 20 =$ _____

c $9 - 4 =$ _____

   $90 - 40 =$ _____

Exercise 20

Complete these subtractions.

a $50 - 10 =$ _____

   $50 - 20 =$ _____

   $50 - 30 =$ _____

   $50 - 40 =$ _____

   $50 - 50 =$ _____

b $80 - 10 =$ _____

   $80 - 20 =$ _____

   $80 - 30 =$ _____

   $80 - 40 =$ _____

   $80 - 50 =$ _____

c $100 - 10 =$ _____

   $100 - 20 =$ _____

   $100 - 30 =$ _____

   $100 - 40 =$ _____

   $100 - 50 =$ _____

Exercise 21

Use $=$ or $\neq$ to complete:

a $5 - 3$ _____ 2

b $6 - 4$ _____ 3

c $10 - 6$ _____ 8

d $9 - 8$ _____ 2

e $11 - 7$ _____ 4

f $10 - 7$ _____ 3

g $13 - 5$ _____ 6

h $16 - 9$ _____ 5

i $15 - 9$ _____ 6

Listening Activity

Click and follow the instructions.

1 _____   **2** _____   **3** _____

4 _____   **5** _____   **6** _____

7 _____   **8** _____   **9** _____

Exercise 22

a Take away 2 each time:

| 14 | → | 12 | → | | → | | → | |

b Take away 2 each time:

| 15 | → | | → | | → | | → | |

c Take away 3 each time:

| 17 | → | | → | | → | | → | |

d Take away 5 each time:

| 20 | → | | → | | → | | → | |

e Take away 10 each time:

| 72 | → | | → | | → | | → | |

Exercise 23

Describe what is happening in each sequence:

a $18 \to 15 \to 12 \to 9 \to 6$

I start with ________ , and subtract ________ each time.

b $19 \to 17 \to 15 \to 13 \to 11$

c $17 \to 13 \to 9 \to 5 \to 1$

d $48 \to 38 \to 28 \to 18 \to 8$

Find the missing number: $11 - \boxed{} = 7$

If I start with 11, how many should I take away, to leave 7?

$11 - 4 = 7$

Exercise 24

Find the missing number.

a $5 - \boxed{} = 3$

b $8 - \boxed{} = 2$

c $10 - \boxed{} = 5$

d $11 - \boxed{} = 3$

e $15 - \boxed{} = 12$

f $16 - \boxed{} = 8$

g $19 - \boxed{} = 10$

h $20 - \boxed{} = 14$

i $18 - \boxed{} = 8$

Exercise 25

Using 6, 7, and 13, I can write:

$$6 + 7 = 13$$
$$7 + 6 = 13$$
$$13 - 6 = 7$$
$$13 - 7 = 6$$

a Use 5, 6, and 11 to complete:

$\boxed{} + \boxed{} = \boxed{}$

$\boxed{} + \boxed{} = \boxed{}$

$\boxed{} - \boxed{} = \boxed{}$

$\boxed{} - \boxed{} = \boxed{}$

b Use 4, 8, and 12 to complete:

$\boxed{} + \boxed{} = \boxed{}$

$\boxed{} + \boxed{} = \boxed{}$

$\boxed{} - \boxed{} = \boxed{}$

$\boxed{} - \boxed{} = \boxed{}$

Find the missing number: $\boxed{} - 4 = 5$

What do I need to start with, so when I take 4 away, I still have 5?

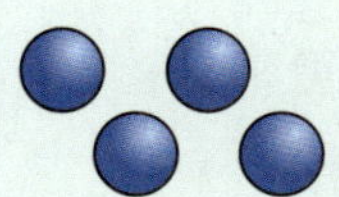

4 taken away

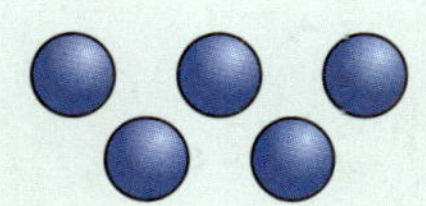

5 left

$9 - 4 = 5$

Exercise 26

Find the missing number.

a $\boxed{} - 4 = 1$ **b** $\boxed{} - 2 = 4$ **c** $\boxed{} - 3 = 5$

d $\boxed{} - 5 = 4$ **e** $\boxed{} - 6 = 2$ **f** $\boxed{} - 7 = 3$

g $\boxed{} - 4 = 8$ **h** $\boxed{} - 8 = 7$ **i** $\boxed{} - 6 = 9$

Exercise 27

Write each word on the appropriate whiteboard:

sum	total	decrease	plus	increase
less	take away	add	minus	

addition   subtraction

Exercise 28

Complete:

a The sum of 9 and 13 is _______ .

b 11 take away 6 = _______

c 18 subtract 10 = _______

d 5 plus 9 = _______

e The total of 6 and 10 is _______ .

f 20 minus 7 = _______

Exercise 29

Emma has 3 children, and Jasmine has 2 children.

In total there are $3 + 2 = 5$ children.

Emma has $3 - 2 = 1$ more child than Jasmine.

Write an addition and a subtraction to solve these problems.

a Linden has 5 red balloons and 3 yellow balloons.

In total, Linden has _______________ = _______ balloons.

Linden has _______________ = _______ fewer yellow balloons than red balloons.

b Isaac has 13 coloured pencils and 6 pens.

In total, Isaac has _______________ = _______ items.

Isaac has _______________ = _______ more coloured pencils than pens.

To subtract numbers with more than one digit, we can use **column subtraction**.

We write the numbers in columns so the place values line up.

Subtract each column, starting with the units.

$28 - 13$

1 Set out your subtraction.

2 Subtract the units:

$8 - 3 = 5$

3 Subtract the tens:

$2 - 1 = 1$

	T	U
	2	8
−	1	3
	1	5

$28 - 13 = 15$

Exercise 30

Complete these subtractions.

a $26 - 5$

	T	U
	2	6
−		5

b $29 - 9$

	T	U
	2	9
−		9

c $24 - 12$

	T	U
	2	4
−	1	2

d $25 - 13$

	T	U
	2	5
−	1	3

e $38 - 16$

	T	U
	3	8
−	1	6

f $36 - 21$

	T	U
	3	6
−	2	1

g $39 - 18$

	T	U
	3	9
−	1	8

h $46 - 24$

	T	U
	4	6
−	2	4

i $48 - 31$

	T	U
	4	8
−	3	1

Exercise 31

Set out these subtractions and find the answer.

a $27 - 11$

	T	U

b $26 - 15$

c $38 - 25$

Exercise 32

a Farmer Giles has 27 sheep in the barn. He asks his dog Shep to move 14 sheep to the field.

How many sheep are left in the barn?

There are ☐ sheep left in the barn.

b Belinda has 47 candles in her cupboard. She uses 24 candles on her daughter's birthday cake.

How many candles does Belinda have left?

Belinda has ☐ candles left.

c In a game of snooker, Xuan scored 68 points and Chu scored 36 points.

By how many points did Xuan win?

Xuan won by ☐ points.

Exercise 33

In a netball match, Tina scored 23 goals and Vy scored 35 goals.

a How many goals did the girls score in total?

The girls scored ☐ goals in total.

b How many fewer goals did Tina score than Vy?

Tina scored ☐ fewer goals than Vy.

Discussion

We *measure* many things, including:

- length
- area
- volume
- capacity
- mass
- time
- temperature

Discuss what each of these things mean.

Exercise 1

Practise your writing.

length

area

volume

capacity

mass

time

temperature

Discussion

Which measurements tell us the *size* of an object?

Exercise 2

If I want to compare the *size* of these objects, what could I measure?

a   I can measure the _______________ of each line.

b   I can measure the _______________ of each square.

c   I can measure the _______________ of each block.

d   I can measure the _______________ of each bucket.

LENGTH

Discussion

Height, *width*, and *depth* are special words we use for _______________ measurements.

Why do we use them?

Hint: What is the height, width, and depth of this cabinet?

Exercise 3

Practise your writing.

height   width   depth

___________________   ___________________   ___________________

Activity

Discuss the meaning of each word. Copy it into the correct group or groups!

thin	fat	shallow	long
short	tall	narrow	deep
wide	thick	slender	skinny

Length words	*Height words*

Width words	*Depth words*

Exercise 4

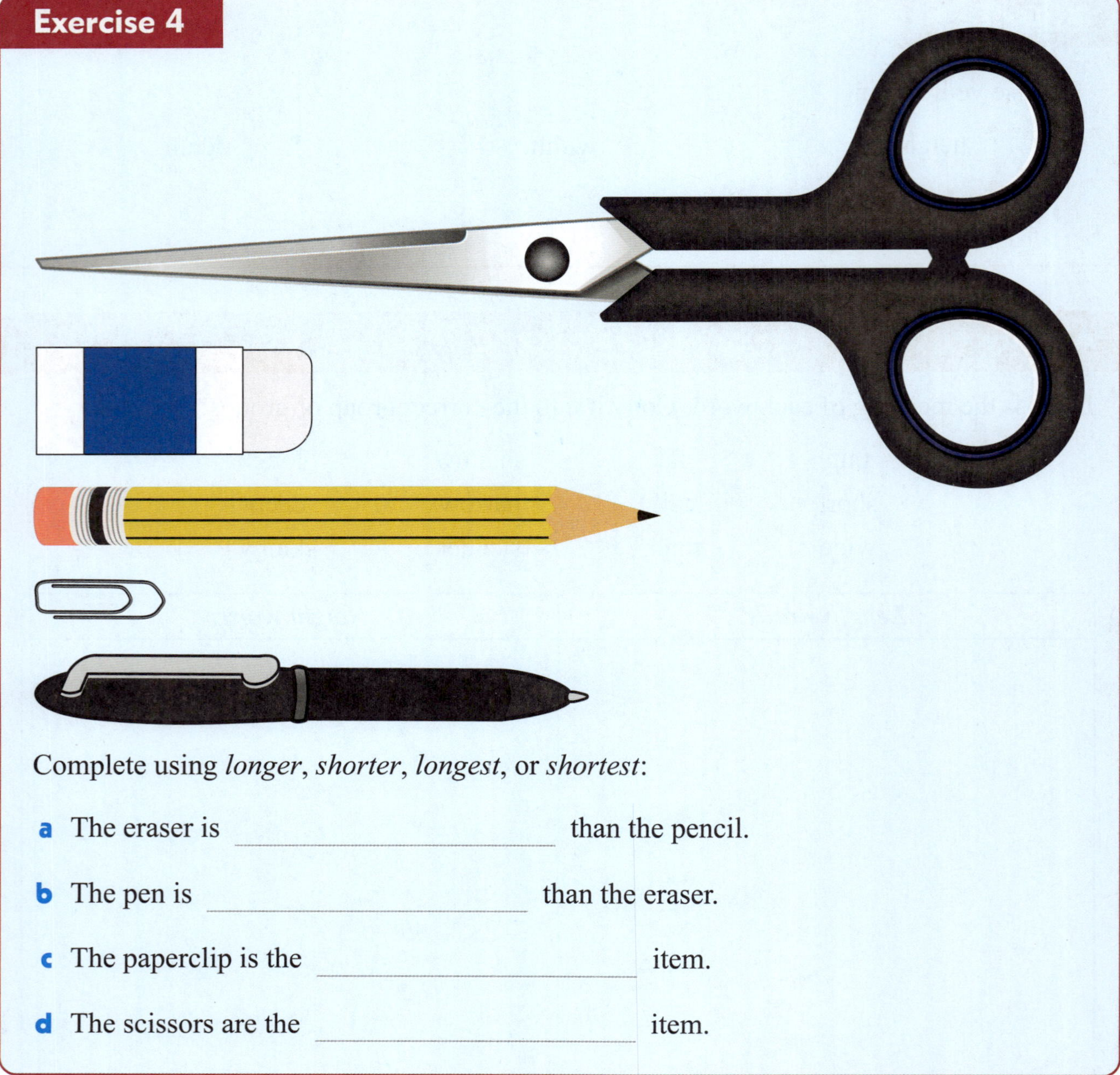

Complete using *longer*, *shorter*, *longest*, or *shortest*:

a The eraser is _________________________ than the pencil.

b The pen is _________________________ than the eraser.

c The paperclip is the _________________________ item.

d The scissors are the _________________________ item.

Activity

a List 4 items longer than your pencil.

1 _________________________ **2** _________________________

3 _________________________ **4** _________________________

b List 4 items shorter than your pencil.

1 _________________________ **2** _________________________

3 _________________________ **4** _________________________

c What is the tallest object in your classroom? _________________________

Exercise 5

puddle

fountain

lake

Complete using *shallower*, *deeper*, or *deepest*:

a The lake is _________________________ .

b The puddle is _________________________ than the lake.

c The fountain is _________________________ than the puddle.

Group Activity

In a small group, work out who is:

a tallest _________________________   **b** shortest _________________________

Write the children in your group in order of height.

____

shortest ⟶ tallest

Activity

Sam, Annie, Paddy, and Lachie want to find out who is the tallest.

Watch the video. Answer the questions when the video pauses.

1 Who is taller, Paddy or Annie? _________________________

2 Is Sam taller or shorter than Lachie? _________________________

3 Was Annie correct about Paddy? _________ . Paddy is _________________ than Sam.

4 Write the children in order of height.

_______________ _______________ _______________ _______________

shortest ⟶ tallest

Discussion

Why is *measuring* useful when we want to compare lengths?

What *tools* can we use to measure lengths?

A **ruler** is used to measure short lengths.

A ruler is a *number line*.

Ask an adult to help you print and cut your ruler.

The marks on this ruler are 1 **centimetre** apart.

We often write centimetre using **cm**.

Exercise 6

Practise your writing.

centimetre

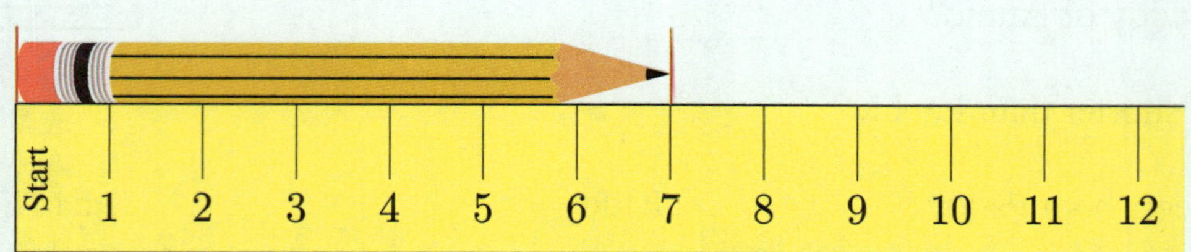

We place the start end of the ruler at one end of the pencil.

The other end of the pencil is at the 7 cm mark.

The pencil is 7 cm long.

Exercise 7

The marker is _________ cm long.

Exercise 8

Use your ruler to measure each line. Write your answer in the space provided.

a ________________________________ _________ cm

b ____________ _________ cm

c ________________________________ _________ cm

d ________________ _________ cm

e __________ _________ cm

f ________________________ _________ cm

Activity

Make a line of 12 centicubes.

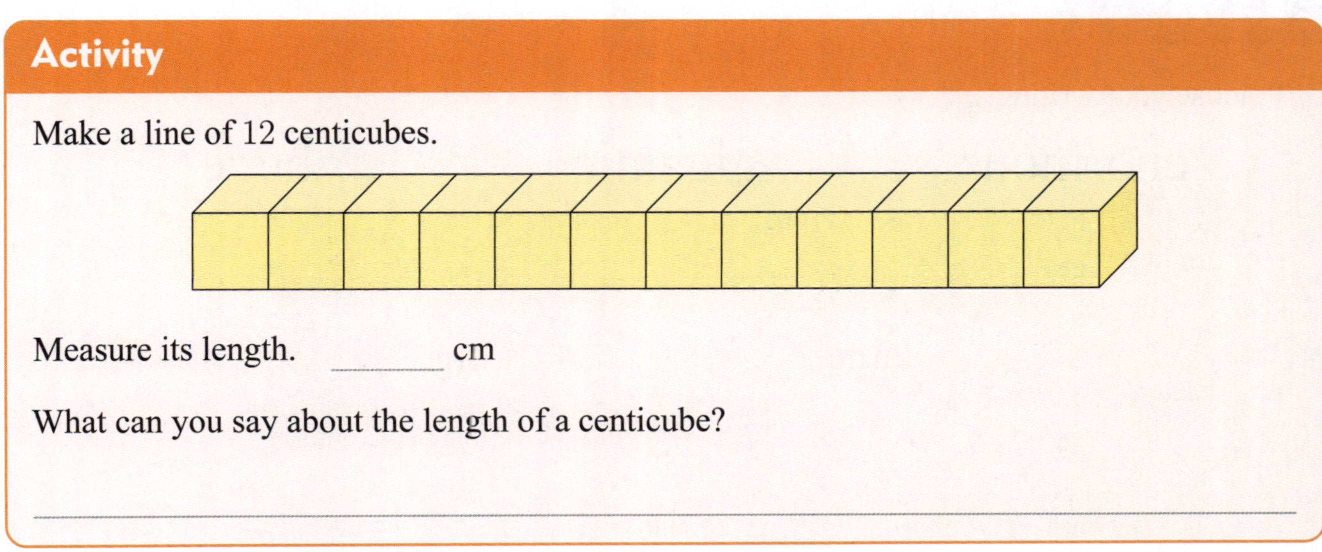

Measure its length. _________ cm

What can you say about the length of a centicube?

____

AREA

The **area** of an object is the amount of *surface* it covers.

Brazil has a *huge* area. It is one of the *largest* countries on Earth.

Bolivia has a *large* area, but it is much *smaller* than Brazil.

Exercise 9

Practise your writing!

enormous   gigantic   huge

_______________   _______________   _______________

large   big

_______________   _______________

small   little   tiny

_______________   _______________   _______________

Exercise 10

Write these areas in order:

your desk

your school

your classroom

largest

smallest

____

____

____

Exercise 11

Write these areas in order:

football field

table tennis table

basketball court

largest

smallest

____

____

____

Exercise 12

Which shape would you describe as:

a big   _______________________

b tiny   _______________________

c little?   _______________________

Activity

Draw a large rectangle.

Draw a tiny circle and a small square inside the rectangle.

Colour the circle purple and the square blue. Colour the rest of the rectangle yellow.

Activity

You will need: 1 or 2 packs of playing cards

What to do:

1   Lay out the cards to cover the surface of each object without overlapping.

   a   a page of this book

   b   the top of your desk

   c   the seat of your chair

2   Record the number of cards for each area in this table.

Area	Number of cards
page	
desk top	
chair seat	

3   Write the areas in order.

________________________   ________________________   ________________________

smallest ⟶ largest

VOLUME

The **volume** of an object is the amount of *space* it takes up.

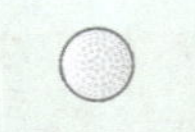

golf ball   tennis ball   softball   volleyball

The volleyball takes up the most space. It has the largest volume.

The golf ball takes up the least space. It has the smallest volume.

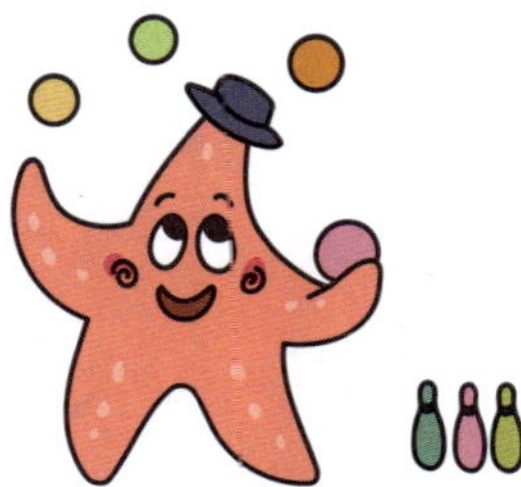

Exercise 13

Look at the balls on the previous page.

Use *smaller* or *larger* to complete each sentence:

a A softball is _________________ than a golf ball.

b A tennis ball is _________________ than a softball.

c A tennis ball is _________________ than a golf ball.

d A softball is _________________ than a volleyball.

Exercise 14

The smallest planet is _________________ .

Uranus is smaller than _________________ and _________________ .

Use *smaller* or *larger* to complete each sentence:

a Venus is _________________ than Neptune.

b Earth is _________________ than Mercury.

Activity

You will need: centicubes

What to do:

1 Your teacher will give you an object.

Arrange centicubes into a solid the same size and shape as the object.

2 Complete:

The volume of _________________ is about the same as _________ centicubes.

Discussion

How can we compare the volumes of objects with different shapes?

Hint: How could we use a fish tank?

Activity

Your teacher will prepare some groups of objects for you to look at.

What to do:

1 Look at the objects carefully.

Try to place each group of objects in order of volume.

Group A	**Group B**
largest ↑ __________	largest ↑ __________
__________	__________
__________	__________
smallest __________	smallest __________

2 Your class will now use a tank of water to compare the volumes of each group of objects.

Listen to your teacher for instructions.

The actual order of volumes is:

Group A	**Group B**
largest ↑ __________	largest ↑ __________
__________	__________
__________	__________
smallest __________	smallest __________

CAPACITY

> The **capacity** of a container is the amount of liquid it can hold.

Exercise 15

Look at these containers:

fishbowl

teaspoon

mug

The container with the smallest capacity is the ________________________.

The container with the largest capacity is the ________________________.

Exercise 16

Mrs Smith has a new fish tank.

She needs to fill it from the tap.

Will it be easier to move water using a teaspoon or a mug?

____

Why? ____

____

Activity

Look around your classroom.

List two containers which have greater capacity than a mug.

1 ________________________   2 ________________________

List two containers which have less capacity than a mug.

1 ________________________   2 ________________________

Activity

You will need: 4 different saucepans, a mug

What to do:

1 Label your saucepans A, B, C, and D.

2 *Guess* how many mugs of water it will take to fill each saucepan. Write your guesses in the table.

3 Use the mug to fill each saucepan. Record the *actual* number of mugs needed.

Saucepan	Guess	Actual
A		
B		
C		
D		

This bottle is **empty**.

It contains no liquid.

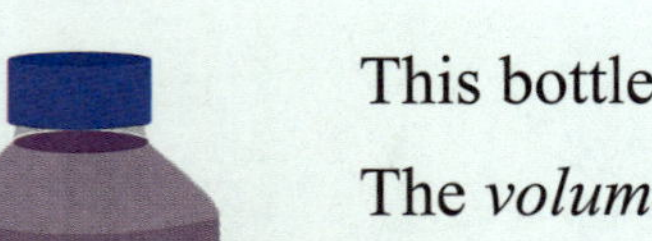

This bottle is **full**.

The *volume* of liquid it contains matches the *capacity* of the bottle.

Exercise 17

Match each bottle to the amount it contains.

| almost full | half full | empty | full | almost empty |

Exercise 18

Complete:

a Bottle _________ has greater capacity.

b Bottle _________ contains a greater volume of liquid.

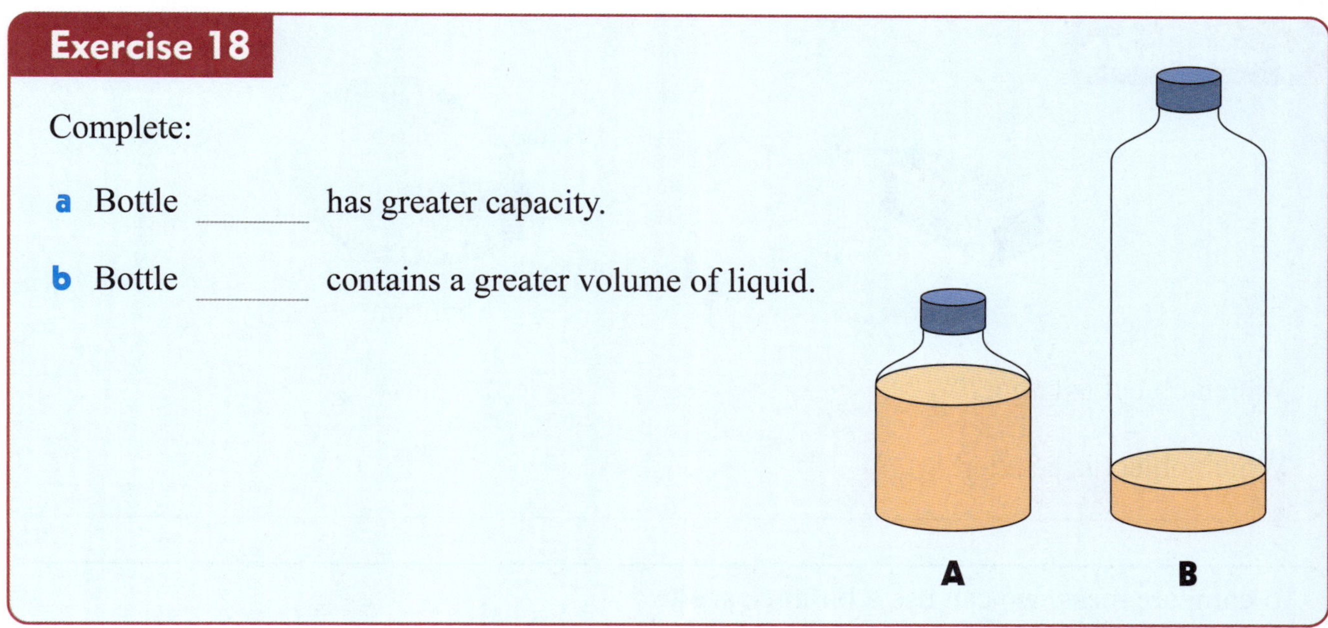

MASS

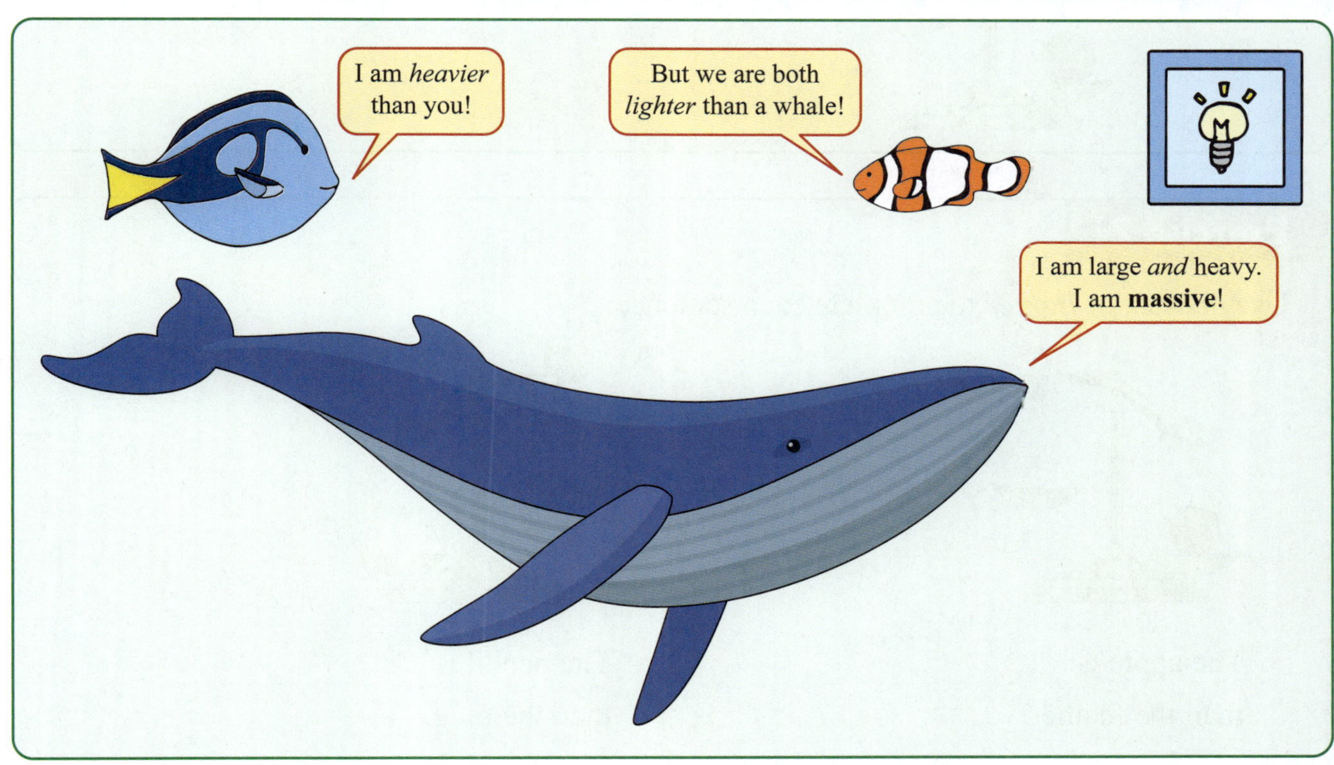

Exercise 19

Practise your writing.

light   heavy   massive

_________________   _________________   _________________

_________________   _________________   _________________

Exercise 20

brick

balloon

Which object is larger?   _______________________

Which object is heavier?   _______________________

To compare mass, we can use a **balance scale**.

The orange is heavier than the strawberry.

Exercise 21

Use *heavier* or *lighter* to complete each sentence:

a

The apple is _______________________
than the feather.

b

The pencil is _______________________
than the mug.

c

The carrot is _______________________
than the orange.

d

The baby is _______________________
than the kitten.

© HAESE MATHEMATICS 2024

Exercise 22

crisps

drink

sandwich

Write these lunch items in order of mass.

__________________   __________________   __________________

lightest ⟶ heaviest

We *weigh* an object to find its mass.

Exercise 23

Complete using *less than*, *more than*, or *the same as*.

a

The acorn weighs

____

2 blocks.

b

The pebble weighs

____

5 blocks.

c

The pinecone weighs

____

13 blocks.

d

The lime weighs

____

10 blocks.

Exercise 24

Write a sentence which *compares* the masses of the objects.

a

b

Activity

Henry has 3 boxes. He used a balance scale to compare their masses:

Which is the lightest box? ___________________

What does Henry need to do next, to be able to order the boxes from lightest to heaviest?

____

____

Activity

You will need: balance scale, centicubes, table tennis ball, pencil, apple, marker, mug

What to do:

1 Find how many cubes balance each item.

Write your answers in the table.

Item	*Mass*
table tennis ball	_______ cubes
pencil	_______ cubes
apple	_______ cubes
marker	_______ cubes
mug	_______ cubes

2 Write the objects in order:

heaviest _______________________

____

____

____

lightest _______________________

Puzzle

Sarah has 3 bags.

Bag 1 is lighter than bag 3.

Bag 2 is heavier than bag 3.

The lightest bag is _______________________ .

The heaviest bag is _______________________ .

TEMPERATURE

The **temperature** of an object is how hot or cold it is.

Exercise 25

Practise your writing.

cold   cool   warm   hot

colder ⟶ hotter

Exercise 26

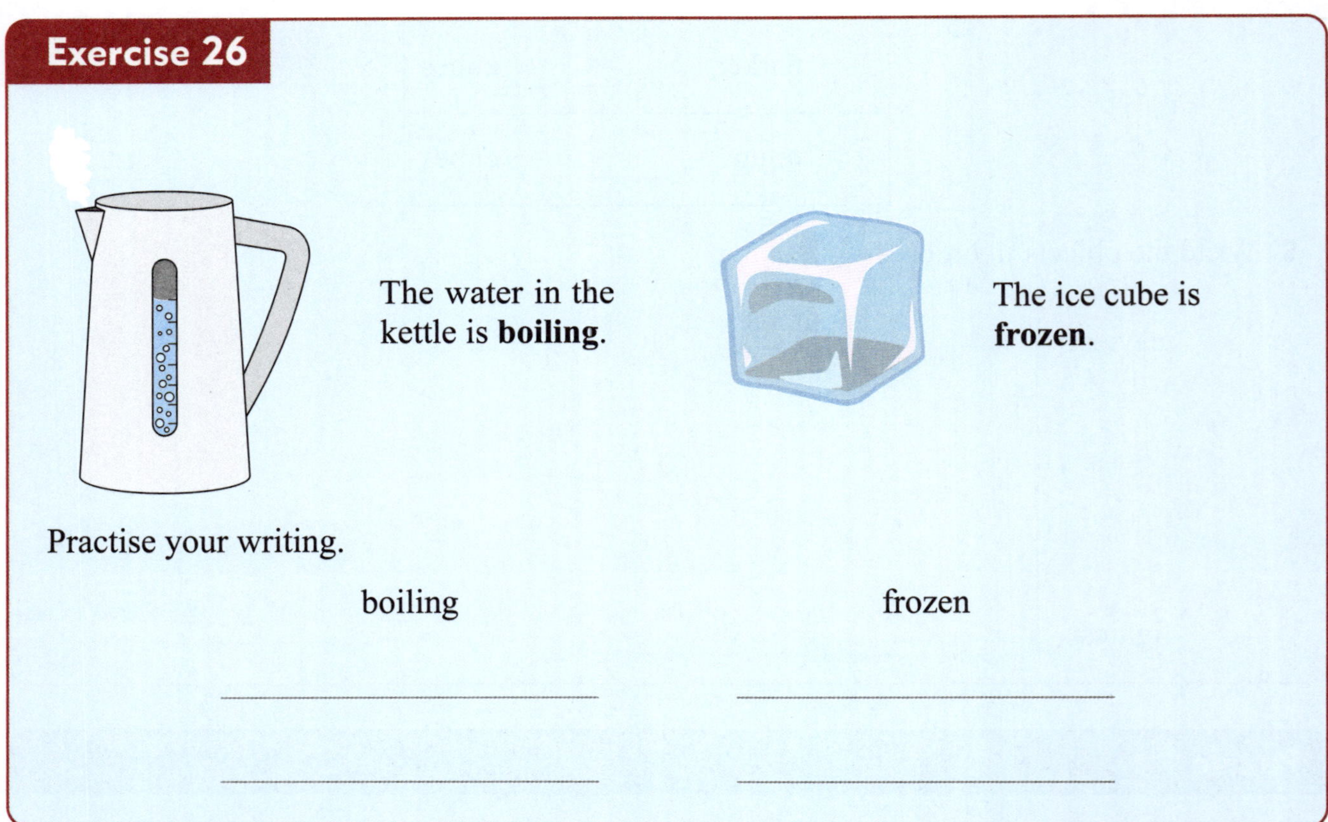

The water in the kettle is **boiling**.

The ice cube is **frozen**.

Practise your writing.

boiling   frozen

Discussion

What do we use to *measure* temperature? ____________________

CHAPTER 7: MULTIPLICATION

Lucy has picked 4 lots of 2 cherries.

In total, she has

$$2 + 2 + 2 + 2 = 8 \text{ cherries}$$

We can also write a **multiplication**:

$$4 \times 2 = 8$$

"Four times two equals eight."

"Four multiplied by two is eight."

Exercise 1

a 

Erin has _______ lots of 2 cherries.

In total, she has

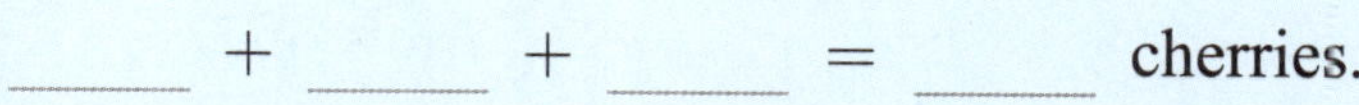

______ + ______ + ______ = ______ cherries.

As a multiplication, we write

______ $\times$ 2 = ______

b 

Adam has ______ lots of 3 bananas.

In total, he has

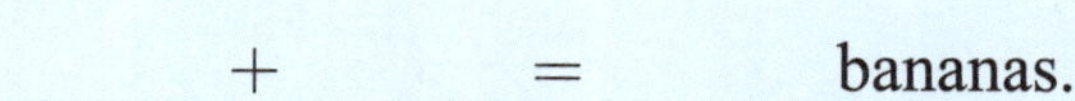

______ + ______ = ______ bananas.

As a multiplication, we write

______ $\times$ 3 = ______

c 

Jess has ______ lots of 2 cherries.

In total, she has

______ + ______ + ______ + ______ + ______ = ______ cherries.

As a multiplication, we write

______ $\times$ 2 = ______

Discussion

What words tell us that we need to multiply?

_______________________   _______________________   _______________________

_______________________   _______________________

Exercise 2

By counting legs, complete these multiplications.

1 lot of 2 =   1   ×   2   = ______   6 lots of 2 = ______ × ______ = ______

2 lots of 2 =   2   ×   2   = ______   7 lots of 2 = ______ × ______ = ______

3 lots of 2 = ______ ×   2   = ______   8 lots of 2 = ______ × ______ = ______

4 lots of 2 = ______ × ______ = ______   9 lots of 2 = ______ × ______ = ______

5 lots of 2 = ______ × ______ = ______   10 lots of 2 = ______ × ______ = ______

Exercise 3

By counting cherries, complete these multiplications.

1 lot of 3 =   1   ×   3   = ______   6 lots of 3 = ______ × ______ = ______

2 lots of 3 =   2   ×   3   = ______   7 lots of 3 = ______ × ______ = ______

3 lots of 3 = ______ ×   3   = ______   8 lots of 3 = ______ × ______ = ______

4 lots of 3 = ______ × ______ = ______   9 lots of 3 = ______ × ______ = ______

5 lots of 3 = ______ × ______ = ______   10 lots of 3 = ______ × ______ = ______

Exercise 4

By counting chair legs, complete these multiplications.

1 lot of 4 =   1   ×   4   = _____   6 lots of 4 = _____ × _____ = _____

2 lots of 4 =   2   ×   4   = _____   7 lots of 4 = _____ × _____ = _____

3 lots of 4 = _____ ×   4   = _____   8 lots of 4 = _____ × _____ = _____

4 lots of 4 = _____ × _____ = _____   9 lots of 4 = _____ × _____ = _____

5 lots of 4 = _____ × _____ = _____   10 lots of 4 = _____ × _____ = _____

Exercise 5

By counting fingers, complete these multiplications.

1 lot of 5 =   1   ×   5   = _____   6 lots of 5 = _____ × _____ = _____

2 lots of 5 =   2   ×   5   = _____   7 lots of 5 = _____ × _____ = _____

3 lots of 5 = _____ ×   5   = _____   8 lots of 5 = _____ × _____ = _____

4 lots of 5 = _____ × _____ = _____   9 lots of 5 = _____ × _____ = _____

5 lots of 5 = _____ × _____ = _____   10 lots of 5 = _____ × _____ = _____

Discussion

What do you notice about the results in the 5 times table?

____

Exercise 6

1	2	3	4	5	6	7	8	9	10
11	12	13	14	15	16	17	18	19	20
21	22	23	24	25	26	27	28	29	30
31	32	33	34	35	36	37	38	39	40
41	42	43	44	45	46	47	48	49	50
51	52	53	54	55	56	57	58	59	60

a Start at zero and add 6 each time.

$0 \rightarrow 6 \rightarrow$ _______ $\rightarrow$ _______ $\rightarrow$ _______ $\rightarrow$ _______ $\rightarrow$ _______

$\rightarrow$ _______ $\rightarrow$ _______ $\rightarrow$ _______ $\rightarrow$ _______

b Complete your 6 times table:

$1 \times 6 =$ _______ $6 \times 6 =$ _______

$2 \times 6 =$ _______ $7 \times 6 =$ _______

$3 \times 6 =$ _______ $8 \times 6 =$ _______

$4 \times 6 =$ _______ $9 \times 6 =$ _______

$5 \times 6 =$ _______ $10 \times 6 =$ _______

Exercise 7

The number 60 means 6 tens or 6×10.

$6 \times 10 = 60$

Complete your 10 times table:

$1 \times 10 =$ _______ $6 \times 10 =$ _______

$2 \times 10 =$ _______ $7 \times 10 =$ _______

$3 \times 10 =$ _______ $8 \times 10 =$ _______

$4 \times 10 =$ _______ $9 \times 10 =$ _______

$5 \times 10 =$ _______ $10 \times 10 =$ _______

Discussion

What do you notice about the results in the 10 times table?

____

Click the icon to practise your **times tables**.

The results of the times tables are called **multiples**.

Exercise 8

The multiples of 2 are:

2, 4, 6, _______, _______, _______, _______, _______, _______, _______, ...

The multiples of 3 are:

3, 6, 9, _______, _______, _______, _______, _______, _______, _______, ...

The multiples of 4 are:

4, 8, _______, _______, _______, _______, _______, _______, _______, ...

The multiples of 5 are:

5, 10, _______, _______, _______, _______, _______, _______, _______, ...

The multiples of 6 are:

6, 12, _______, _______, _______, _______, _______, _______, _______, ...

The multiples of 10 are:

10, 20, _______, _______, _______, _______, _______, _______, _______, ...

Exercise 9

a List the multiples of 2 on the snake by adding 2 each time.

b Colour the multiples of 2 on the chart.

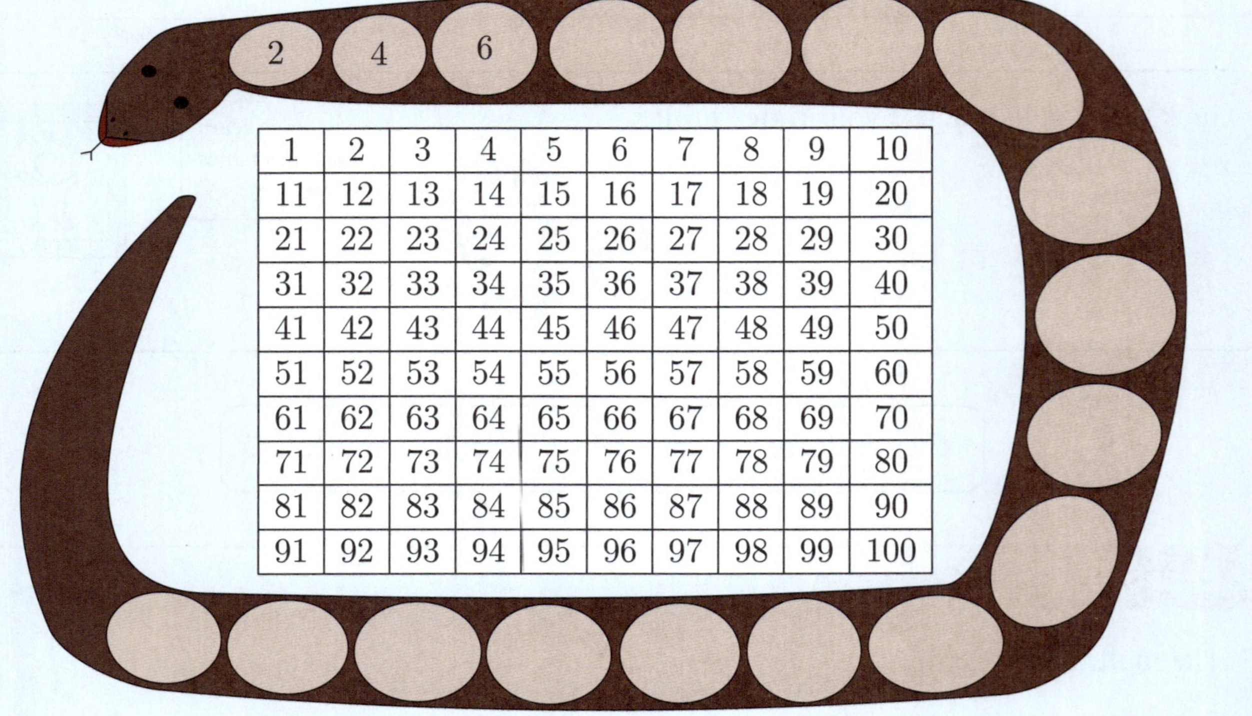

1	2	3	4	5	6	7	8	9	10
11	12	13	14	15	16	17	18	19	20
21	22	23	24	25	26	27	28	29	30
31	32	33	34	35	36	37	38	39	40
41	42	43	44	45	46	47	48	49	50
51	52	53	54	55	56	57	58	59	60
61	62	63	64	65	66	67	68	69	70
71	72	73	74	75	76	77	78	79	80
81	82	83	84	85	86	87	88	89	90
91	92	93	94	95	96	97	98	99	100

Exercise 10

a List the multiples of 3 on the snake by adding 3 each time.

b Colour the multiples of 3 on the chart.

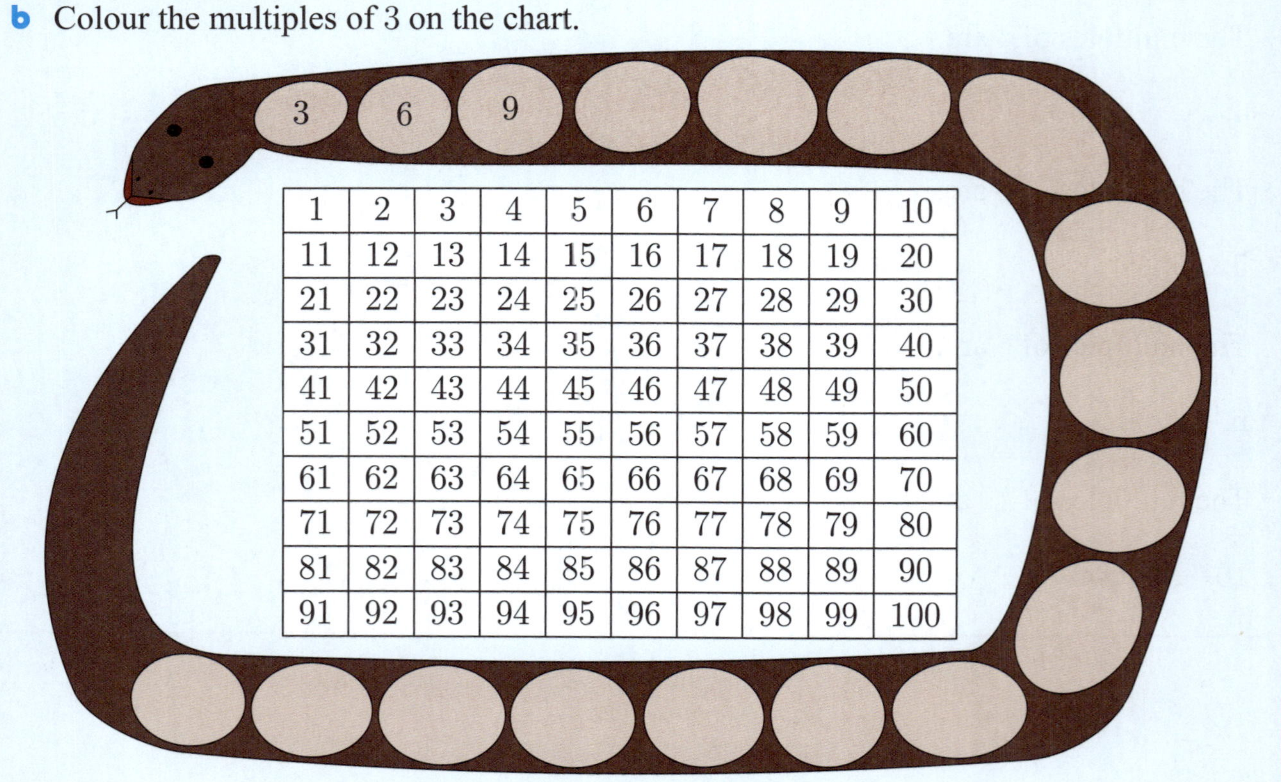

1	2	3	4	5	6	7	8	9	10
11	12	13	14	15	16	17	18	19	20
21	22	23	24	25	26	27	28	29	30
31	32	33	34	35	36	37	38	39	40
41	42	43	44	45	46	47	48	49	50
51	52	53	54	55	56	57	58	59	60
61	62	63	64	65	66	67	68	69	70
71	72	73	74	75	76	77	78	79	80
81	82	83	84	85	86	87	88	89	90
91	92	93	94	95	96	97	98	99	100

We can use an **array** to help us find a multiplication.

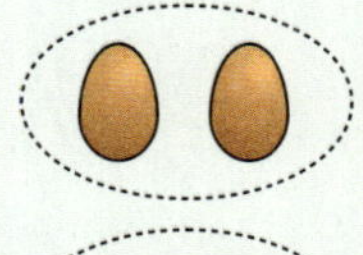
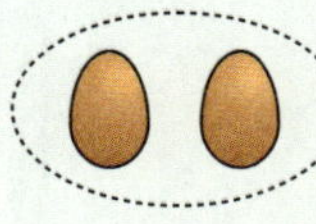

$$2 + 2 + 2 = 6$$
$$3 \text{ lots of } 2 = 6$$
$$3 \times 2 = 6$$

Exercise 11

a

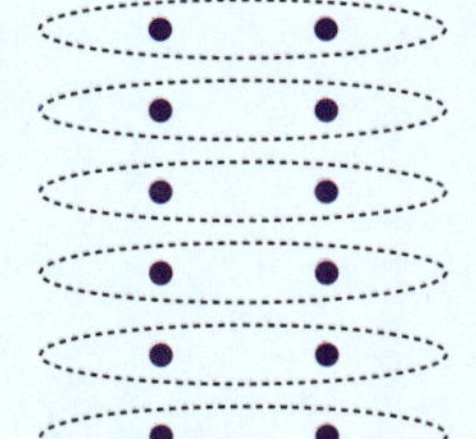

$$2 \quad + \quad 2 \quad + \quad 2 \quad + \underline{\quad} + \underline{\quad} + \underline{\quad} = \underline{\quad}$$
$$\underline{\quad} \text{ lots of } \quad 2 \quad = \underline{\quad}$$
$$\underline{\quad} \times \quad 2 \quad = \underline{\quad}$$

b

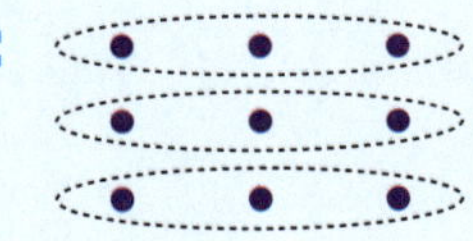

$$\underline{\quad} + \underline{\quad} = \underline{\quad}$$
$$\underline{\quad} \text{ lots of } \underline{\quad} = \underline{\quad}$$
$$\underline{\quad} \times \quad 3 \quad = \underline{\quad}$$

c

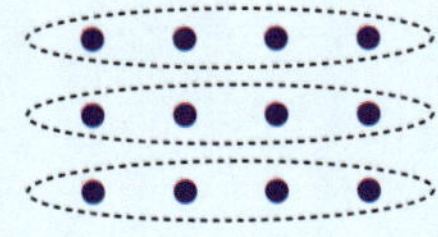

$$\underline{\quad} + \underline{\quad} + \underline{\quad} = \underline{\quad}$$
$$\underline{\quad} \text{ lots of } \underline{\quad} = \underline{\quad}$$
$$\underline{\quad} \times \underline{\quad} = \underline{\quad}$$

d

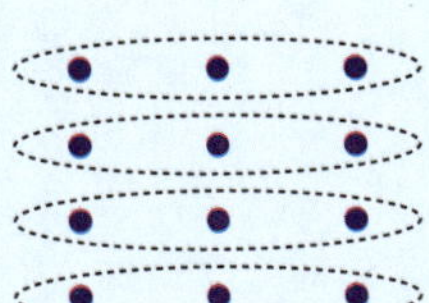

$$\underline{\quad} + \underline{\quad} + \underline{\quad} = \underline{\quad}$$
$$\underline{\quad} \text{ lots of } \underline{\quad} = \underline{\quad}$$
$$\underline{\quad} \times \underline{\quad} = \underline{\quad}$$

e

$$\underline{\quad} + \underline{\quad} + \underline{\quad} + \underline{\quad} = \underline{\quad}$$
$$\underline{\quad} \text{ lots of } \underline{\quad} = \underline{\quad}$$
$$\underline{\quad} \times \underline{\quad} = \underline{\quad}$$

Exercise 12

Draw an array to help answer each multiplication:

a

$2 + 2 + 2 + 2 =$ _______

$4 \times 2 =$ _______

b

$3 + 3 =$ _______

$2 \times 3 =$ _______

c

$4 + 4 + 4 + 4 + 4 =$ _______

$5 \times 4 =$ _______

d

$5 + 5 + 5 + 5 + 5 + 5 =$ _______

$6 \times 5 =$ _______

Exercise 13

Write each addition as a multiplication, and find the answer:

a $2 + 2 + 2 + 2 + 2 + 2 + 2 = $ _______ $\times$ _______ $= $ _______

b $3 + 3 + 3 + 3 = $ _______ $\times$ _______ $= $ _______

c $5 + 5 + 5 = $ _______ $\times$ _______ $= $ _______

d $3 + 3 + 3 + 3 + 3 + 3 = $ _______ $\times$ _______ $= $ _______

e $5 + 5 + 5 + 5 + 5 = $ _______ $\times$ _______ $= $ _______

f $6 + 6 + 6 = $ _______ $\times$ _______ $= $ _______

g $10 + 10 + 10 + 10 + 10 + 10 + 10 = $ _______ $\times$ _______ $= $ _______

h $4 + 4 + 4 + 4 + 4 + 4 = $ _______ $\times$ _______ $= $ _______

Exercise 14

Complete these times tables.

$1 \times 1 = $ _____	$1 \times 2 = $ _____	$1 \times 3 = $ _____	$1 \times 4 = $ _____
$2 \times 1 = $ _____	$2 \times 2 = $ _____	$2 \times 3 = $ _____	$2 \times 4 = $ _____
$3 \times 1 = $ _____	$3 \times 2 = $ _____	$3 \times 3 = $ _____	$3 \times 4 = $ _____
$4 \times 1 = $ _____	$4 \times 2 = $ _____	$4 \times 3 = $ _____	$4 \times 4 = $ _____
$5 \times 1 = $ _____	$5 \times 2 = $ _____	$5 \times 3 = $ _____	$5 \times 4 = $ _____
$6 \times 1 = $ _____	$6 \times 2 = $ _____	$6 \times 3 = $ _____	$6 \times 4 = $ _____
$7 \times 1 = $ _____	$7 \times 2 = $ _____	$7 \times 3 = $ _____	$7 \times 4 = $ _____
$8 \times 1 = $ _____	$8 \times 2 = $ _____	$8 \times 3 = $ _____	$8 \times 4 = $ _____
$9 \times 1 = $ _____	$9 \times 2 = $ _____	$9 \times 3 = $ _____	$9 \times 4 = $ _____
$10 \times 1 = $ _____	$10 \times 2 = $ _____	$10 \times 3 = $ _____	$10 \times 4 = $ _____

Exercise 15

Complete these times tables.

$1 \times 5 =$ _____	$1 \times 6 =$ _____	$1 \times 10 =$ _____
$2 \times 5 =$ _____	$2 \times 6 =$ _____	$2 \times 10 =$ _____
$3 \times 5 =$ _____	$3 \times 6 =$ _____	$3 \times 10 =$ _____
$4 \times 5 =$ _____	$4 \times 6 =$ _____	$4 \times 10 =$ _____
$5 \times 5 =$ _____	$5 \times 6 =$ _____	$5 \times 10 =$ _____
$6 \times 5 =$ _____	$6 \times 6 =$ _____	$6 \times 10 =$ _____
$7 \times 5 =$ _____	$7 \times 6 =$ _____	$7 \times 10 =$ _____
$8 \times 5 =$ _____	$8 \times 6 =$ _____	$8 \times 10 =$ _____
$9 \times 5 =$ _____	$9 \times 6 =$ _____	$9 \times 10 =$ _____
$10 \times 5 =$ _____	$10 \times 6 =$ _____	$10 \times 10 =$ _____

Exercise 16

a Sarah has 4 boxes with 2 lightbulbs in each.

In total she has _____ $\times$ _____ = _____ lightbulbs.

b Riley has 7 layers with 5 bricks in each.

In total he has _____ $\times$ _____ = _____ bricks.

c Philip has 8 ladybirds with 3 dots on each.

In total there are _____ $\times$ _____ = _____ dots.

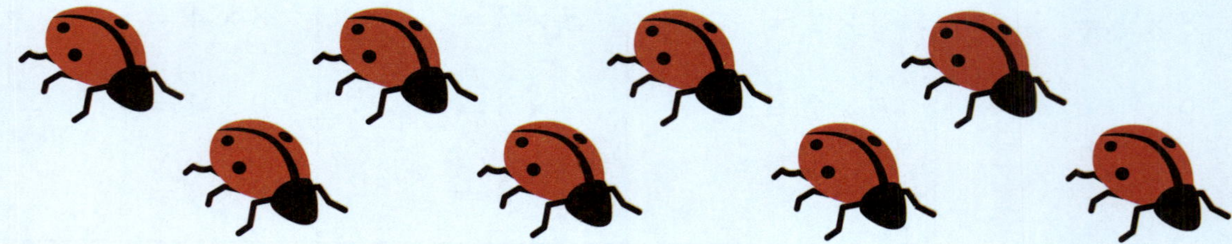

Activity

What multiplications can you write for 12 eggs in a carton?

Illustrate each answer.

1   ×   12   = _______

2   × _______ = _______

_______ × _______ = _______

_______ × _______ = _______

_______ × _______ = _______

_______ × _______ = _______

Game

This is a game for 2 players.

You will need: 2 dice, 8 red counters, 8 blue counters.

Setting up:

- Give one player the red counters, and the other the blue counters.
- Both players will play on the same grid. Decide whose grid will be used.
- Decide who will start the game.

Playing the game:

- Players take turns.
- On your turn:

Roll the 2 dice.

Multiply the numbers together.

$$3 \times 5 = 15$$

If you can see the result on the grid, place one of your counters over it. This square cannot be used again.

6	20	2	30	4	16
25	5	15	10	18	1
3	12	8	36	9	24

- Keep playing until one player has used all of their counters.

That player is the winner!

CHAPTER 8: DATA HANDLING

When we look at the world around us, we often **count** things.
The information we collect is called **data**.

Exercise 1

When Benita went for a hike, she counted the animals she saw.

a How many foxes did Benita see? ....

b Which animal did Benita see the most of?

____

c How many more mice than deer did Benita see? ....

Animal	Number
squirrel	5
fox	2
mouse	3
deer	1

Exercise 2

John's vegetable garden is very neat.

a John is growing:

☐ cabbages,

☐ carrots,

and ☐ onions.

b John is growing ☐ more carrots than cabbages.

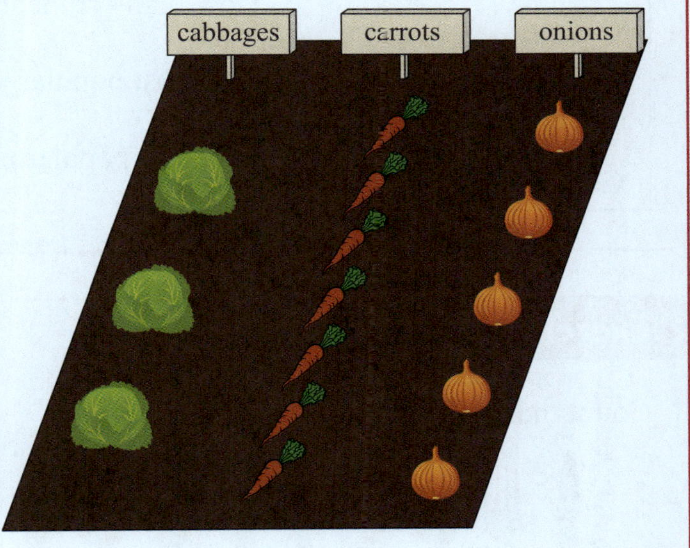

Discussion

The vegetables in John's garden are easy to count because they are planted neatly in lines.
How can we count objects which are all mixed together?

Helena counts the books on her mother's shelf according to their type.

She uses a **tally**.

She draws one **tally mark** for each item.

| represents 1 item.

|||| represents 5 items.

Book type	Tally
art	|||
cooking	|||||
craft	||||| ||
history	||

Helena counted:

3 art books

5 cooking books

7 craft books

2 history books

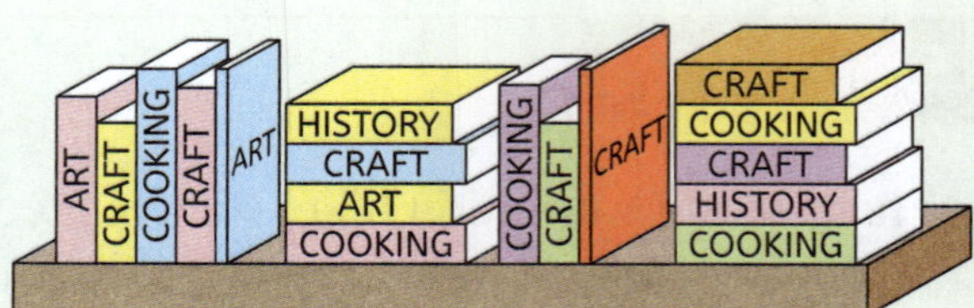

Exercise 3

Mrs Henry asked her students "Where should we go for our excursion?"

She gave them 5 places to choose from.

Place	Tally
Beach	||||| ||
Museum	||||
Theatre	||||| |||
Park	|||||
Gallery	|||

a How many students wanted to go to the museum? _______

b How many students wanted to go to the beach? _______

c The most popular place was the _____________________.

d The least popular place was the _____________________.

Exercise 4

Count these items using a tally.

Item	Tally
fork	
knife	
spoon	

There are [ ] forks, [ ] knives, and [ ] spoons.

There are [ ] items in total.

Exercise 5

Count these items using a tally.

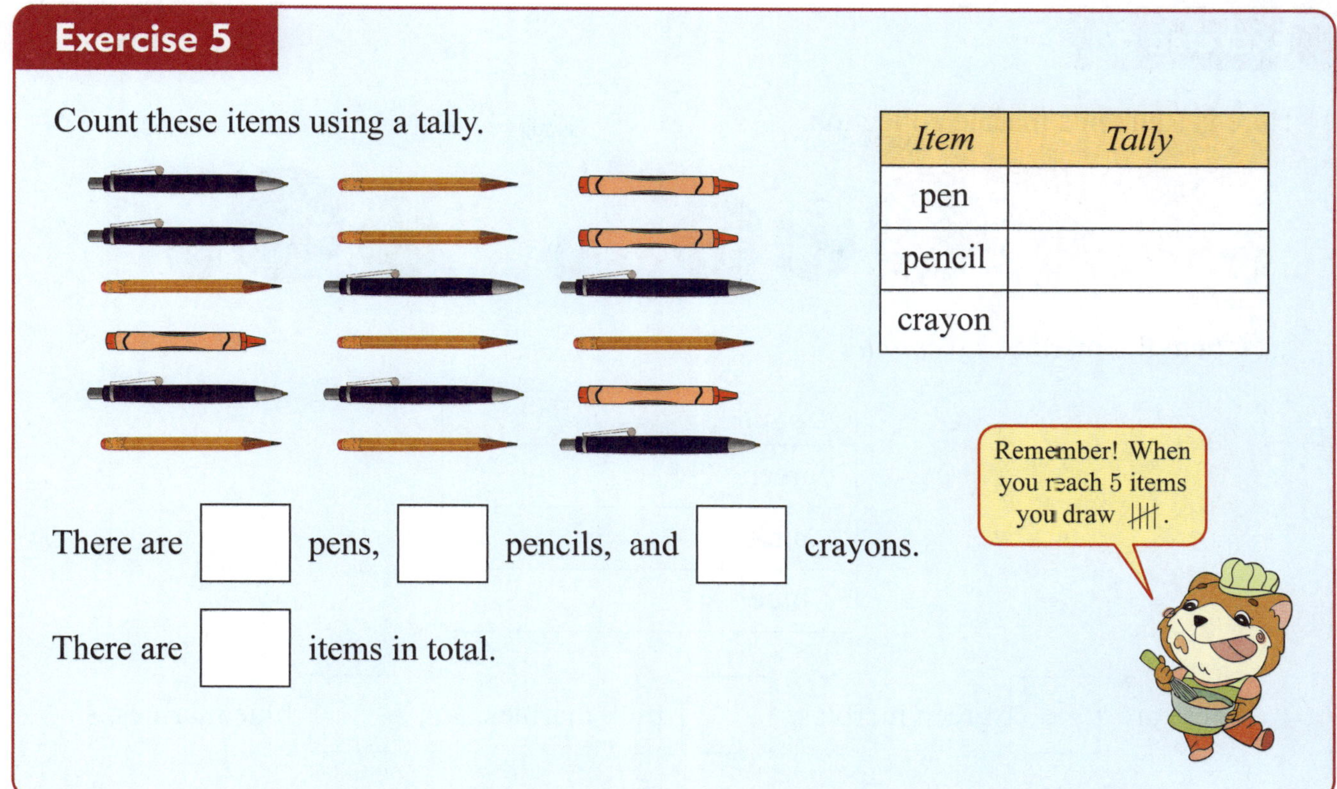

Item	Tally
pen	
pencil	
crayon	

There are ☐ pens, ☐ pencils, and ☐ crayons.

There are ☐ items in total.

Look at the magnets on the fridge.

Luca counts the magnets by *colour*.

Colour	Tally
blue	‖‖‖	
red	||
yellow	||||

Luca counted:

    6 blue magnets

    2 red magnets

    4 yellow magnets

Grace counts the magnets by *shape*.

Shape	Tally
square	|||
triangle	||||
circle	‖‖‖

Grace counted:

    3 square magnets

    4 triangle magnets

    5 circle magnets

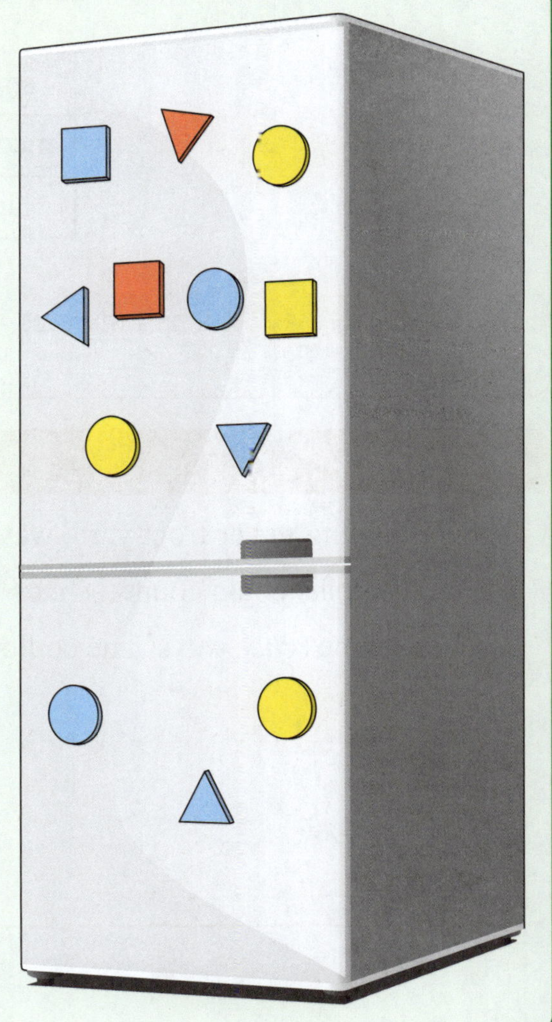

Exercise 6

Here is Reuben's marble collection.

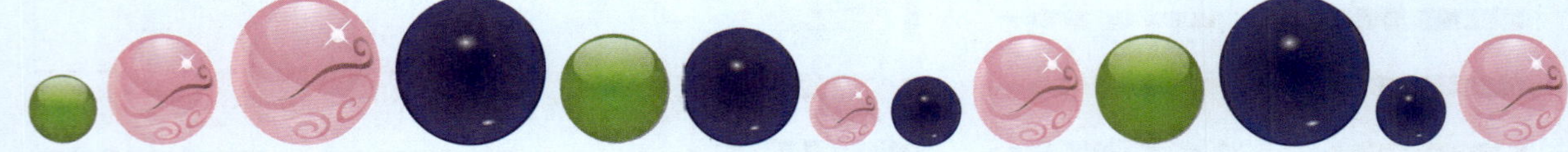

a Count the marbles by *colour*.

Colour	Tally
green	
pink	
blue	

There are ☐ green marbles, ☐ pink marbles, and ☐ blue marbles.

b Count the marbles by *size*.

Size	Tally
small	
medium	
large	

There are ☐ small marbles, ☐ medium marbles, and ☐ large marbles.

Discussion

Jane is standing in her front yard, watching cars as they go past her house.

Jane realises she could count the cars by their *colour*.

What are some *other* ways Jane could count the cars?

____

____

____

Group Activity

You will need:

- a printout of a horse from your teacher
- coloured pencils including black, grey, brown, and tan.

What to do:

1   Choose *one* colour you would like your horse to be.

   Colour your horse in that colour.

2   Colour the rest of the picture as you wish.

3   Give your picture to your teacher.

4   Your teacher will show the pictures to the class one at a time.

   Count the horse colours using a tally:

Colour	Tally
black	
grey	
brown	
palomino	

5   The most popular colour chosen was ___________________ .

6   The least popular colour chosen was ___________________ .

Activity

In this Activity you will count skydiving animals!

What to do:

1   Click the icon, and press the WATCH button.

2   Count the animals you see using a tally.

   I counted [ ] cows, [ ] bears,

   [ ] moose, and [ ] elephants.

Animal	Tally
cow	
bear	
moose	
elephant	

Exercise 7

Leon counted the number of eggs his chickens laid this week.

He drew a **pictograph** to show his data.

Number of eggs

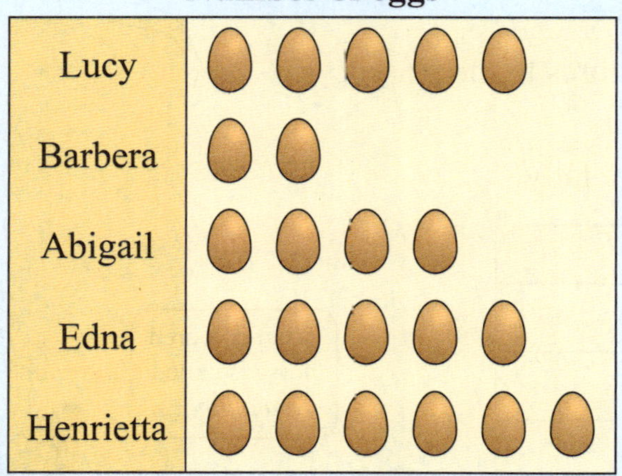

a Which chicken laid the most eggs? _______________________

b How many eggs did Abigail lay? _________

c Which chicken laid the fewest eggs? _______________________

d Which chickens laid the same number of eggs? _______________________

e How many more eggs did Edna lay than Barbera? _________

f In total, how many eggs did Leon's chickens lay this week? _________

Discussion

What features of the pictograph above help us understand what it means?

____

____

____

Exercise 8

Harry counted the apples in a fruit bowl.

Draw a pictograph to show this information.

Use 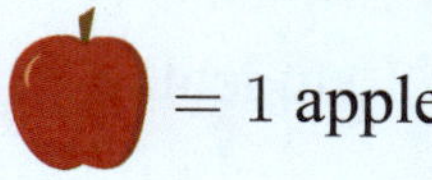 = 1 apple

Apples

Gala	3
Red Delicious	6
Fuji	2
Pink Lady	5

Number of apples

Gala	
Red Delicious	
Fuji	
Pink Lady	

Exercise 9

Evelyn played a car racing game with 3 friends. She recorded her result for each race in a tally.

Draw a pictograph to show this information.

Use 🚗 = 1 race

Result	*Tally*
first	IIII I
second	IIII
third	IIII II
fourth	II

Race results

first	
second	
third	
fourth	

Group Activity

What is your favourite season?

In this Activity you will create a "living pictograph" with the other students in your class!

You should then draw a pictograph to show the results, using $\big\uparrow$ = 1 student.

Favourite season

Summer	
Autumn or Fall	
Winter	
Spring	

The most popular season is ________________ .

The least popular season is ________________ .

Nick has 12 cherries. He wants to share them **equally** with Harry.

Nick's bowl

Harry's bowl

Each person will have **6** cherries.

Nick has **divided** the cherries **equally** into 2 groups.

Discussion

What do we mean by sharing *equally*?

Exercise 1

Describe the division:

a

6 has been divided into _______ equal groups of _______ .

b

_______ has been divided into _______ equal groups of _______ .

c

_______ has been divided into _______ equal groups of _______ .

d

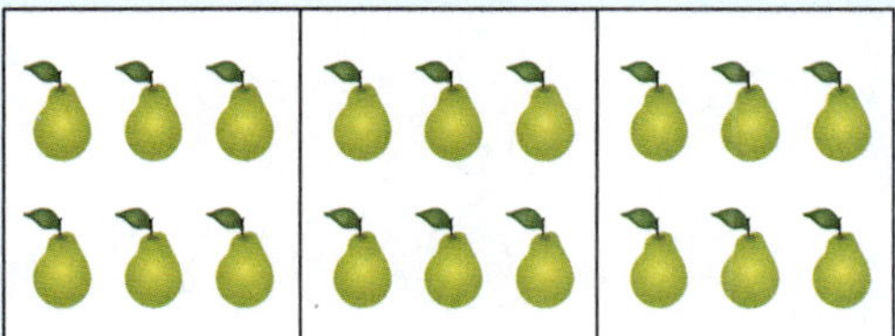

_______ has been divided into _______ equal groups of _______ .

Exercise 2

Describe the division:

__________ has been __________________ into

__________ __________ groups of _______ .

Exercise 3

Use counters to share the amounts, then draw your answer.

a Share 2 into 2 equal groups.

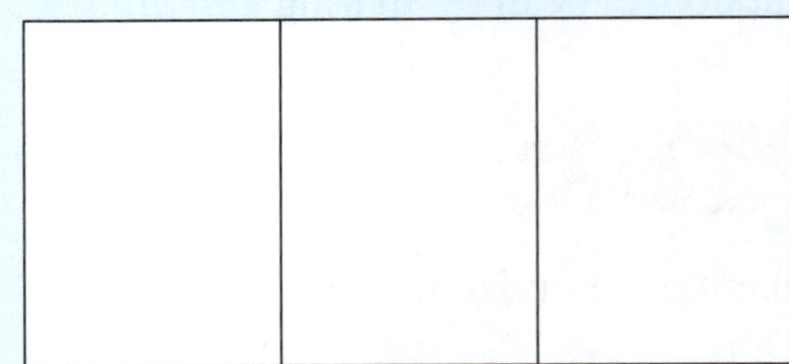

b Share 6 into 3 equal groups.

c Share 8 into 2 equal groups.

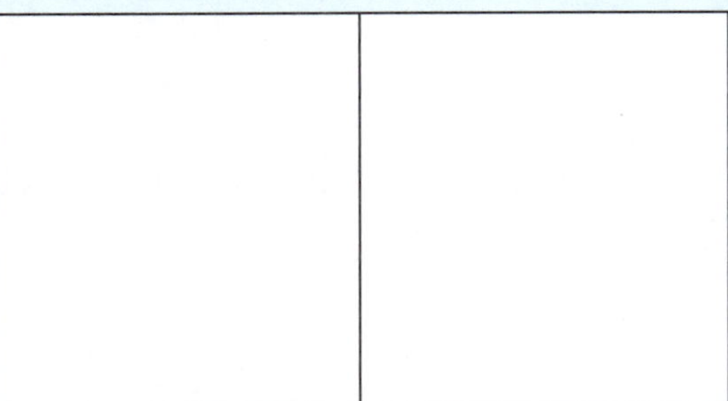

d Share 9 into 3 equal groups.

e Share 15 into 3 equal groups.

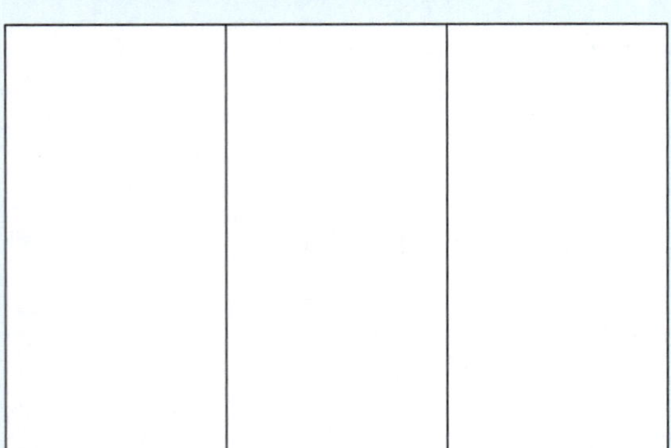

f Share 14 into 2 equal groups.

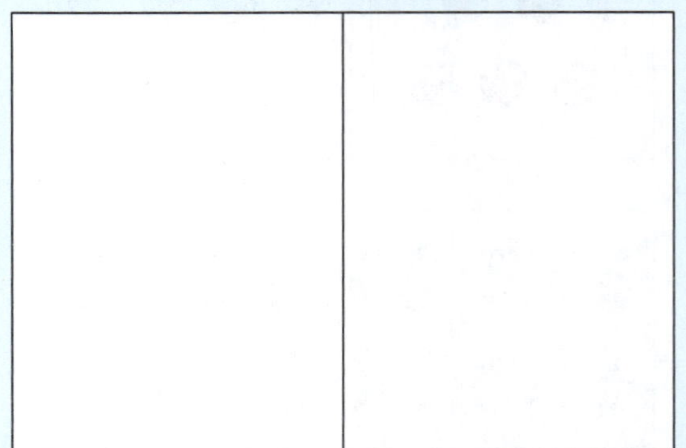

Exercise 4

In how many ways can you divide 12 equally?

12 can be divided into ....

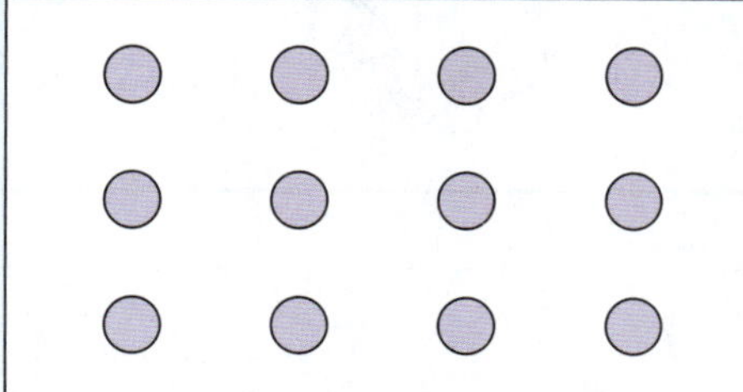

1 group of _______ .

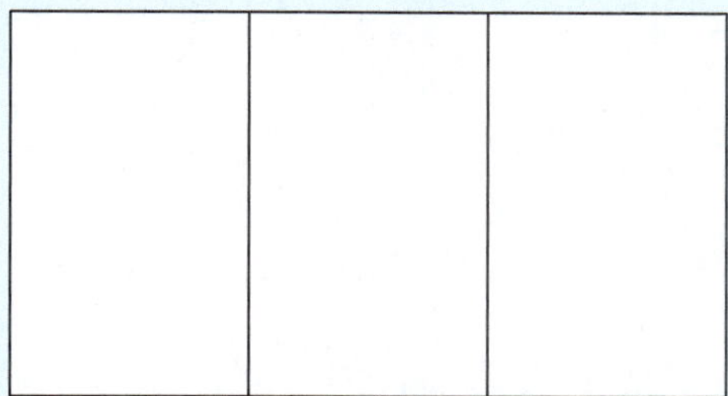

_______ equal groups of _______ .

_______ equal groups of _______ .

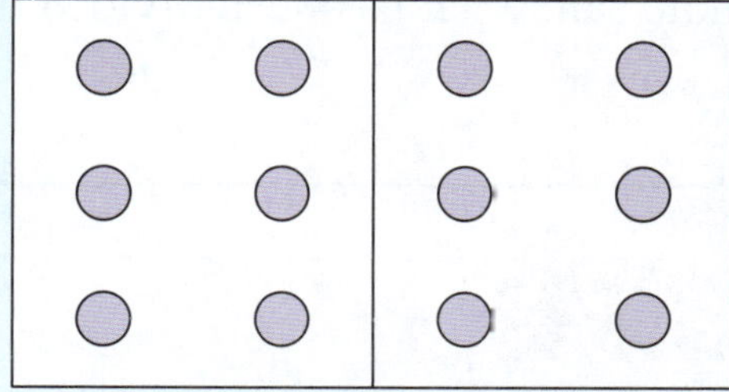

2 equal groups of _______ .

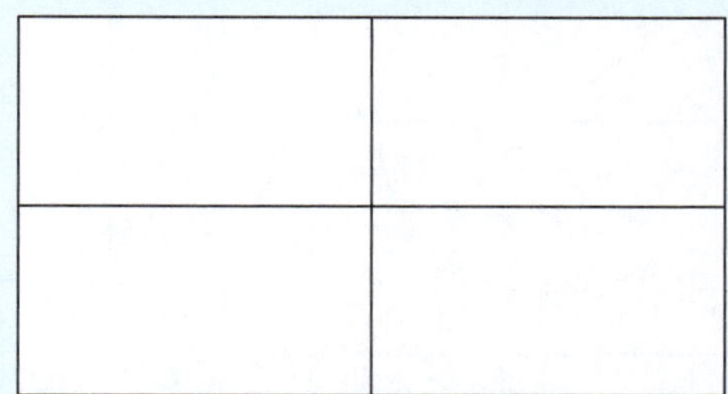

_______ equal groups of _______ .

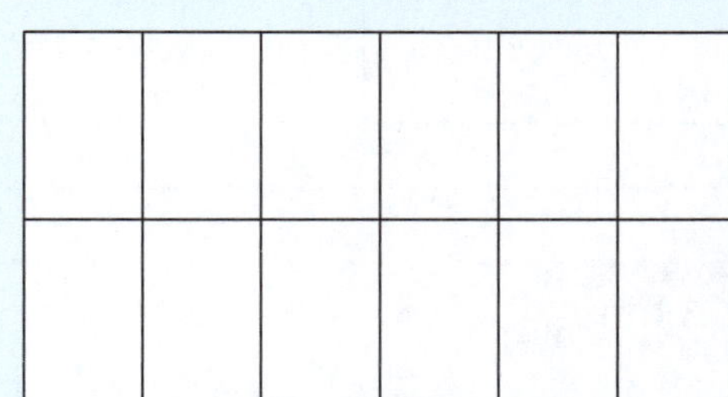

_______ equal groups of _______ .

Exercise 5

Draw a diagram to show each division:

a Divide 8 into 4 equal groups.

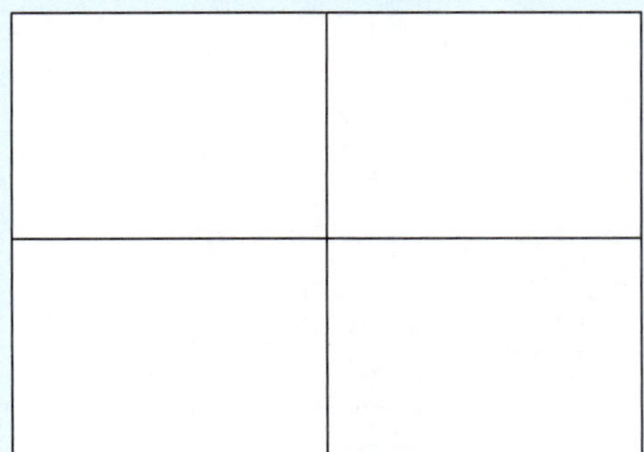

b Divide 15 into 5 equal groups.

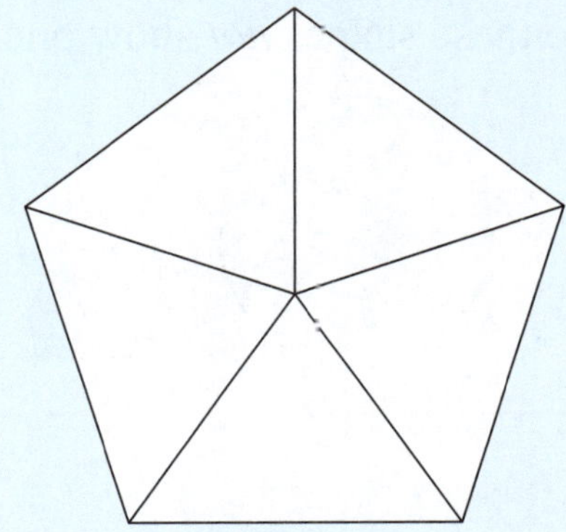

Jack cut his sandwich into 2 equal sized pieces.

He has **divided** his sandwich into 2.

Each piece is **one half** of the sandwich.

The whole sandwich has been divided into 2 halves.

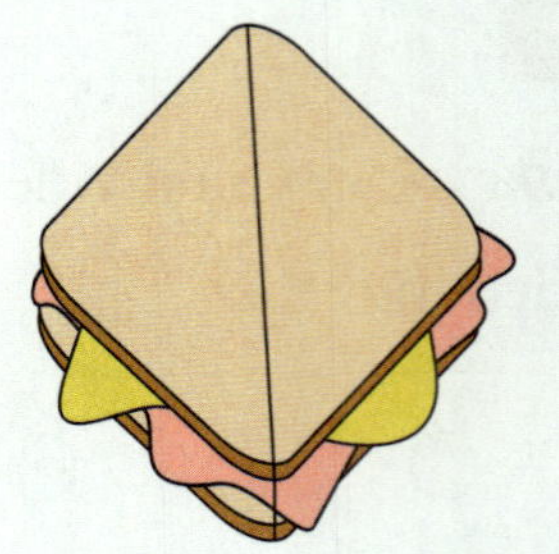

Exercise 6

Shade one half of each shape:

a
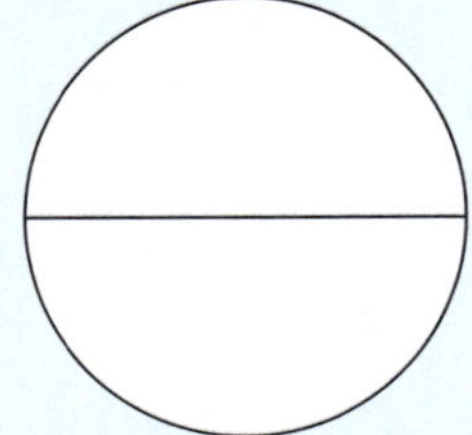

b
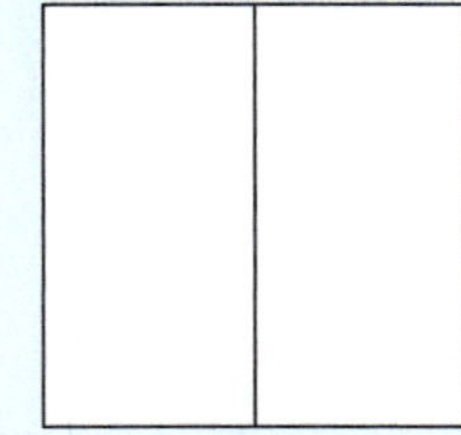

c
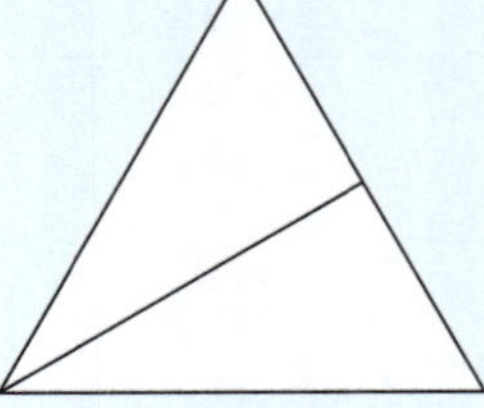

d
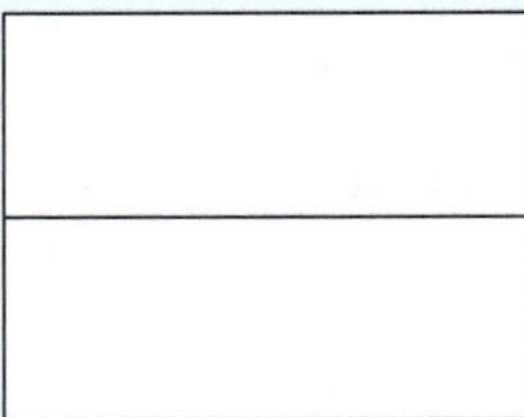

e
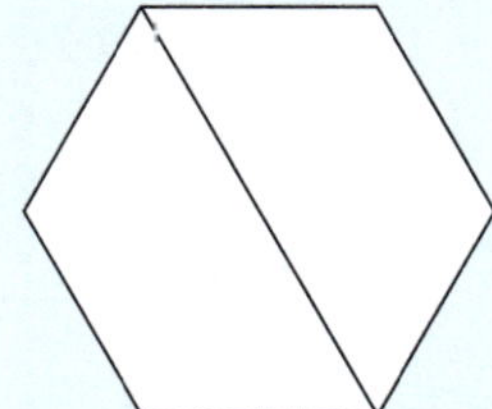

f
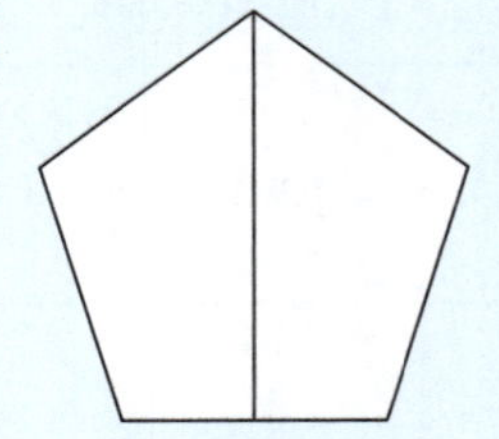

Exercise 7

Complete:

To divide a shape into *halves*, we divide it into ________ ________________ pieces.

There are ______ halves in one whole.

Discussion

Why do these shapes *not* show one half?

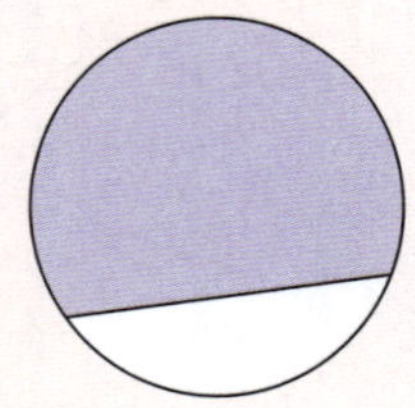
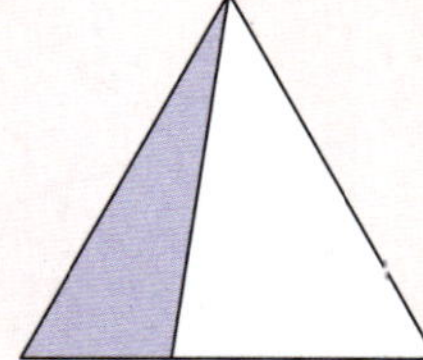
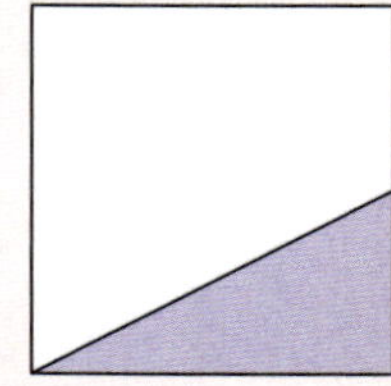
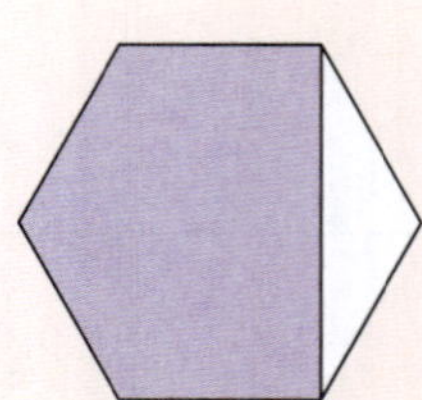

Exercise 8

Shade one half of each shape:

a   b   c

To find **one half** of a number, we divide it into **two equal groups**.

One half of 14 is 7.

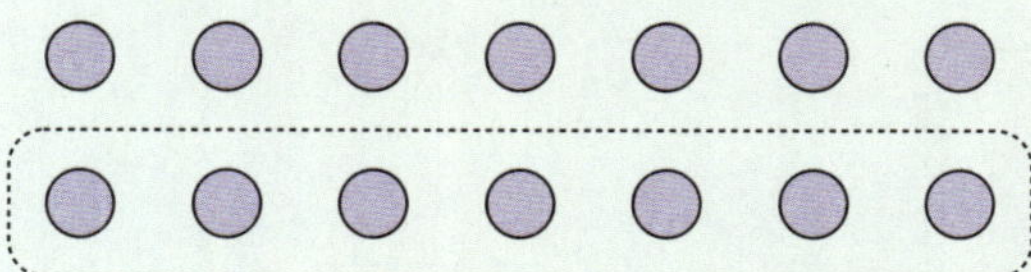

Exercise 9

Circle one half of the objects.

a 

One half of _________ is _________ .

b

One half of _________ is _________ .

c

One half of _________ is _________ .

d

One half of _________ is _________ .

Exercise 10

Dion has 16 lollies. He gives one half of them to Meg.

How many lollies does Dion give to Meg?

Dion gives _________ lollies to Meg.

Ella cut her melon into 4 equal pieces.

She has **divided** her melon into 4.

Each piece is **one quarter** of the melon.

The whole melon has been divided into 4 quarters.

Exercise 11

Shade the correct amount.

a one quarter

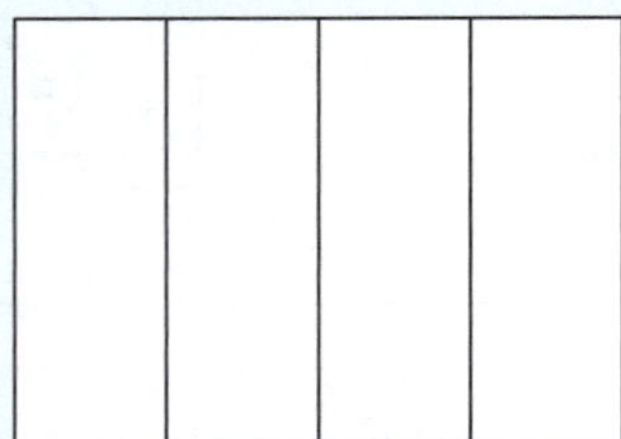

b one quarter

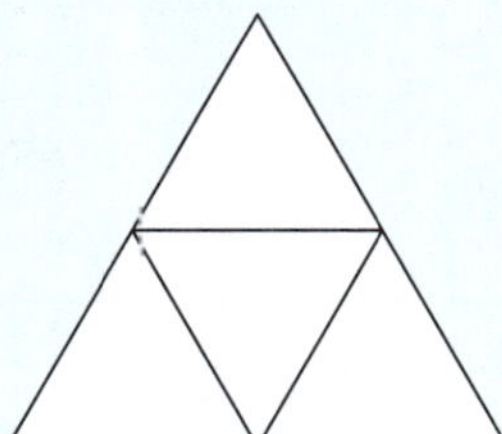

c two quarters

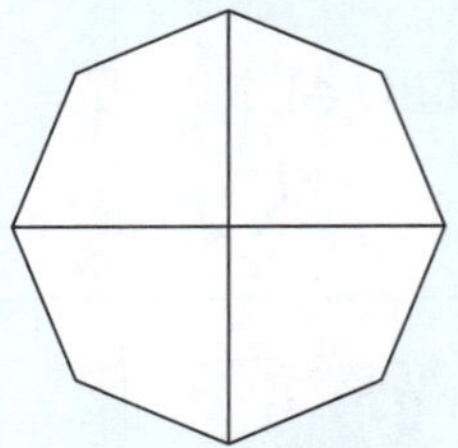

d three quarters

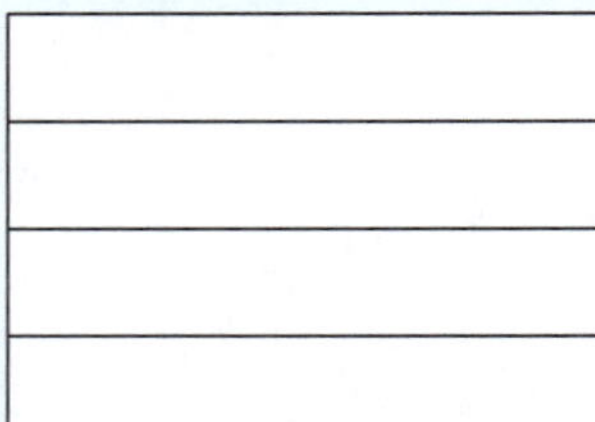

e one quarter

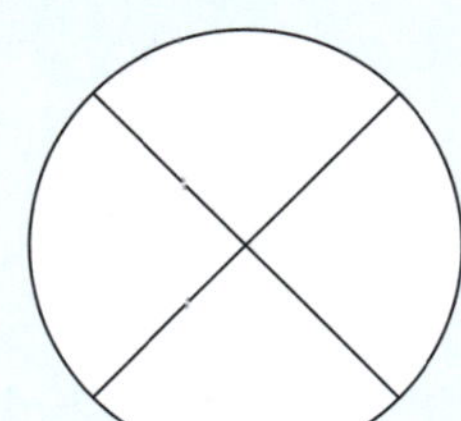

f four quarters

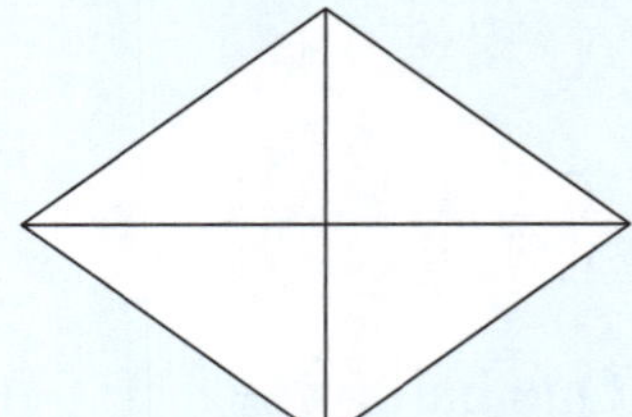

Exercise 12

Complete:

To divide a shape into *quarters* we divide it into _______ _______________ pieces.

There are _______ quarters in one whole.

This figure does *not* show one quarter because the pieces are _______ _______________ .

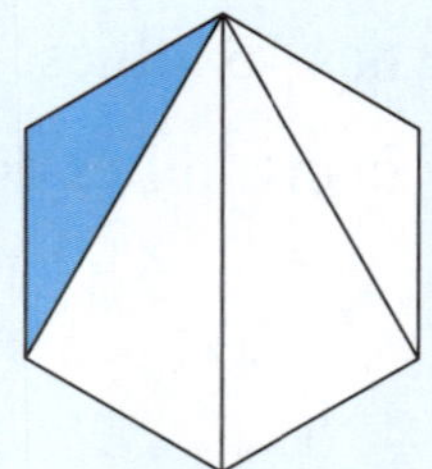

Exercise 13

How many quarters are shaded?

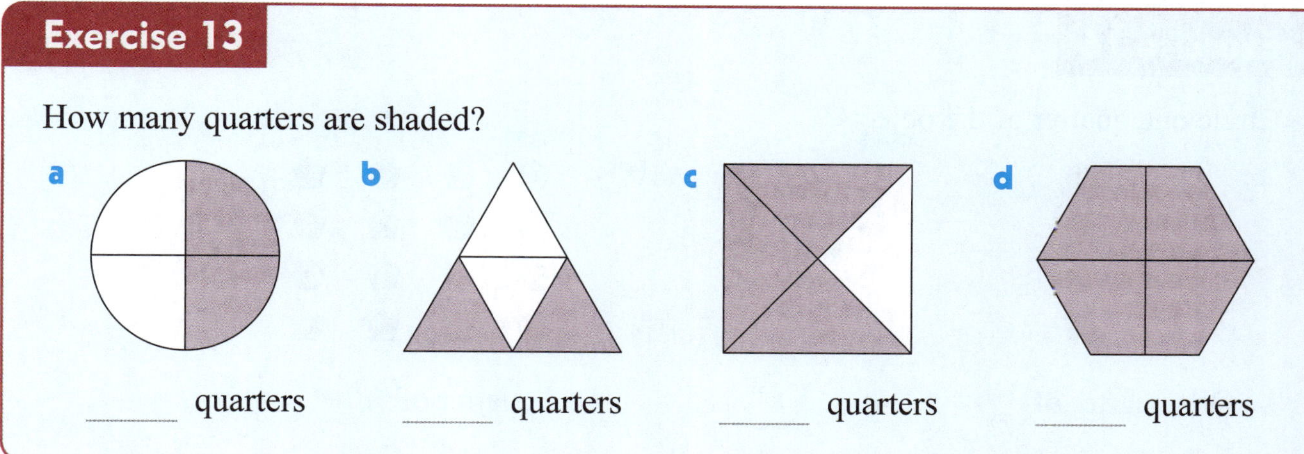

a ________ quarters   b ________ quarters   c ________ quarters   d ________ quarters

Exercise 14

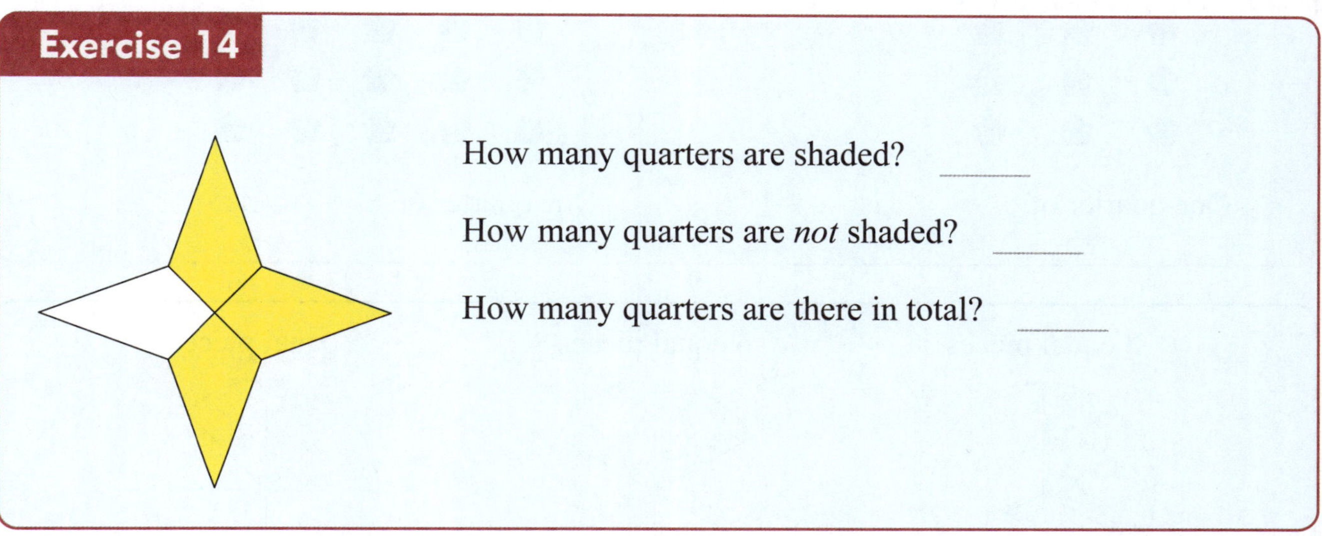

How many quarters are shaded? ________

How many quarters are *not* shaded? ________

How many quarters are there in total? ________

Exercise 15

Shade the correct amount.

a one quarter   b two quarters   c three quarters

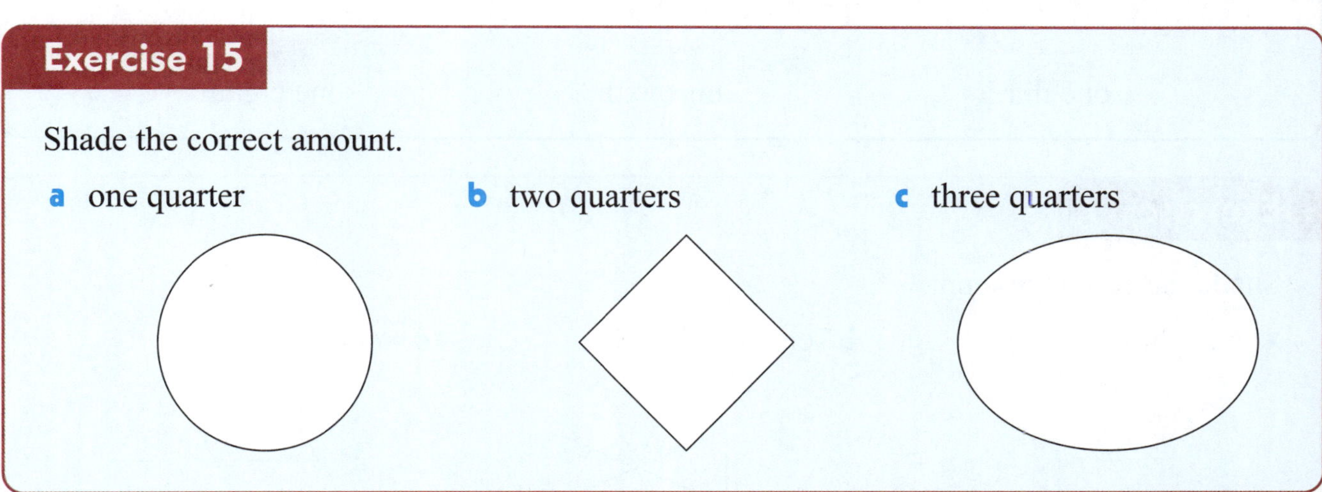

Discussion

Giorgio loves chocolate cake.

Would he rather have *one half* or *two quarters*?

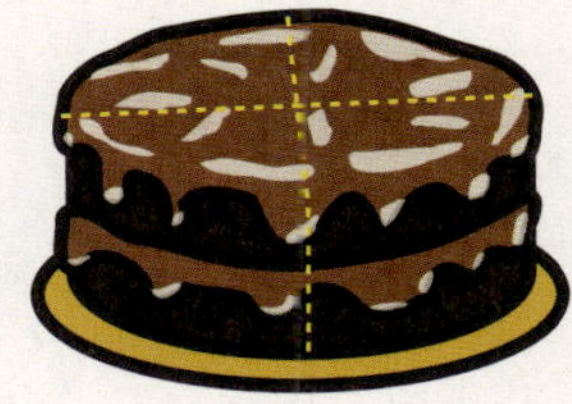

Exercise 16

Circle one quarter of the objects.

a

One quarter of _______ is _______ .

b

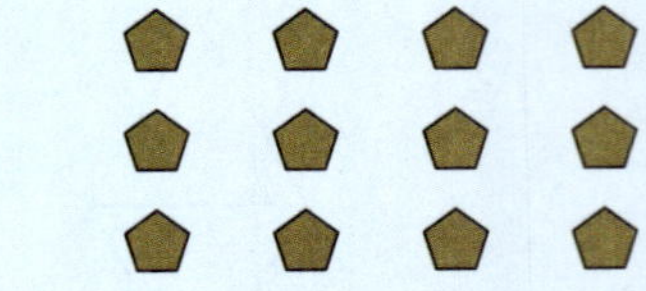

One quarter of _______ is _______ .

c

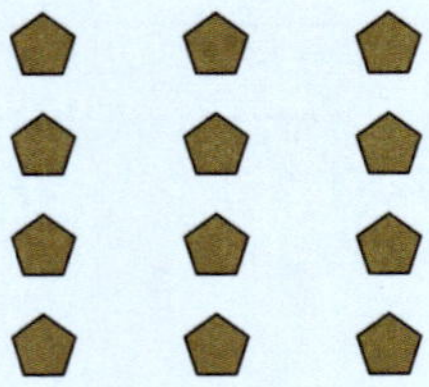

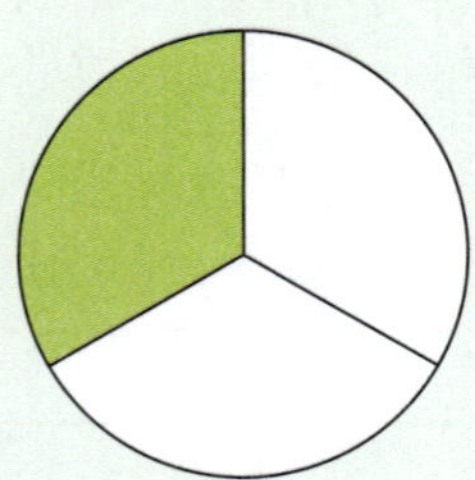

One quarter of _______ is _______ .

d

One quarter of _______ is _______ .

---

3 equal pieces	6 equal pieces	8 equal pieces
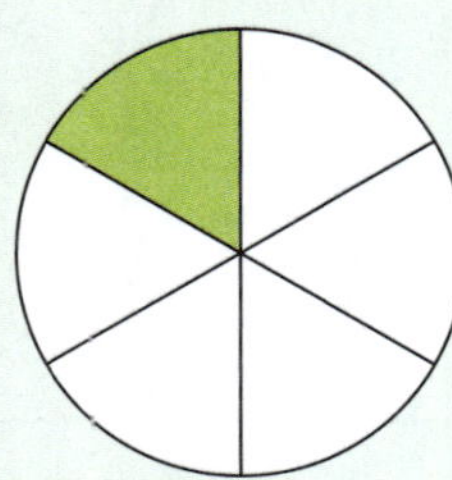		
one third	one sixth	one eighth

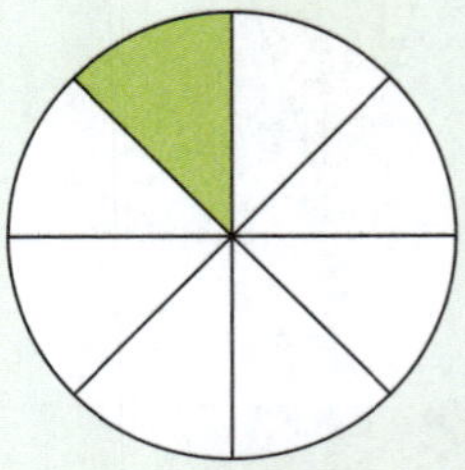

Exercise 17

Shade the correct amount.

a one third

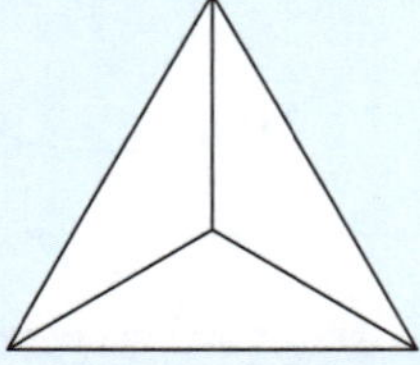

b one eighth

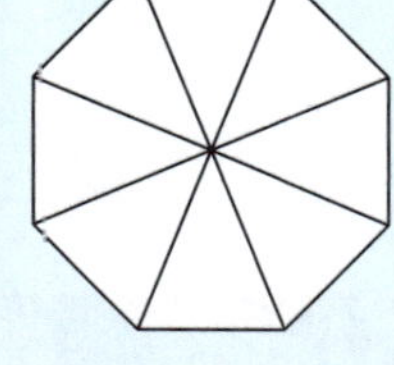

c one sixth

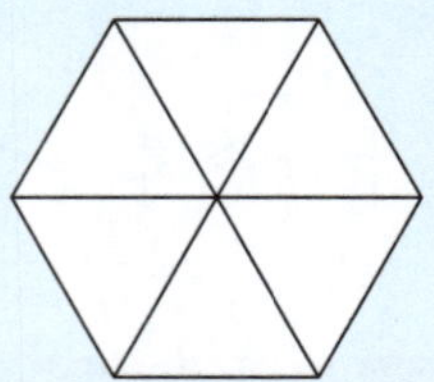

d four sixths

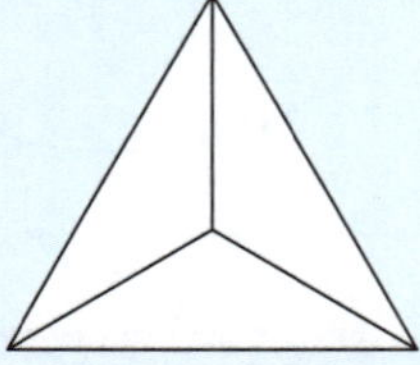

e two thirds

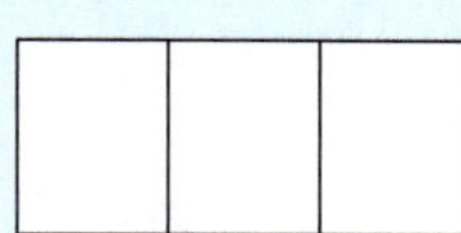

f five eighths

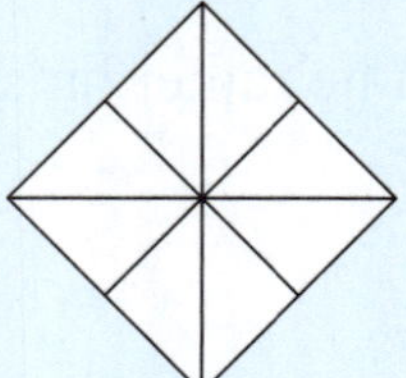

Exercise 18

How much is shaded?

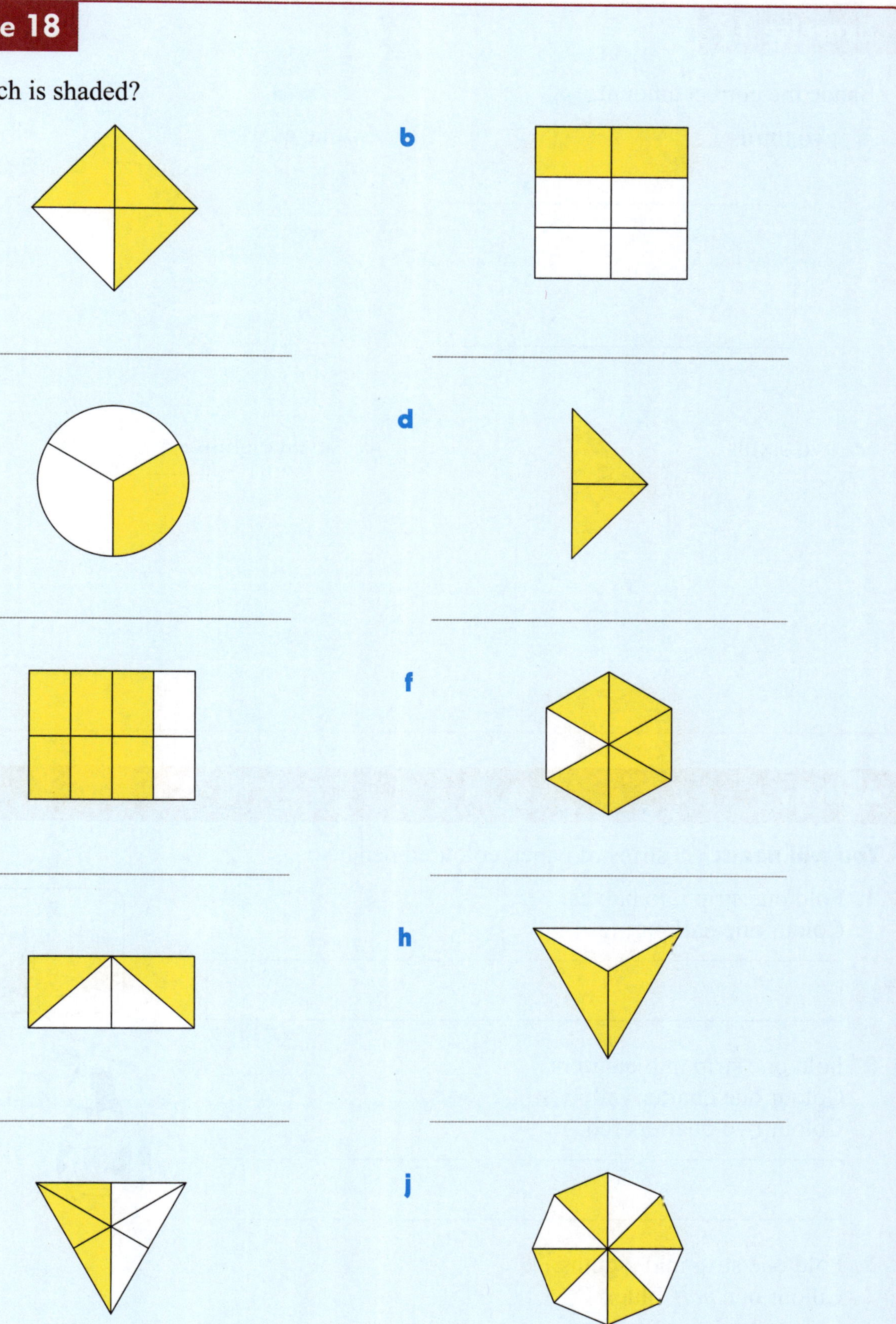

Exercise 19

Shade the correct amount.

a two thirds

b one quarter

c two sixths

d seven eighths

Activity

You will need: 3 strips of paper, coloured pencils

1 Fold one strip into halves.
Colour one half purple.

2 Fold one strip into quarters.
Colour one quarter yellow.
Colour two quarters red.

3 Fold one strip into eighths.
Colour one *half* blue.
Colour one *quarter* green.
Colour one *eighth* pink.

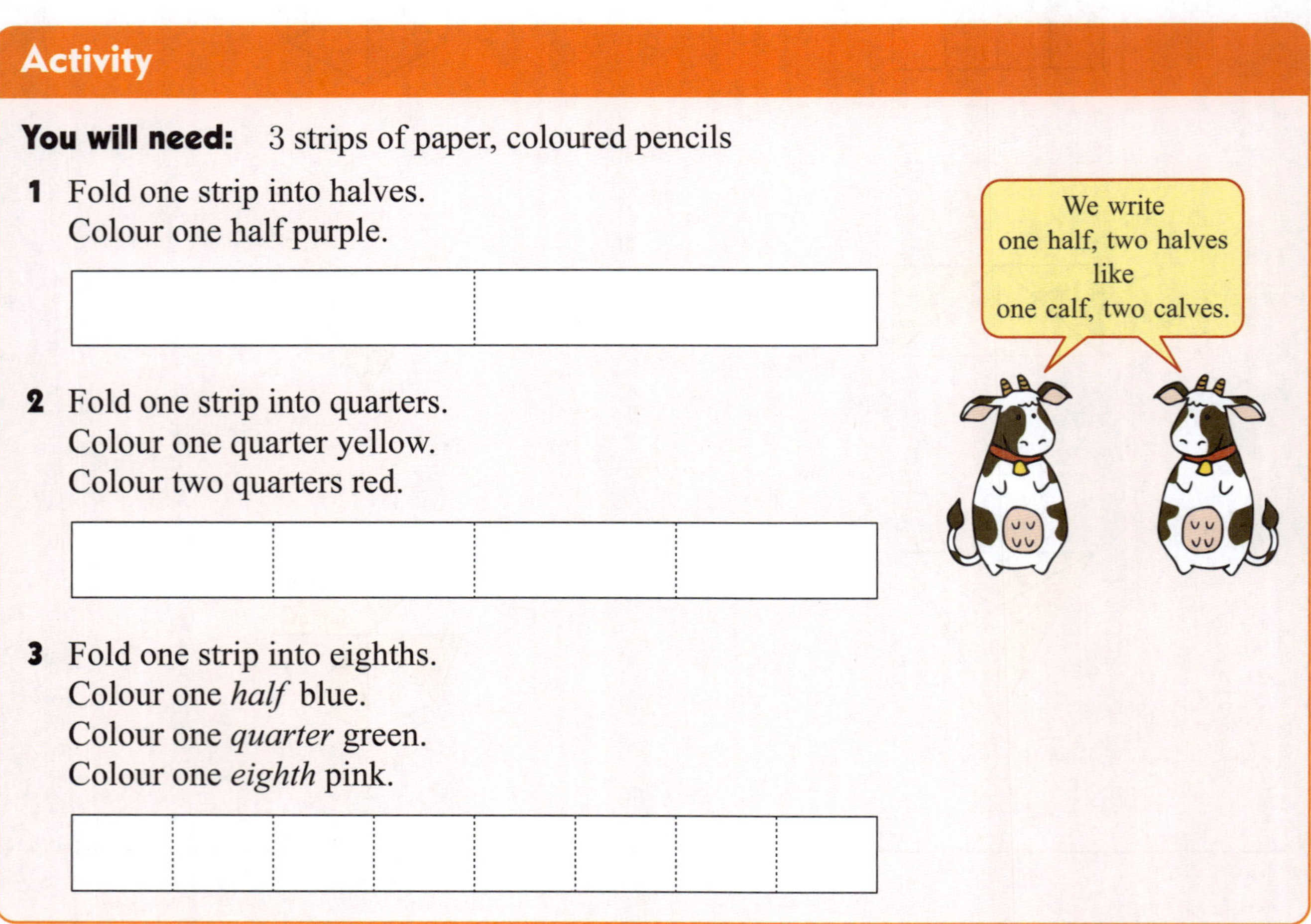

> **Time** measures *when* things happen.

Discussion

1 Discuss each sentence:

- We wash our hands *before* we prepare food.
- We can go home *after* we tidy the classroom.
- I always wake up *early*.
- Igor is often *late* for school.

2 What do we mean by "*on time*"? Why is it important to be "on time"?

Exercise 1

earliest   earlier   early   "on time"   late   later   latest

Practise your writing.

earliest	earlier	early
___________	___________	___________
___________	___________	___________
	on time	
	___________	
	___________	
late	later	latest
___________	___________	___________
___________	___________	___________

Exercise 2

Four girls arrived at a party in this order:

Isla,   Jess,   Ebony,   Felicity

Complete using *earliest*, *earlier*, *later*, or *latest*:

a Isla was ___________________________ .

b Felicity arrived ___________________________ than Jess.

c Jess was ___________________________ than Ebony.

Today is the day right now.

Yesterday is the day *before* today.

Tomorrow is the day *after* today.

Exercise 3

Practise your writing.

before	after	today
_________________	_________________	_________________
_________________	_________________	_________________

yesterday	tomorrow
_________________	_________________
_________________	_________________

Activity

1 List two things you did *yesterday*.

____

____

2 List two different things you did *today*.

____

____

3 What would you like to do *tomorrow*?

____

____

We use a **clock** to tell the time.

The **minute hand** is longer. It is pointing to 12.

The **hour hand** is shorter. It is pointing to 3.

The time is 3 o'clock.

The hands on a clock always turn in the same direction.

The hands move **clockwise**.

Exercise 4

What time is shown on the clock?

a

___________ o'clock

b

____

c

____

d

____

e

____

f

____

Exercise 5

Practise your writing.

minute   hour   o'clock

______________________   ______________________   ______________________

______________________   ______________________   ______________________

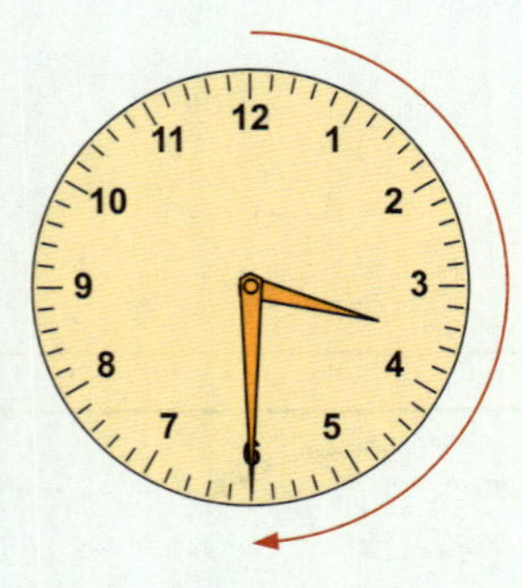

The minute hand has turned from 12 to 6.

The minute hand is **half way** around the clock.

The hour hand is **half way** between 3 and 4.

The time is half past 3.

Exercise 6

The minute hand is ______________________ around the clock.

The hour hand is ______________________ between

__________ and __________ .

The time is ______________________ .

Exercise 7

a Label the hands of the clock.

b Draw the clockwise direction.

c The time is ______________________ .

____

____

Exercise 8

What time is shown on the clock?

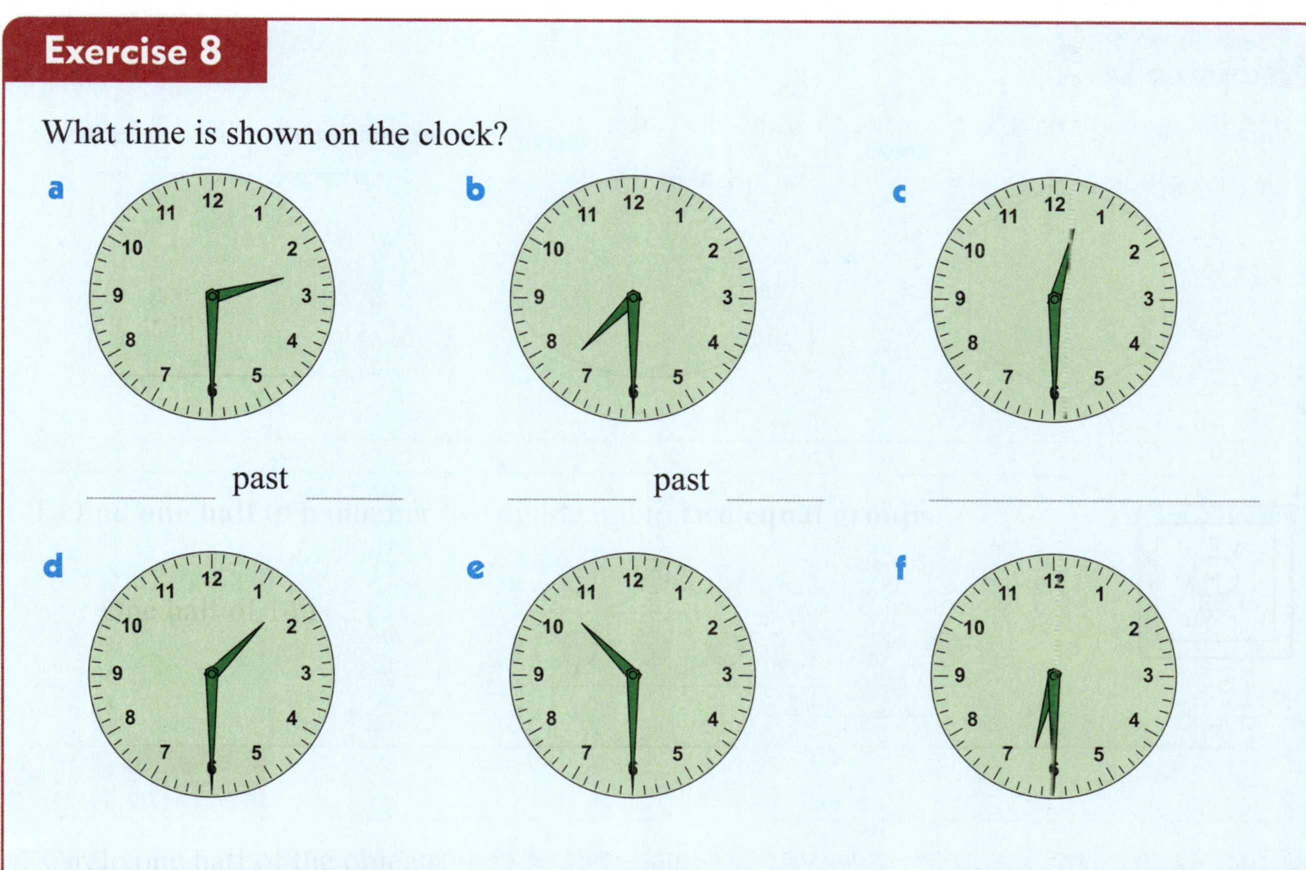

a _________ past _________

b _________ past _________

c _________

d _________

e _________

f _________

Exercise 9

Draw the missing hand.

a 2 o'clock

b half past 4

c half past 8

d 11 o'clock

e half past 9

f half past 5

Exercise 10

Draw hands on each clock face to show the time:

a 6 o'clock

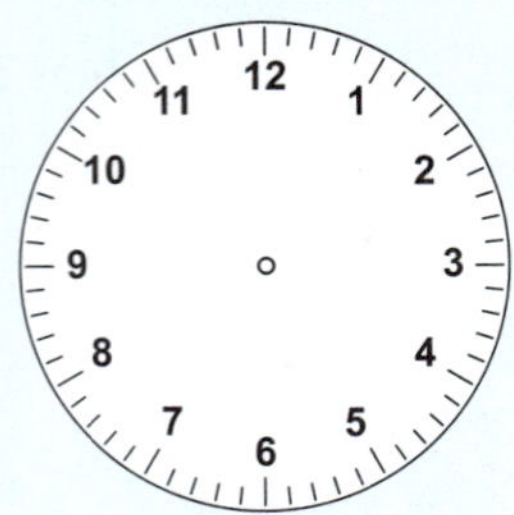

b half past 7

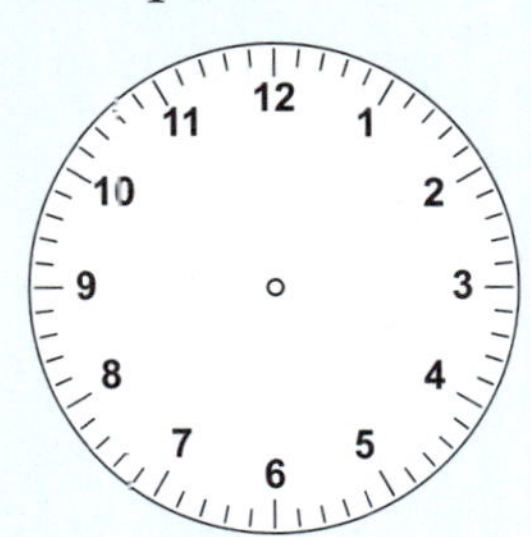

c half past 1

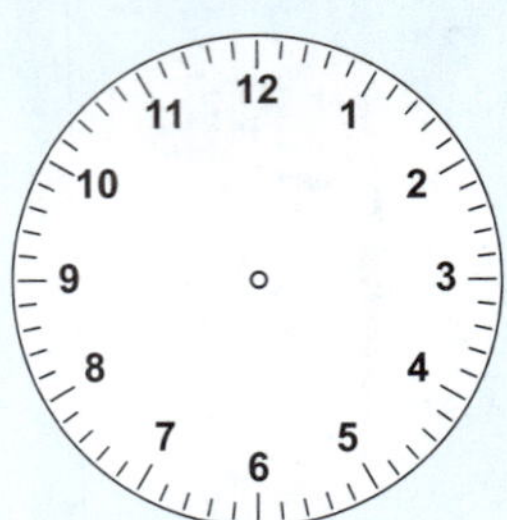

d 8 o'clock

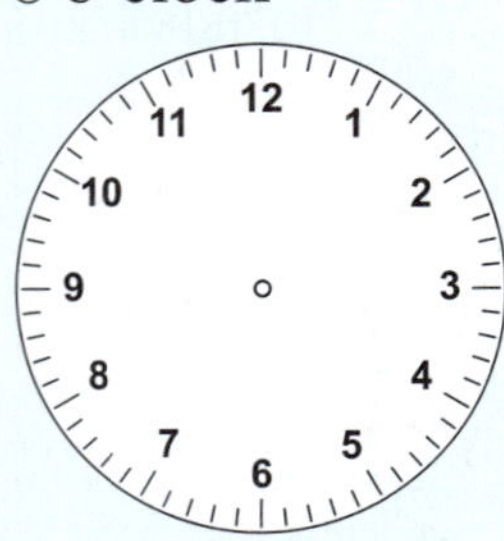

e 10 o'clock

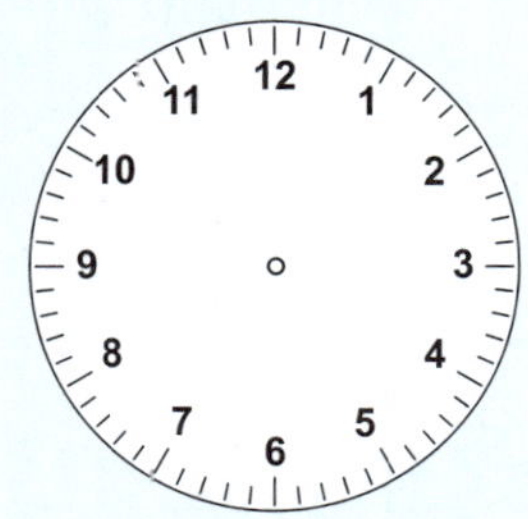

f half past 3

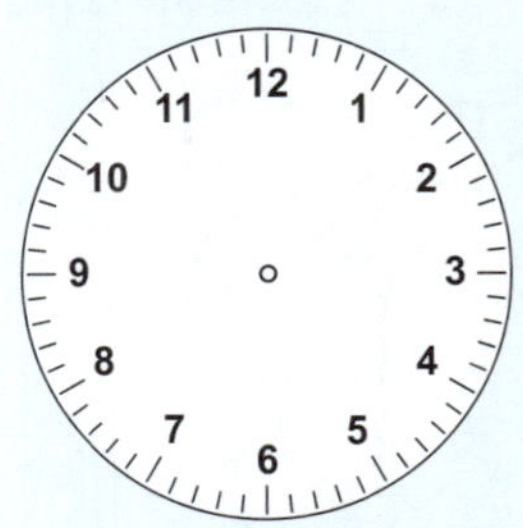

Discussion

"My clock shows half past 9," says Molly.

Is Molly's clock working properly?

The minute hand has turned from 12 to 3.

The minute hand is a **quarter** of the way around the clock.

The hour hand is between 1 and 2.

The time is quarter past 1.

Exercise 11

The minute hand is a _________________________ of the way around the clock.

The hour hand is between _________ and _________ .

The time is _________________________________ .

Exercise 12

What time is shown?

a

quarter past ____________

b

____________ past ____________

c

____________ past ____________

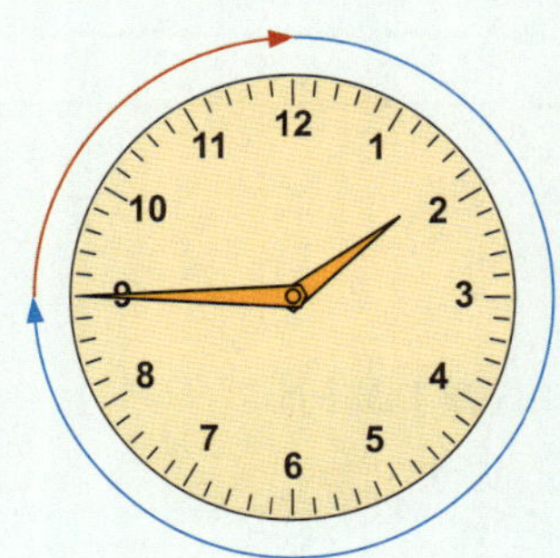

The minute hand has turned from 12 to 9.

The minute hand still has a **quarter** of the way to go, to get back to 12.

The hour hand is between 1 and 2.

The time is quarter to 2.

Exercise 13

The minute hand still has a ____________________ of the way to go, to get back to 12.

The hour hand is between ________ and ________.

The time is ____________________.

Exercise 14

What time is shown?

a

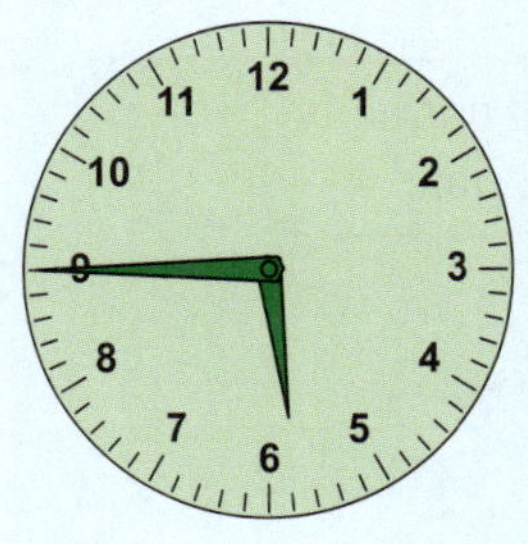

quarter to ____________

b

____________ to ____________

c

____________ to ____________

Exercise 15

Complete the labels.

The time is ______________________ .

Exercise 16

Draw hands on each clock to show the time:

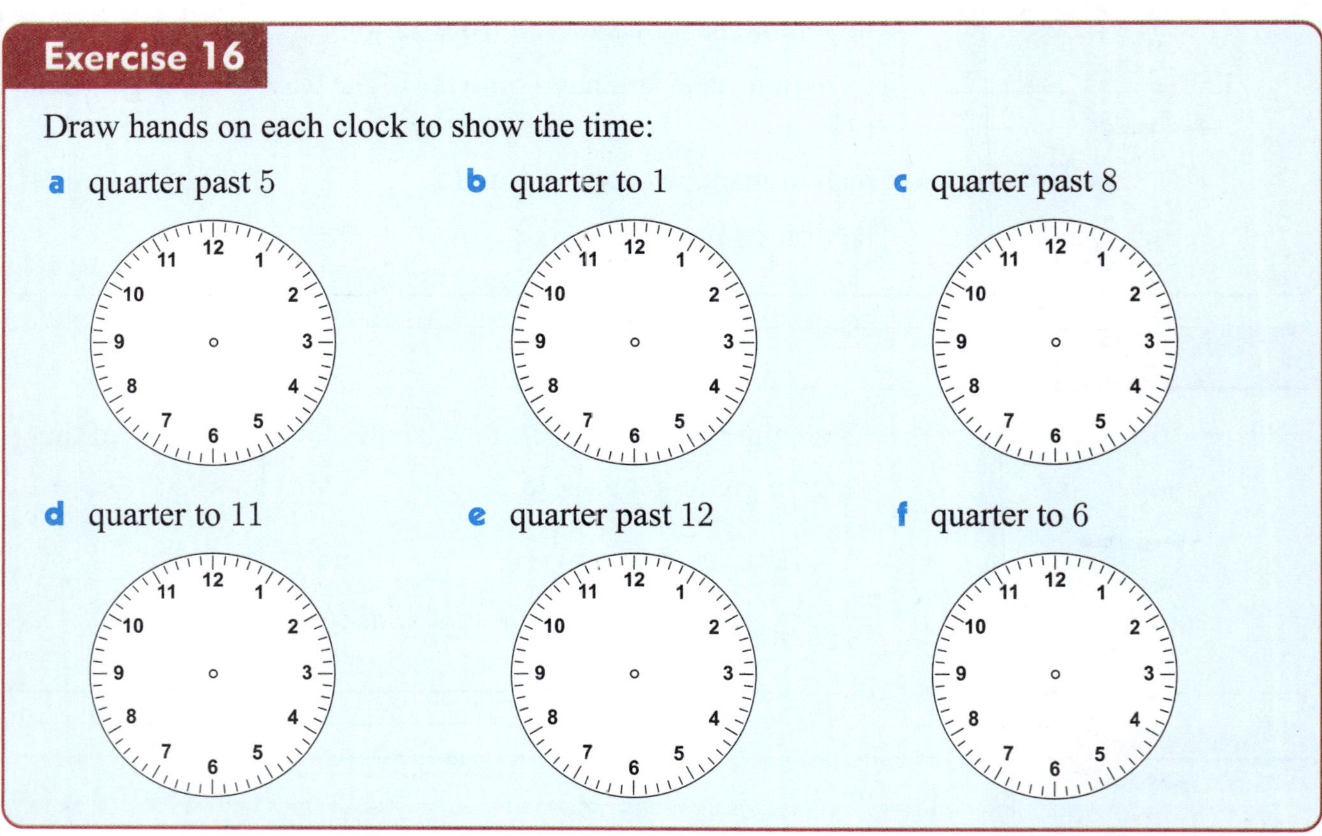

a quarter past 5

b quarter to 1

c quarter past 8

d quarter to 11

e quarter past 12

f quarter to 6

There are sixty minutes in **one hour**.

There are twenty four hours in **one day**.

There are seven days in **one week**.

There are twelve months in **one year**.

Exercise 17

Complete:

a 1 day = __________ hours

b 1 hour = __________ minutes

c 1 year = __________ months

d 1 week = __________ days

Exercise 18

These are the days of the week, written in order.

Practise your writing.

Sunday

Monday

Tuesday

Wednesday

Thursday

Friday

Saturday

Discussion

Does everyone around the world have the same "first" day for their week?

Which days are the **weekend**?

__________________________ and __________________________

Exercise 19

How many days are there in 2 weeks?

In 2 weeks there are ________ + ________ = ________ days.

Exercise 20

How many hours are there in 2 days?

In 2 days, there are [ ] hours.

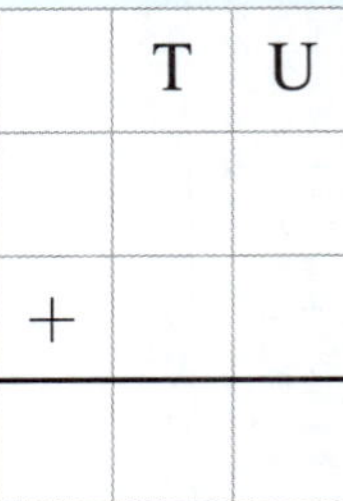

Exercise 21

These are the months of the year, written in order.
Practise your writing.

January

February

March

April

May

June

July

August

September

October

November

December

Exercise 22

Write down the day *after*:

a Tuesday _______________________

b Sunday _______________________

c Friday _______________________

Exercise 23

Write down the day *before*:

a Thursday _______________________

b Saturday _______________________

c Wednesday _______________________

A **calendar** shows the months and days of the year.

Some months have more days than others.

Calendar 2025

January

Su	M	T	W	Th	F	Sa
			1	2	3	4
5	6	7	8	9	10	11
12	13	14	15	16	17	18
19	20	21	22	23	24	25
26	27	28	29	30	31	

February

Su	M	T	W	Th	F	Sa
						1
2	3	4	5	6	7	8
9	10	11	12	13	14	15
16	17	18	19	20	21	22
23	24	25	26	27	28	

March

Su	M	T	W	Th	F	Sa
						1
2	3	4	5	6	7	8
9	10	11	12	13	14	15
16	17	18	19	20	21	22
23	24	25	26	27	28	29
30	31					

April

Su	M	T	W	Th	F	Sa
		1	2	3	4	5
6	7	8	9	10	11	12
13	14	15	16	17	18	19
20	21	22	23	24	25	26
27	28	29	30			

May

Su	M	T	W	Th	F	Sa
				1	2	3
4	5	6	7	8	9	10
11	12	13	14	15	16	17
18	19	20	21	22	23	24
25	26	27	28	29	30	31

June

Su	M	T	W	Th	F	Sa
1	2	3	4	5	6	7
8	9	10	11	12	13	14
15	16	17	18	19	20	21
22	23	24	25	26	27	28
29	30					

July

Su	M	T	W	Th	F	Sa
		1	2	3	4	5
6	7	8	9	10	11	12
13	14	15	16	17	18	19
20	21	22	23	24	25	26
27	28	29	30	31		

August

Su	M	T	W	Th	F	Sa
					1	2
3	4	5	6	7	8	9
10	11	12	13	14	15	16
17	18	19	20	21	22	23
24	25	26	27	28	29	30
31						

September

Su	M	T	W	Th	F	Sa
	1	2	3	4	5	6
7	8	9	10	11	12	13
14	15	16	17	18	19	20
21	22	23	24	25	26	27
28	29	30				

October

Su	M	T	W	Th	F	Sa
		1	2	3	4	
5	6	7	8	9	10	11
12	13	14	15	16	17	18
19	20	21	22	23	24	25
26	27	28	29	30	31	

November

Su	M	T	W	Th	F	Sa
						1
2	3	4	5	6	7	8
9	10	11	12	13	14	15
16	17	18	19	20	21	22
23	24	25	26	27	28	29
30						

December

Su	M	T	W	Th	F	Sa
	1	2	3	4	5	6
7	8	9	10	11	12	13
14	15	16	17	18	19	20
21	22	23	24	25	26	27
28	29	30	31			

Exercise 24

Practise your writing.   calendar

____

____

Be careful! "Calendar" is often misspelt!

Exercise 25

The months with 30 days are:

___________________   ___________________   ___________________

____

The months with 31 days are:

___________________   ___________________   ___________________

___________________   ___________________   ___________________

____

How many days were in February 2025?   ___________

Discussion

Does February always have the same number of days?

February could have _________ or _________ days.

Exercise 26

Write down the month *after*:

a   April   _______________________

b   October   _______________________

c   December   _______________________

Exercise 27

Write down the month *before*:

a   June   _______________________

b   September   _______________________

c   January   _______________________

Exercise 28

September 2025						
Su	M	T	W	Th	F	Sa
	1	2	3	4	5	6
7	(8)	9	10	11	12	13
14	15	16	17	18	19	20
21	22	23	24	25	26	27
28	29	30				

This calendar shows the days of the week in the month of September, 2025.

The **date** of the day circled in red is Monday, 8th September.

We say "Monday, the eighth of September".

a Use the calendar to fill in the day of the week:

__________________ , 10th September

__________________ , 26th September

__________________ , 4th September

__________________ , 23rd September

b The day before Saturday, 20th September is

____ .

c The day after Tuesday, 2nd September is

____ .

Exercise 29

a Use the calendar to fill in the day of the week:

__________________ , 7th December

__________________ , 24th December

__________________ , 13th December

__________________ , 29th December

December 2025						
Su	M	T	W	Th	F	Sa
	1	2	3	4	5	6
7	8	9	10	11	12	13
14	15	16	17	18	19	20
21	22	23	24	25	26	27
28	29	30	31			

b The day after Thursday, 4th December is

____ .

c The day before Sunday, 21st December is

____ .

Exercise 30

Write the date of the day *after*:

a Monday, 4th July

b Friday, 13th February

c Saturday, 23rd August

d Wednesday, 31st March

Discussion

What **seasons** do you experience?

☐ I do not experience seasons. The weather is the same all year.

☐ I experience a wet season and a dry season.

The wet season is usually from ______________________ until ______________________ .

The dry season is usually from ______________________ until ______________________ .

☐ I experience 4 seasons each year. They are:

Summer	Autumn or Fall
Winter	**Spring**

Exercise 31

Write the date of the day *before*:

a Thursday, 16th June

b Monday, 7th December

c Sunday, 26th October

d Friday, 1st May

Discussion

Match each season with a picture. Discuss *why* the picture connects with that season.

Summer

Autumn or Fall

Winter

Spring

Activity

Collect a list of birthdays.

My birthday is ____ .

My mum's birthday is ____ .

My dad's birthday is ____ .

Other family members:

____

____

____

Friends:

____

____

____

Print a calendar for *this year*.

Mark each of the birthdays on your calendar.

Discussion

Is your birthday always on the same day of the week? _________

Click on the icon to find which day of the week your birthday is on.

- This year, my birthday is on a _____________________ .

- Next year, my birthday will be on a _____________________ .

CHAPTER 11: MONEY

How much is shaded?

Exercise 1

To make sure you are ready to work with money, practise your counting.

a Add 5 each time:

$5 \rightarrow 10 \rightarrow$ _______ $\rightarrow 20 \rightarrow$ _______ $\rightarrow$ _______ $\rightarrow 35$

b Add 10 each time:

$10 \rightarrow 20 \rightarrow$ _______ $\rightarrow 40 \rightarrow$ _______ $\rightarrow$ _______ $\rightarrow 70$

c Add 5 each time:

$50 \rightarrow 55 \rightarrow$ _______ $\rightarrow$ _______ $\rightarrow$ _______ $\rightarrow$ _______ $\rightarrow 80$

d Add 20 each time:

$20 \rightarrow$ _______ $\rightarrow$ _______ $\rightarrow$ _______ $\rightarrow$ _______

We use **money** to buy items.

Activity

Make a list of items in your classroom, and write down where each item could be bought.

Item	Bought at
desk	furniture shop
pencil	

Different types of money are used in different countries.

The money used in a country is called its **currency**.

In most countries, the currency includes **coins** and **notes**.

coins   notes

Exercise 2

Practise your writing.

money   ____________________________   coins   ____________________________

currency   ____________________________   notes   ____________________________

Activity

Match each country with its currency.

      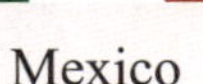  

Peru   Denmark   Mexico   Thailand   Germany

Mexican peso   Peruvian sol   Thai baht   Euro   Danish krone

Click on the icon to print some Exercises for your currency.

CHAPTER 12: CHANCE

The **chance** of something happening describes how *likely* it is to happen.

Exercise 1

Who do you think is *more likely* to win this race?

____

Exercise 2

Who do you think is *less likely* to get hurt?

____

Exercise 3

Erica is blindfolded. She has been asked to pick a ball out of this bag.

Use *more* or *less* to complete each statement:

a Erica is _______________ likely to choose a blue ball than a black ball.

b Erica is _______________ likely to choose a blue ball than a red ball.

Exercise 4

Cody will spin this spinner once.

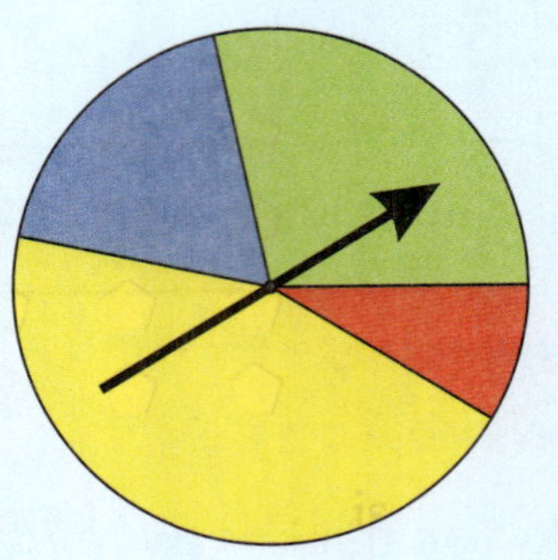

a Which colour is Cody *most likely* to spin?

____

b Which colour is Cody *least likely* to spin?

____

c Use *more* or *less* to complete:

Cody is _________________ likely to spin blue than green.

Discussion

When you roll a die:

- Are you more likely to roll a 1 than a 6?   _______

- Are you less likely to roll a 2 than a 5?   _______

When you roll a die, the possible results are _________________ _________________ .

Activity   What will happen tomorrow?

For each pair of activities, place a tick ✓ next to the activity that is *more likely* to happen tomorrow.

| ☐ You will ride an elephant. | ☐ You will ride in a car. |

| ☐ You will eat an apple. | ☐ You will eat a biscuit. |

| ☐ You will wear a hat. | ☐ You will wear a coat. |

| ☐ You will watch TV. | ☐ You will read a book. |

Activity

Circle the words that are used to *describe* the chance of something happening.

If something will *definitely* happen, we say it is **certain**.

If something will *definitely not* happen, we say it is **impossible**.

If something is as likely to happen as not happen, we say it has a **50-50 chance** of happening.

Class Activity

As a class, write these words in order.

| likely |
| impossible |
| unlikely |
| 50-50 chance |
| certain |

↑ most likely

least likely

Exercise 5

Match each spinner to the chance it will stop on red.

impossible   unlikely   50-50 chance   likely   certain

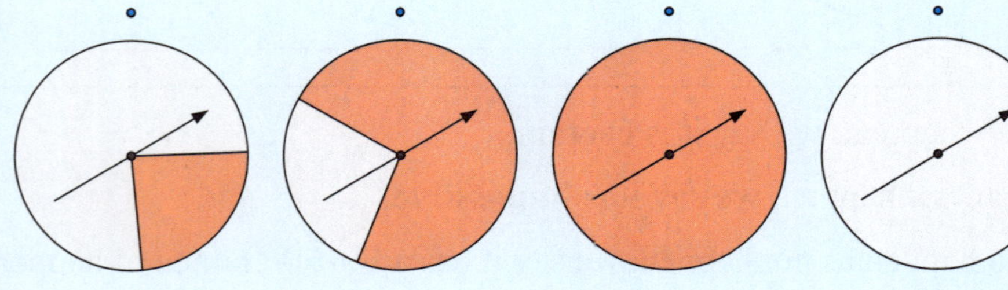

Exercise 6

When a coin is tossed, the two possible **outcomes** are "head" and "tail".

There is a ______________ chance of tossing a "head".

head   tail

Exercise 7

Jo is blindfolded. He will pick a ball from this bag.

Use *certain*, *likely*, *50-50 chance*, *unlikely*, or *impossible* to describe the chance that Jo will:

a choose a blue ball

b choose a red ball

c choose a yellow ball

Exercise 8

Maria is asked to spin this spinner.

Describe the chance that Maria will:

a spin black

b spin white

c spin green

d *not* spin black

e *not* spin orange

Exercise 9

When I roll a die, the six possible outcomes are "1", "2", "3", "4", "5", and "6".

Describe the chance that I will:

a roll a "5"

b roll a number less than 7

c roll a "7"

d roll a "4", "5", or "6"

e *not* roll a "1"

Exercise 10

Describe the chance that:

a you will wash your hands today ______________________________

b you will climb a tree tomorrow ______________________________

c you will read a book this weekend ______________________________

d you will drink water today ______________________________

e a rhinoceros will crash through the classroom ceiling

____

Discussion

In the Exercise above, will everyone give the same answers?

Activity

Write some sentences of your own which use the language of chance.

____

____

____

____

Revision   Number

1 Write each number in words.

1	11	30
2	12	40
3	13	50
4	14	60
5	15	70
6	16	80
7	17	90
8	18	100
9	19	
10	20	

2 Complete the chart.

Diagram	Numeral	Word
		four
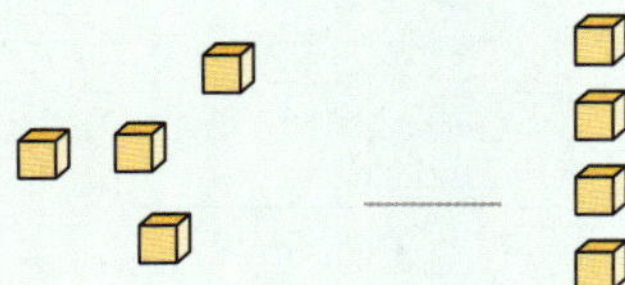		

3 Use $=$ or $\neq$.

a 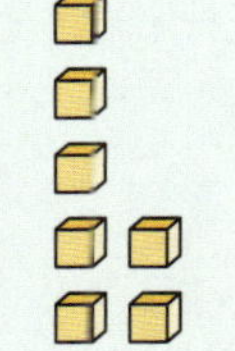

b

c 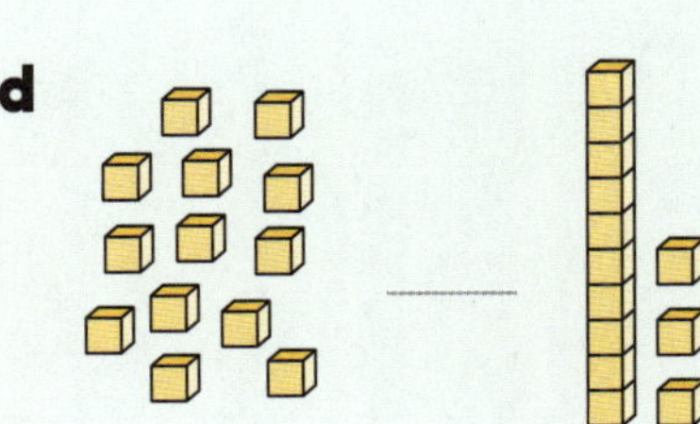

d

4 What numbers are shown?

a 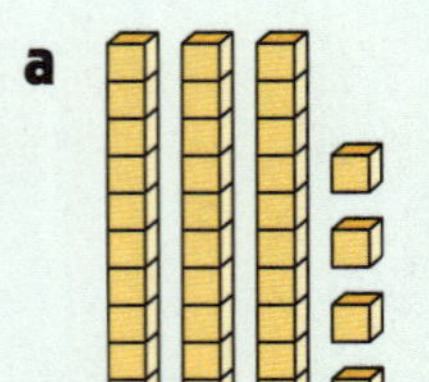 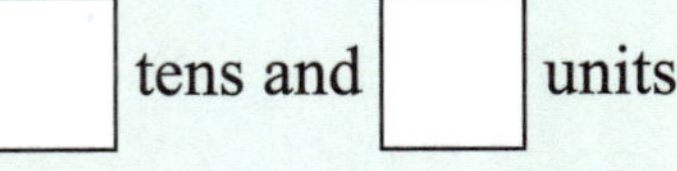 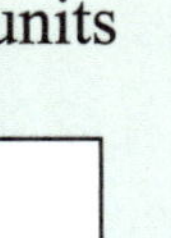 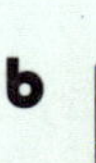 tens and ☐ units

☐ + ☐ = ☐

b 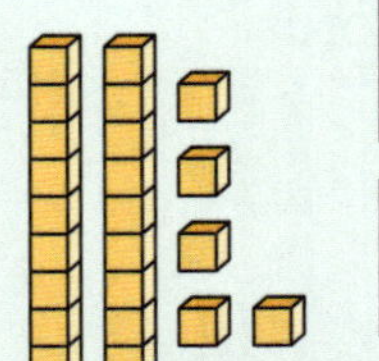 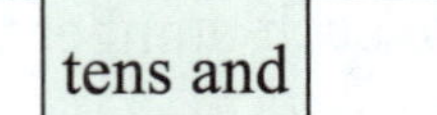 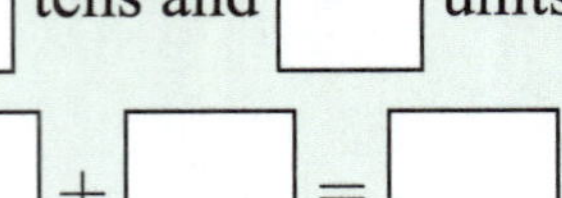 tens and ☐ units

☐ + ☐ = ☐

c 

☐ hundred ☐ tens ☐ units

The number is ☐ or ☐ .

5 Partition each number using place values:

a 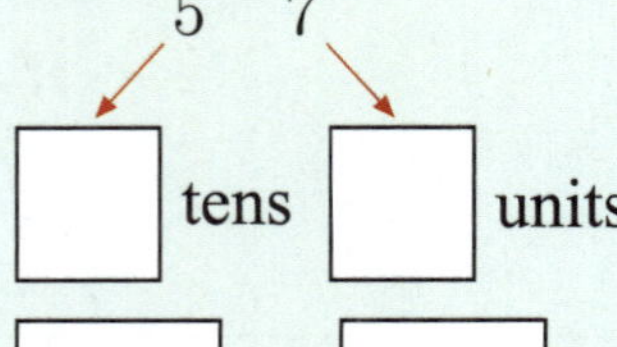 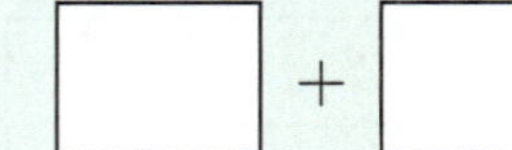   5   7

☐ tens ☐ units

☐ + ☐

b 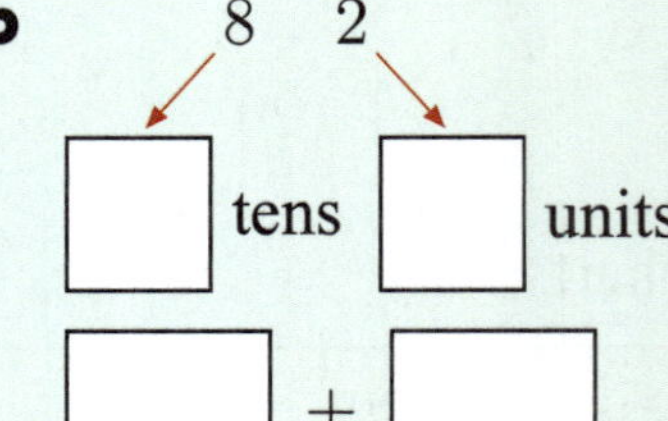   8   2

☐ tens ☐ units

☐ + ☐

c 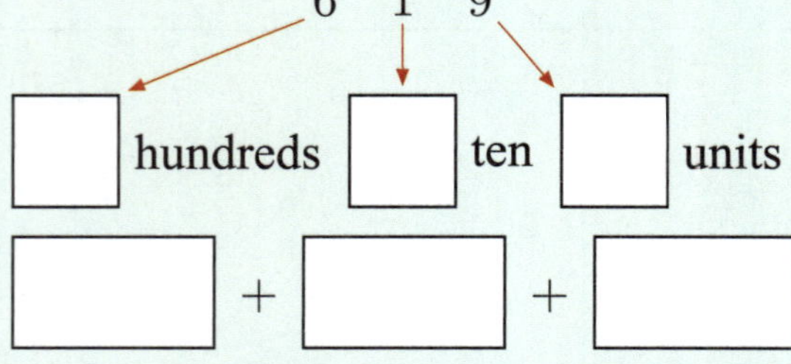   6   1   9

☐ hundreds ☐ ten ☐ units

☐ + ☐ + ☐

d 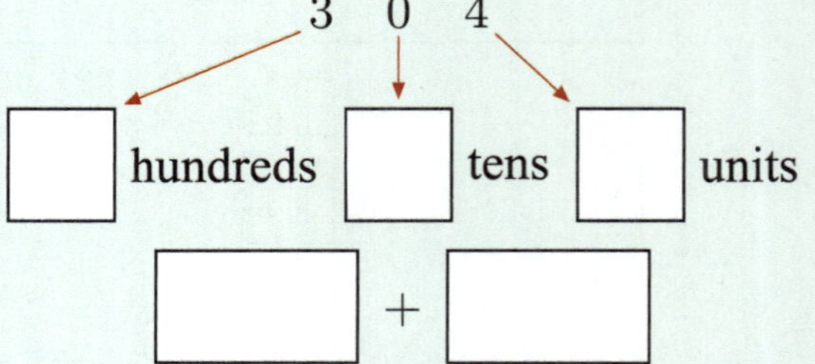   3   0   4

☐ hundreds ☐ tens ☐ units

☐ + ☐

6 Write these numbers in words.

a 46 _______________________________

b 83 _______________________________

c 295 _______________________________

d 712 _______________________________

e 360 _______________________________

f 509 _______________________________

7 What is the value of the red digit?

   a 547 ________________________   **b** 486 ________________________

   c 913 ________________________   **d** 205 ________________________

8 What number does each abacus show?

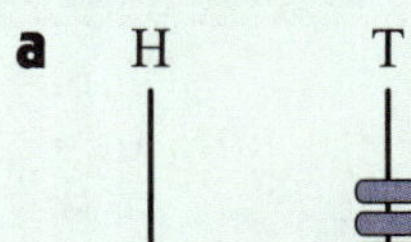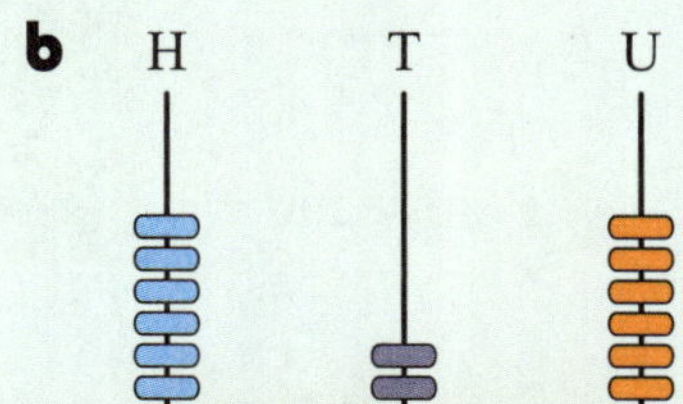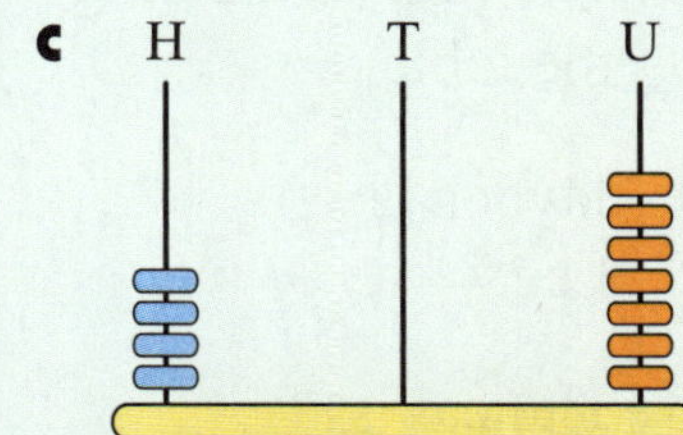

   a ________   **b** ________   **c** ________

9 Show the number on the abacus:

   a 514   **b** 843   **c** 490

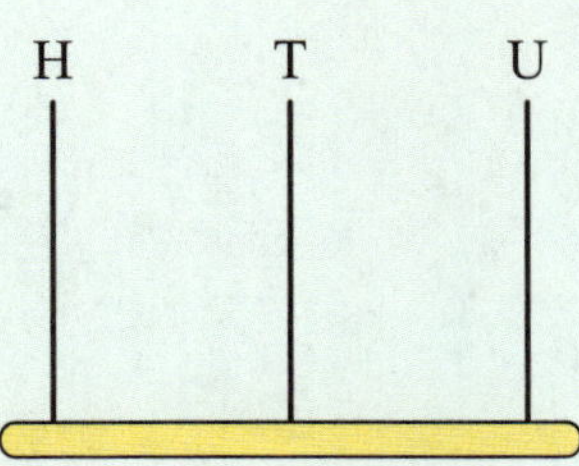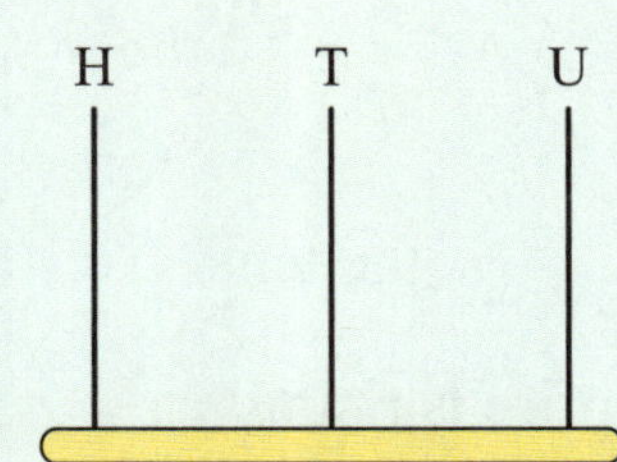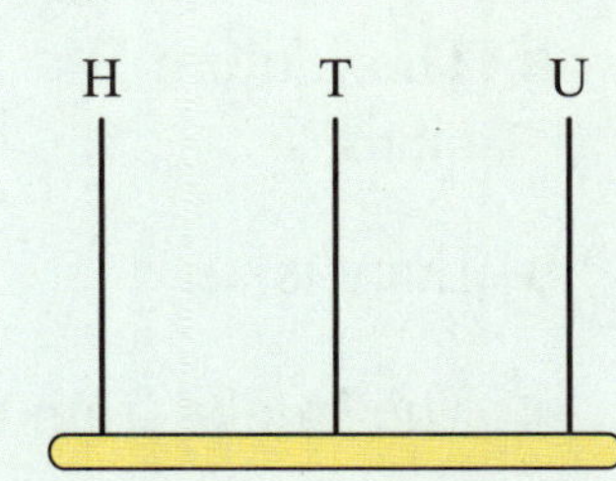

10 Use *greater* or *less*.

   a 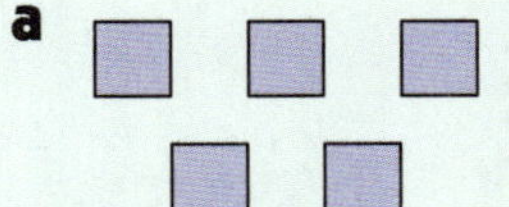   **b** 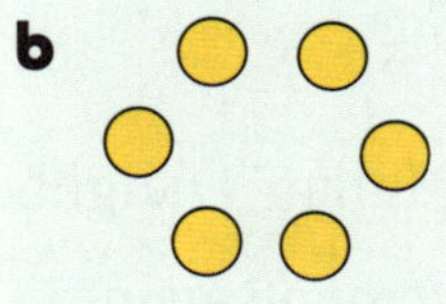

   5 is ________________ than 3.   6 is ________________ than 8.

11 Complete each number line:

   a

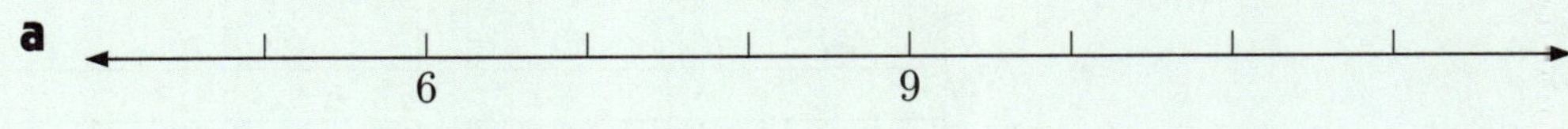

   b

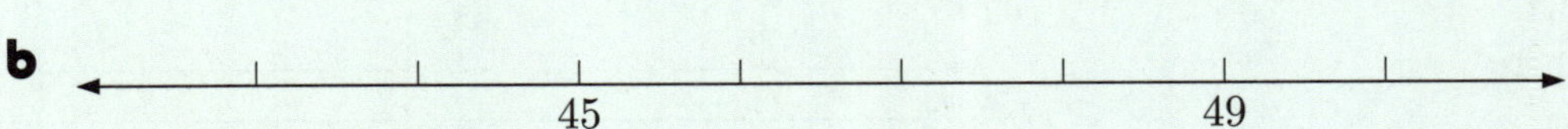

12

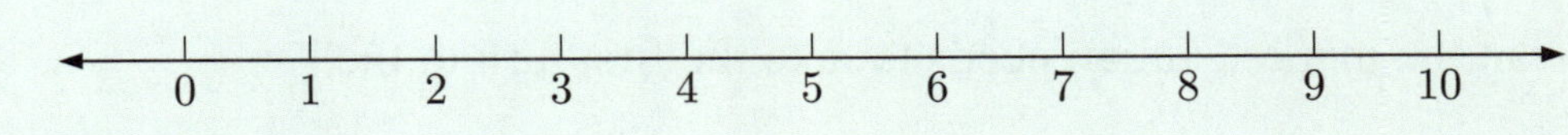

   Use < or >.

   a 4 ____ 7   **b** 10 ____ 8   **c** 5 ____ 9

Revision — Position and direction

1 Complete each statement using *over*, *under*, *in front of*, *behind*, and *between*.

a Lee is ________________ Penny.

b The light is hanging ________________ the table.

c Jayden is ________________ Penny and Lee.

d The cat is ________________ the table.

e Penny is ________________ Jayden and Lee.

2 Complete each statement using *inside*, *outside*, *upwards*, and *next to*.

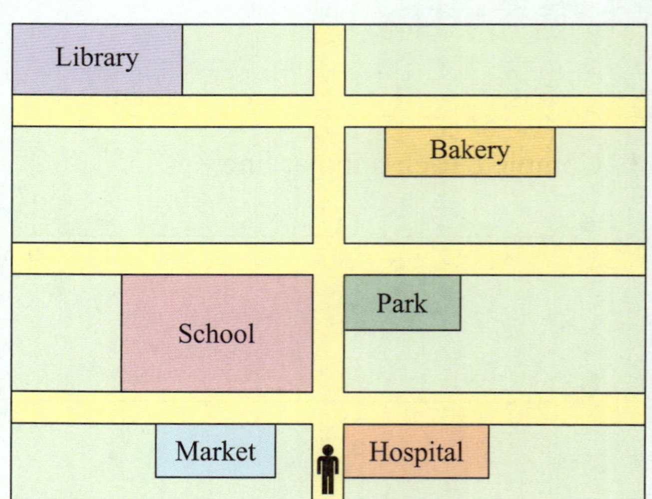

a The lantern is ________________ the chair.

b Lucy is ________________ the tent.

c The smoke from the campfire is rising

________________ .

d Wendy is ________________ the tent.

3 Here are some streets in Julian's neighbourhood.

a Complete each statement using *near* or *far from*.

Julian is:

- ________________ the hospital

- ________________ the library.

b To visit the market, Julian needs to take the first turn to the ________________ .

c To visit the bakery, Julian needs to take the ________________ turn to the ________________ .

Revision   Addition

1 Use the number line to help you add:

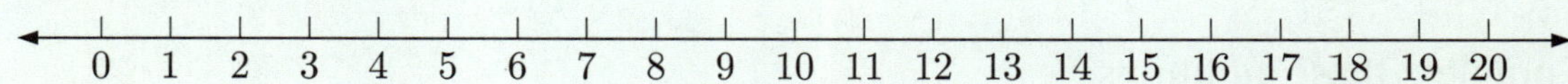

a $3 + 2 =$ _______

b 6 plus $3 =$ _______

c $4 + 4 =$ _______

d $5 + 6 =$ _______

e 9 add $4 =$ _______

f $8 + 7 =$ _______

2 Write pairs of numbers that total 7.

$$7 = 0 + \square \qquad 7 = \square + \square \qquad 7 = \square + \square$$

$$7 = \square + \square \qquad 7 = \square + \square \qquad 7 = \square + \square$$

$$7 = \square + \square \qquad 7 = \square + \square$$

3 Colour the objects to match the sum.

a $4 + 5 = 9$

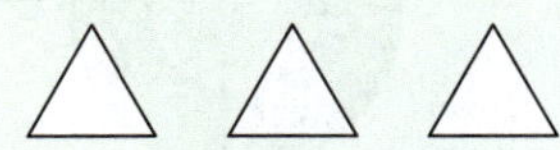
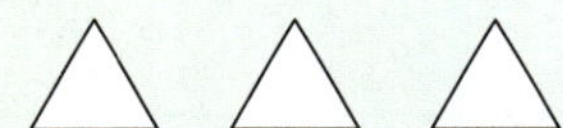

b $5 + 9 = 14$

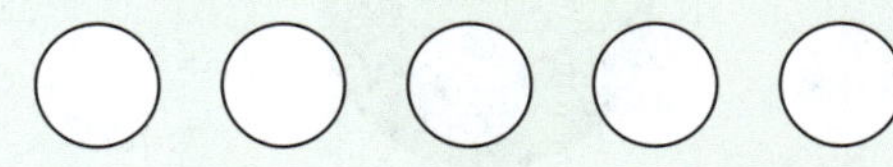
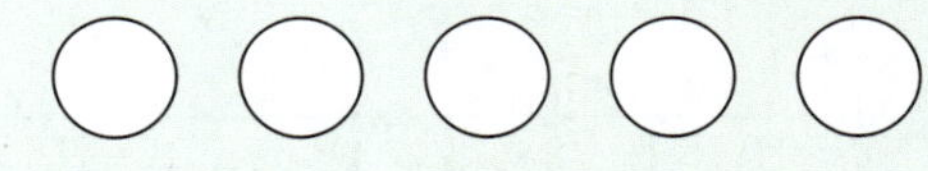
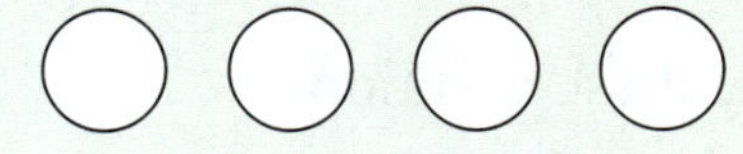

4 a Add 3 each time:

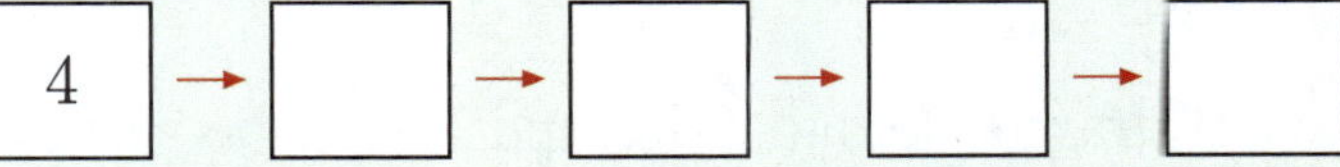

b Add 4 each time:

5 Describe what is happening in this sequence.

I start with _______ , and __________________ each time.

6 Complete these pairs of additions.

a $6 + 2 =$ _______

$2 + 6 =$ _______

b $5 + 8 =$ _______

$8 + 5 =$ _______

c $0 + 9 =$ _______

$9 + 0 =$ _______

7 Use $=$ or $\neq$.

   a $1+8$ _____ 9 **b** $6+6$ _____ 10 **c** $9+8$ _____ 17

8 Complete these additions.

   a $5+$ _____ $=7$ **b** _____ $+6=9$ **c** $9+$ _____ $=10$

   d $7+$ _____ $=14$ **e** _____ $+4=12$ **f** $6+$ _____ $=15$

9 Complete these additions.

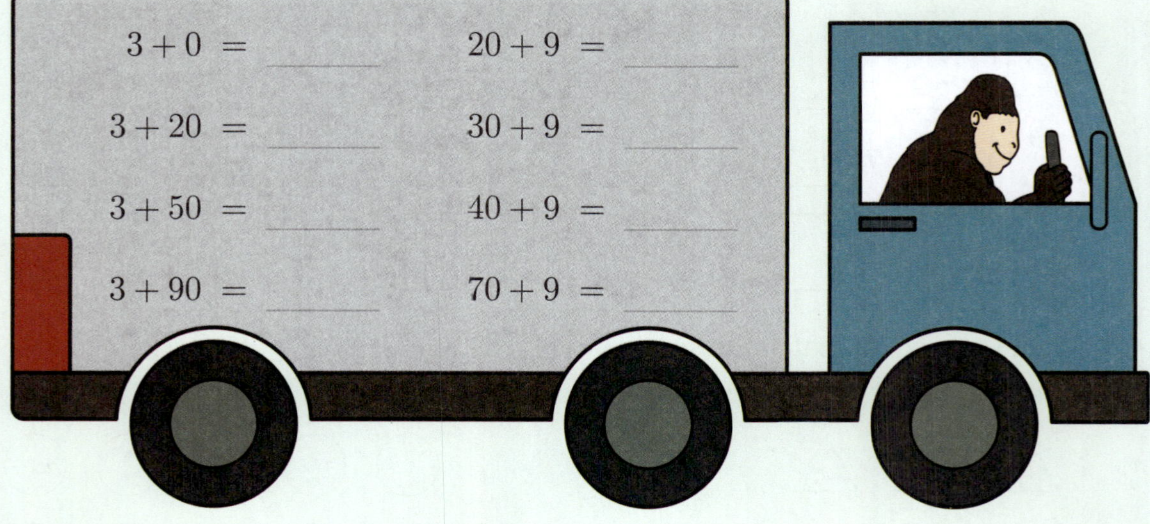

$$3+0 = \rule{2cm}{0.4pt} \qquad 20+9 = \rule{2cm}{0.4pt}$$
$$3+20 = \rule{2cm}{0.4pt} \qquad 30+9 = \rule{2cm}{0.4pt}$$
$$3+50 = \rule{2cm}{0.4pt} \qquad 40+9 = \rule{2cm}{0.4pt}$$
$$3+90 = \rule{2cm}{0.4pt} \qquad 70+9 = \rule{2cm}{0.4pt}$$

10

$$0\ \ 1\ \ 2\ \ 3\ \ 4\ \ 5\ \ 6\ \ 7\ \ 8\ \ 9\ \ 10\ \ 11\ \ 12\ \ 13\ \ 14\ \ 15\ \ 16\ \ 17\ \ 18\ \ 19\ \ 20$$

Complete these additions.

   a $2+3+4=$ _____   **b** $3+5+5=$ _____   **c** $6+4+2=$ _____

11 Amelia gave 9 strawberries to Erica and 7 strawberries to Fernando.

   In total, Amelia gave away

   _________________ $=$ _____ strawberries.

12 Write an addition to find the total number of:

   a tulips _____

   b daisies _____

   c roses _____

   d pink flowers _____

   e white flowers _____

Tulips   Daisies   Roses

13 Complete these additions.

   a $4 + 2 = $ _______

       $40 + 20 = $ _______

   b $5 + 3 = $ _______

       $50 + 30 = $ _______

   c $6 + 4 = $ _______

       $60 + 40 = $ _______

14 Complete each addition:

a

	T	U
	2	2
+		5

b

	T	U
	3	5
+	1	4

c

	T	U
	2	3
+	5	2

15 Set out each addition and find the answer:

   a $23 + 12$

   b $30 + 17$

   c $47 + 21$

16 Along the street there are 32 trees and 26 shrubs.

How many plants are there in total?

There are ☐ plants in total.

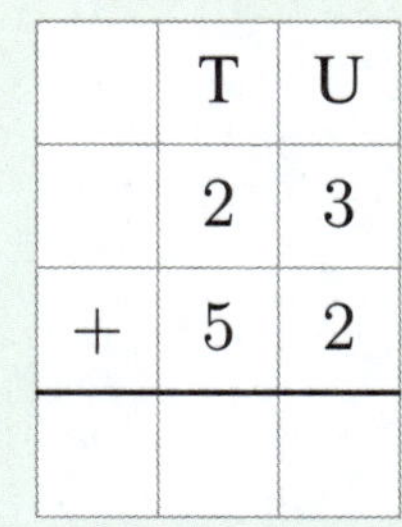

17 Complete:

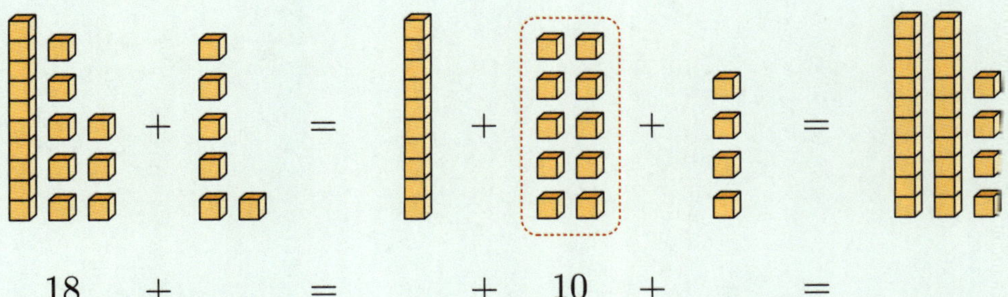

   18 $+$ _____ $=$ _____ $+$ 10 $+$ _____ $=$ _____

18 Use blocks to help you add:

   a $17 + 5 = $ _______

   b $23 + 7 = $ _______

   c $18 + 13 = $ _______

Revision   Shape

1 Name each shape:

_______________   _______________   _______________

_______________   _______________   _______________

2 Complete:

  a An octagon has _______ edges and _______ corners.

  b A hexagon has _______ more edges than a triangle.

  c A triangle has _______ fewer corners than a pentagon.

3 Draw each shape. Use a ruler if needed.

  a a square   **b** a hexagon

4 How would you *describe* a quadrilateral to someone?

____

____

5 Name each solid:

a 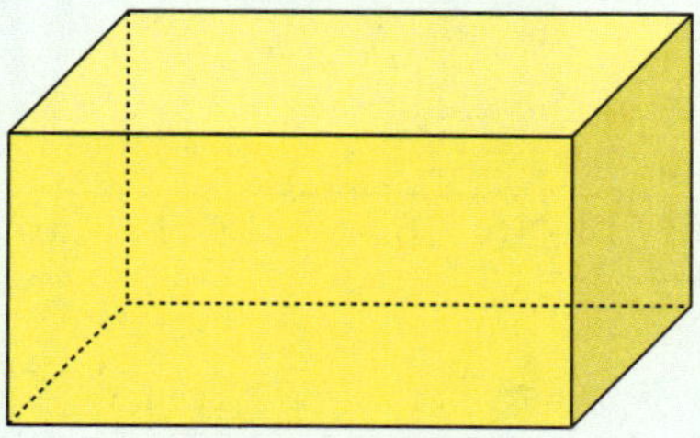   **b** 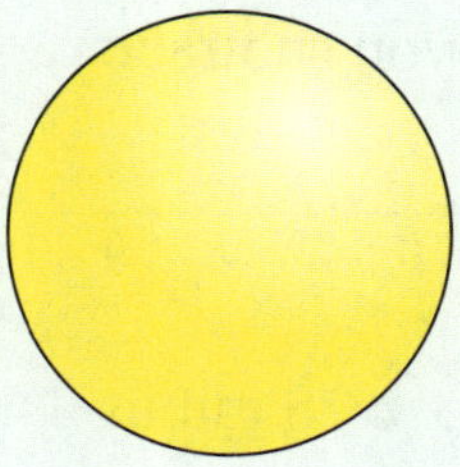   **c**

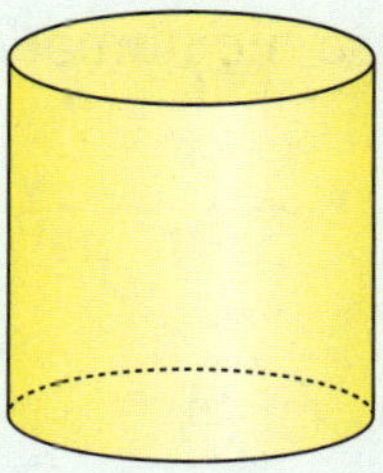

____

____

6 Complete:

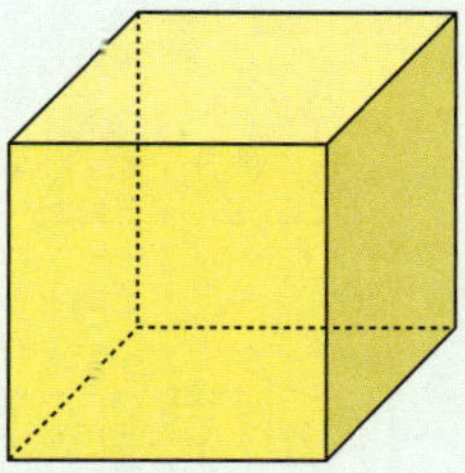

a This is a _______________ .

It has _______ faces, _______ edges, and _______ corners.

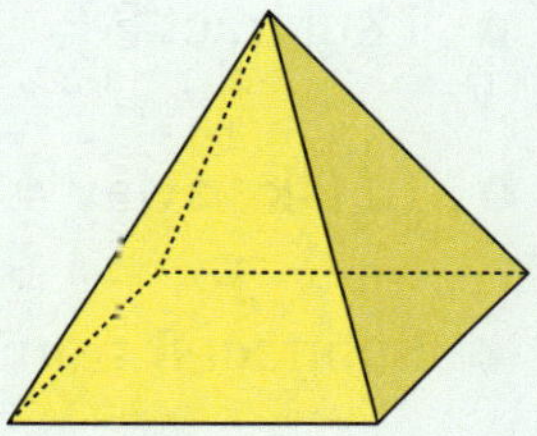

b This is a ____ .

_________ face is a square.

The other _______ faces are triangles.

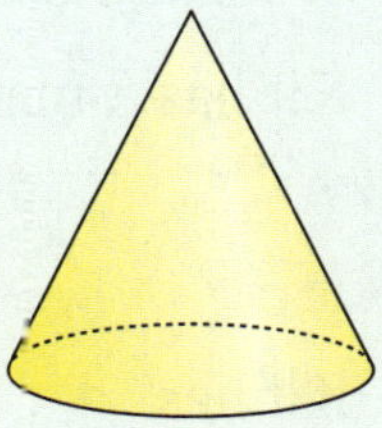

c This is a _______________ .

It has _______ flat face and _______ curved surface.

7 **a** The faces of a rectangular prism are all

_________________________ .

b Sketch a rectangular prism in the space
alongside.

Revision — Subtraction

1 Use the number line to help you subtract:

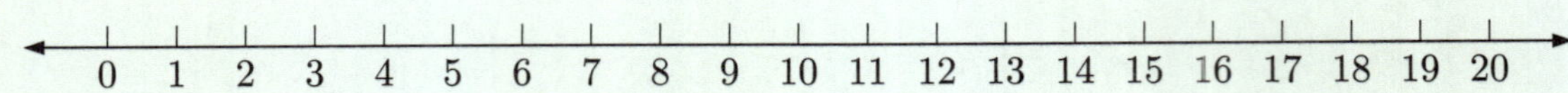

a $7 - 4 = $ _______

b 8 minus $3 = $ _______

c $9 - 5 = $ _______

d $11 - 2 = $ _______

e 14 minus $7 = $ _______

f $18 - 5 = $ _______

2 Cross out pictures to help you subtract.

a

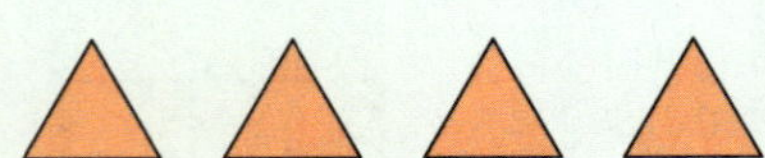

$8 - 5 = $ _______

b

$9 - 7 = $ _______

3 **a** 7 subtract $2 = $ _______

b 10 take away $4 = $ _______

c Subtract 5 from 12:   _______ $-$ _______ $=$ _______

4 Write a subtraction to solve each problem:

a Kit has 8 marbles. He gives 2 marbles to his friend.

_______ $-$ _______ $=$ _______

Kit has _______ marbles left.

b A team has 13 players. 4 players are injured.

_______ $-$ _______ $=$ _______

_______ players are not injured.

5 Use the addition to write *two* subtractions.

a $6 + 3 = 9$

so $9 - $ _______ $=$ _______

and $9 - $ _______ $=$ _______

b $8 + 7 = $ _______

so   _______ $- 8 = $ _______

and   _______ $- 7 = $ _______

6 Complete these subtractions.

a $7 - 0 =$ _______   **b** $4 - 4 =$ _______   **c** $9 - 0 =$ _______

7 Complete each pattern of subtractions:

a

$13 - 10 =$ _______

$23 - 10 =$ _______

$33 - 10 =$ _______

$43 - 10 =$ _______

$53 - 10 =$ _______

$63 - 10 =$ _______

$73 - 10 =$ _______

$83 - 10 =$ _______

$93 - 10 =$ _______

b

$6 - 4 =$ _______

$16 - 4 =$ _______

$26 - 4 =$ _______

$36 - 4 =$ _______

$46 - 4 =$ _______

$56 - 4 =$ _______

$66 - 4 =$ _______

$76 - 4 =$ _______

$86 - 4 =$ _______

8 Complete these subtractions.

a $3 - 1 =$ _______   **b** $7 - 3 =$ _______   **c** $9 - 2 =$ _______

$30 - 10 =$ _______ $70 - 30 =$ _______ $90 - 20 =$ _______

9 Use $=$ or $\neq$ to complete:

a $8 - 4$ _____ 4 **b** $14 - 6$ _____ 9 **c** $16 - 9$ _____ 7

10 **a** Take away 4 each time:

$\boxed{19} \rightarrow \boxed{} \rightarrow \boxed{} \rightarrow \boxed{} \rightarrow \boxed{}$

b Take away 10 each time:

$\boxed{67} \rightarrow \boxed{} \rightarrow \boxed{} \rightarrow \boxed{} \rightarrow \boxed{}$

11 Describe what is happening in this sequence.

I start with _______ , and subtract _______ each time.

12 Find the missing number.

a $7 - \boxed{} = 6$ **b** $11 - \boxed{} = 5$ **c** $19 - \boxed{} = 9$

13 Find the missing number.

a $\boxed{} - 3 = 3$ **b** $\boxed{} - 4 = 6$ **c** $\boxed{} - 5 = 8$

14 Write an addition and a subtraction to solve this problem.

There are 12 woodpeckers and 4 owls in a tree.

In total, there are ____________ = _______ birds.

There are ____________ = _______ more woodpeckers than owls.

15 Complete these subtractions.

a

T	U
3	8
− 3	4

b

T	U
2	7
− 1	3

c

T	U
4	6
− 2	6

16 Set out these subtractions and find the answer.

a $28 - 12$ **b** $37 - 26$ **c** $44 - 31$

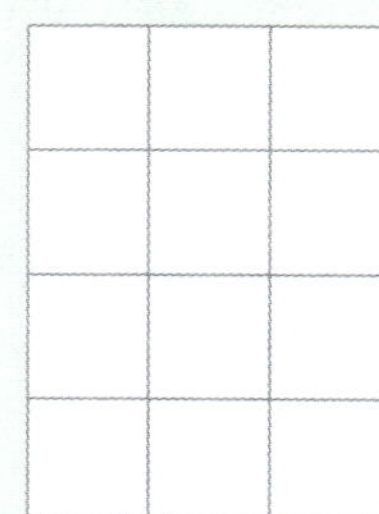

17 Joan baked 36 cupcakes for a party.

23 cupcakes were eaten.

How many cupcakes were left?

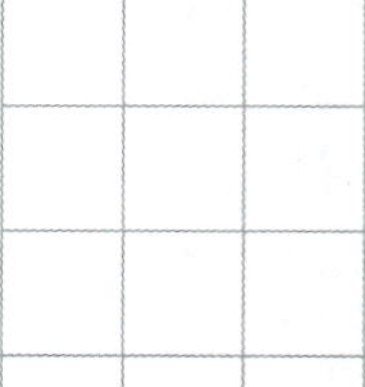

There were $\boxed{}$ cupcakes left.

Revision   Measurement

1

Complete using *taller*, *shorter*, *tallest* or *shortest*:

a Tree **B** is ________________ than tree **A**.

b Tree **A** is ________________ than tree **C**.

c Tree **D** is the ______________ tree.

d Tree **C** is the ______________ tree.

2 Use your ruler to measure each line. Write your answer in the space provided.

a ________________ ________ cm

b ____ ________ cm

c ________________________ ________ cm

3 Write these areas in order:

tennis court

picnic rug

golf course

largest

smallest

4 Complete each statement:

a The block is smaller than the ______________ .

b The box is larger than the ______________
and the ______________ .

c The smallest object is the ______________ .

d The largest object is the ______________ .

block   die   box

5

pool   bucket   bathtub

a The object with the smallest capacity is the ______________ .

b The object with the largest capacity is the ______________ .

c The bathtub has a larger capacity than the ______________ .

6

netball   bowling ball   tennis ball

Write the balls in order of mass.

______________   ______________   ______________

lightest ⟶ heaviest

7 Complete using *less than*, *more than*, or *the same as*:

a

The phone weighs

____

5 blocks.

b

The watermelon weighs

____

9 blocks.

8 Write the words *warm*, *cold*, *hot*, and *cool* in order.

______________   ______________   ______________   ______________

colder ⟶ hotter

1 Complete these times tables.

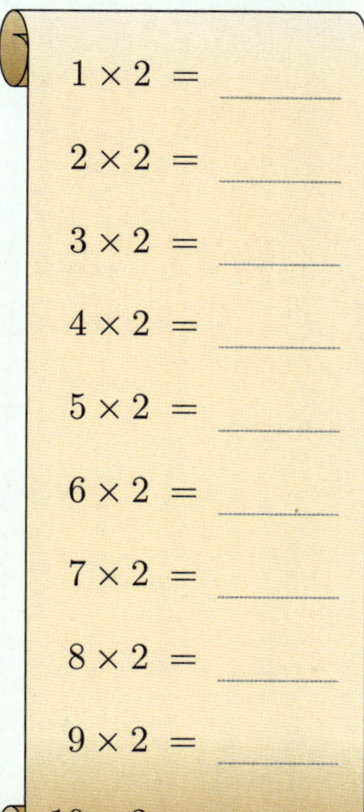

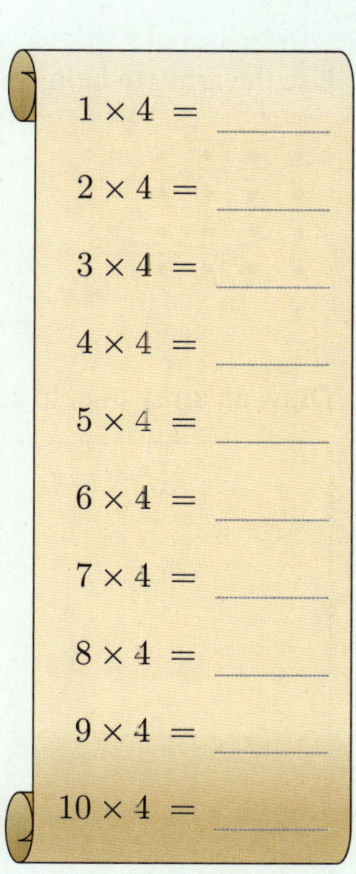

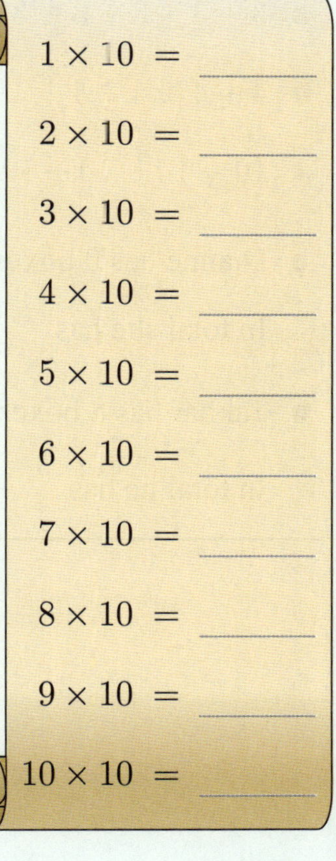

2 The multiples of 10 are:

10, _______ , _______ , 40, _______ , _______ , 70, _______ , _______ , _______ , ...

3 Use the array to help find the multiplication.

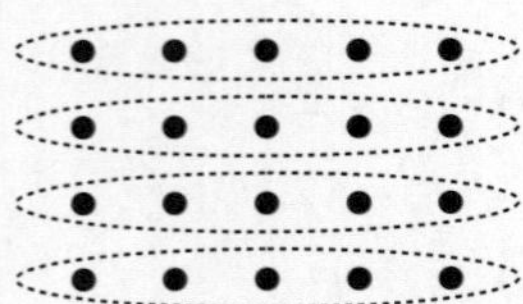

$5 + 5 +$ _______ $+$ _______ $=$ _______

_______ lots of _______ $=$ _______

_______ $\times$ 5 $=$ _______

4 Draw an array to help answer the multiplication.

$6 + 6 + 6 =$ _______

$3 \times 6 =$ _______

5 Write each addition as a multiplication, and find the answer:

a $3 + 3 + 3 + 3 + 3 =$ _______ $\times$ _______ $=$ _______

b $4 + 4 + 4 + 4 + 4 + 4 + 4 =$ _______ $\times$ _______ $=$ _______

c $10 + 10 + 10 + 10 + 10 + 10 =$ _______ $\times$ _______ $=$ _______

6 a Dianne has 9 boxes with 2 shoes in each.

In total she has _______ $\times$ _______ $=$ _______ shoes.

b Takhar has 8 boxes with 6 glasses in each.

In total he has _______ $\times$ _______ $=$ _______ glasses.

Revision — Data handling

1 Mr Farrer asked his class which game they liked most on Sports Day.

Game	Tally
Tunnel ball	卌 ||||
Tug 'o' war	||||
Running relay	卌 |||
Tennis ball throw	卌 ||

a How many students chose the running relay? ______

b 7 students chose the

____ .

c The most popular game was

____ .

2 Count these items using a tally.

Fruit	Tally
blueberry	
raspberry	
strawberry	

There are ☐ blueberries, ☐ raspberries, and ☐ strawberries.

There are ☐ berries in total.

3 This pictograph shows the flavours of soup sold at a restaurant.

Soup flavours sold

Pumpkin	🥣 🥣 🥣 🥣 🥣
Sweet potato	🥣 🥣 🥣
Vegetable	🥣 🥣 🥣 🥣 🥣 🥣
Chicken	🥣 🥣 🥣 🥣

🥣 = 1 bowl of soup

a How many bowls of chicken soup were sold? ______________

b 5 bowls of ____________________ soup were sold.

c Which soup flavour was the most popular? ____________________

d How many more bowls of vegetable soup were sold than sweet potato soup?

____

4 Baby Ava plays with this collection of building blocks.

a Count the blocks by *colour*.

There are ⬚ blue blocks, ⬚ red blocks,

and ⬚ yellow blocks.

Colour	Tally
blue	
red	
yellow	

b Count the blocks by *shape*.

There are ⬚ cubes,

⬚ square-based pyramids,

⬚ cylinders, and

⬚ cones.

Shape	Tally
cube	
square-based pyramid	
cylinder	
cone	

5 Jennifer recorded the favourite meal of each student in her class.

Draw a pictograph to show this information.

Use 🧍 = 1 student.

Favourite meal	Tally
breakfast	ⵌ|||
lunch	ⵌ|
dinner	ⵌ||||

Favourite meal

breakfast	
lunch	
dinner	

Revision — Division

1 Describe the division:

10 has been divided into _______ equal groups of _______ .

2 **a** Share 6 into 2 equal groups.

 b Share 18 into 3 equal groups.

3 Draw diagrams to show how 9 can be divided equally.

1 group of _______

_______ equal groups of _______

_______ equal groups of _______

4 Shade one half of each shape:

a

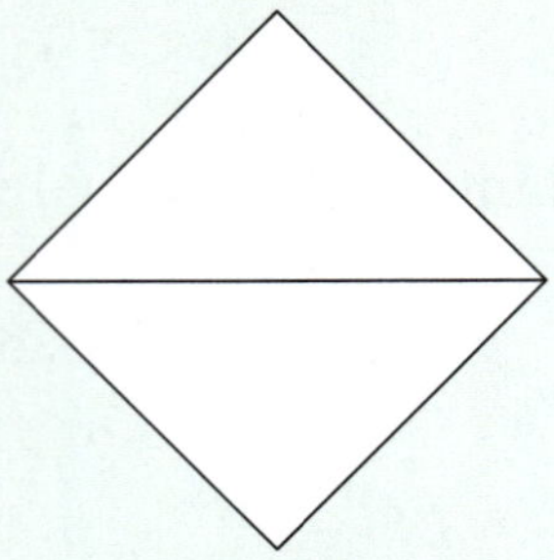

b

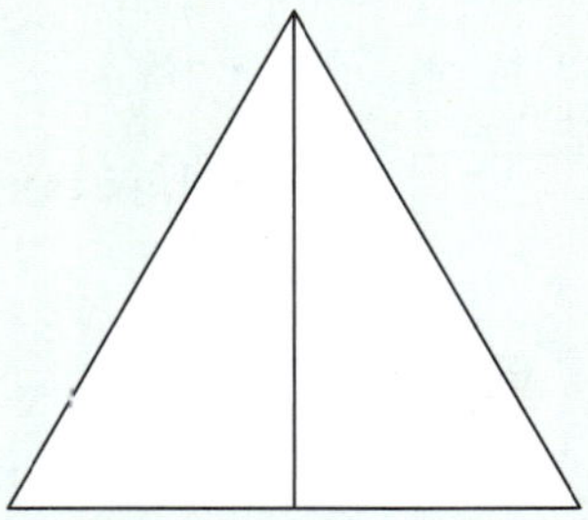

c

5 Circle one half of the objects.

a

b

One half of _______ is _______ .   One half of _______ is _______ .

6 Shade the correct amount.

a one quarter

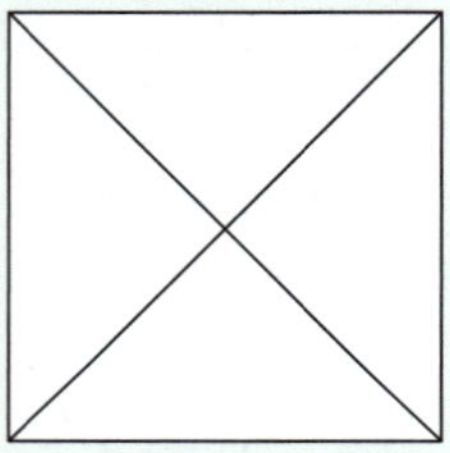

b three quarters

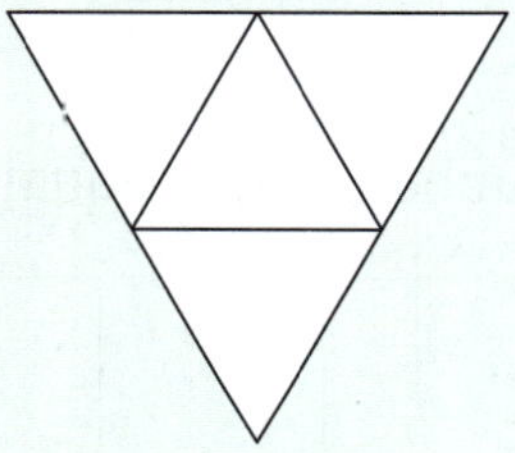

c four quarters

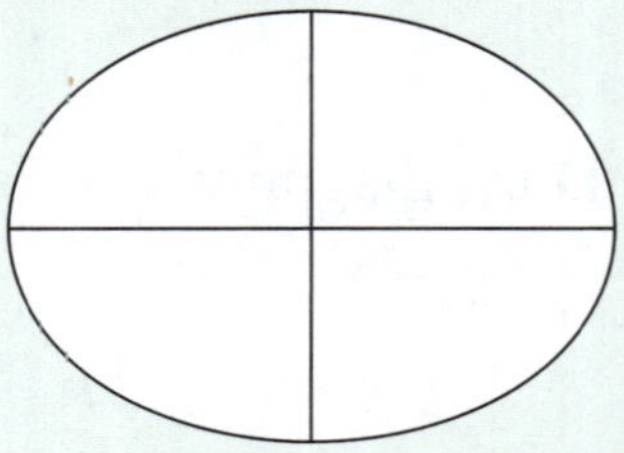

d one quarter

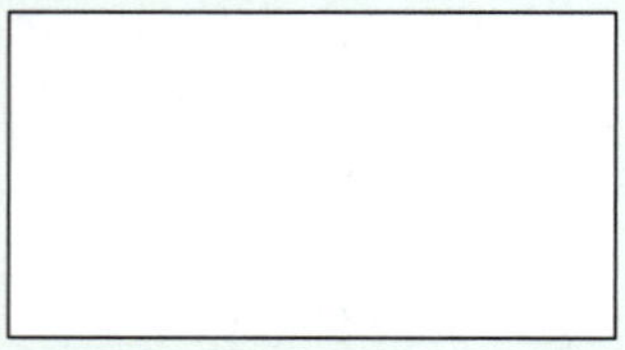

e two quarters

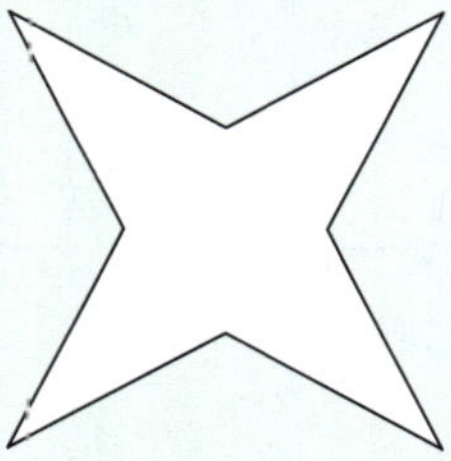

f three quarters

7 Circle one quarter of the objects.

a

b

One quarter of _______ is _______ .

One quarter of _______ is _______ .

8 Shade the correct amount.

a one third

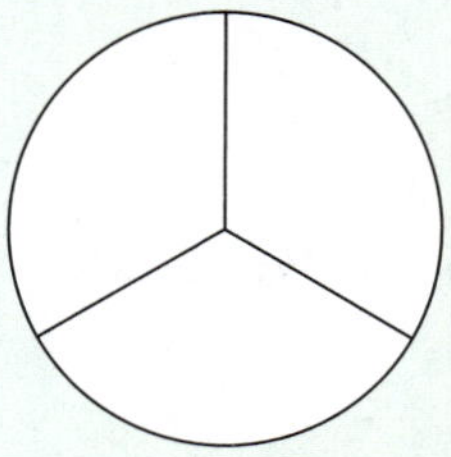

b one sixth

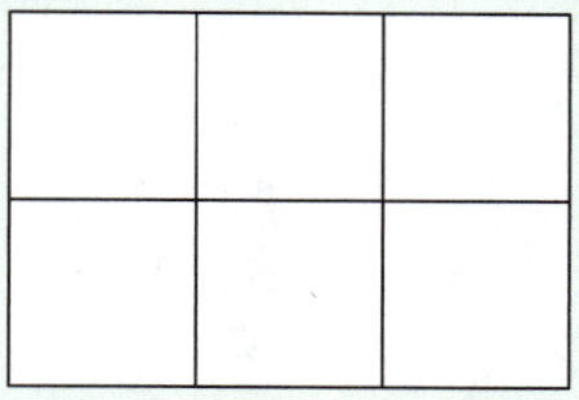

c three eighths

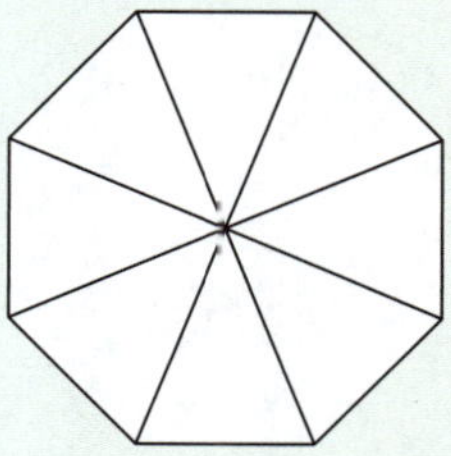

d two thirds

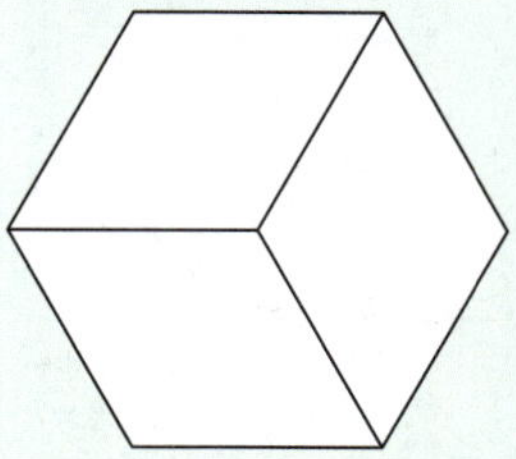

e five eighths

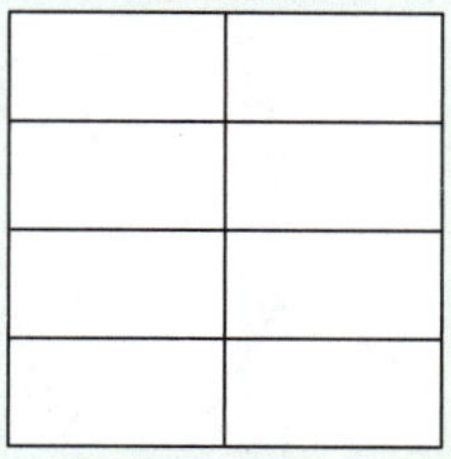

f two sixths

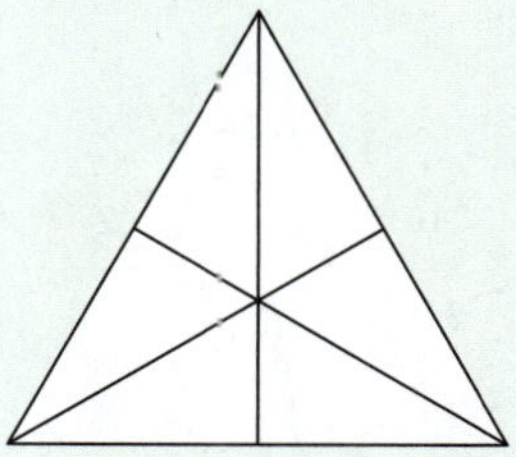

g one third

h five sixths

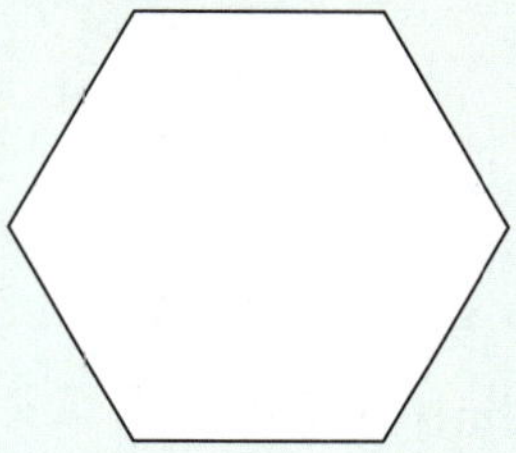

i two eighths

9 How much is shaded?

a

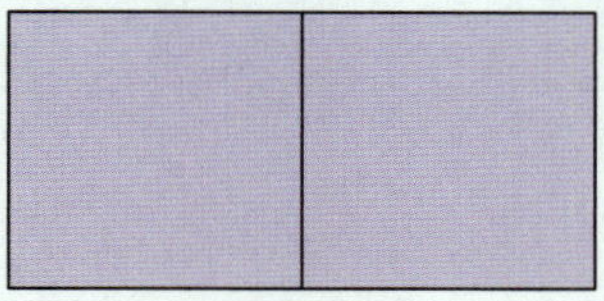

b

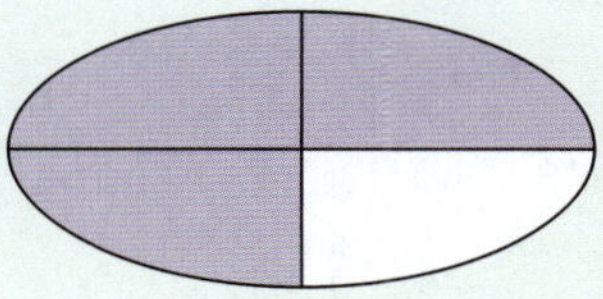

____

c

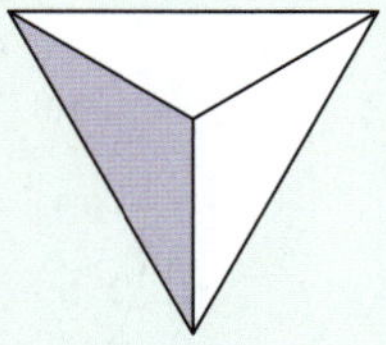

d

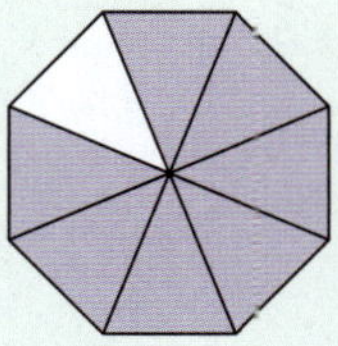

____

e

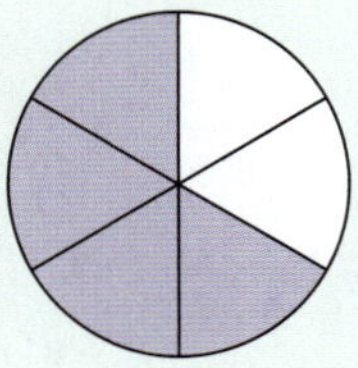

f

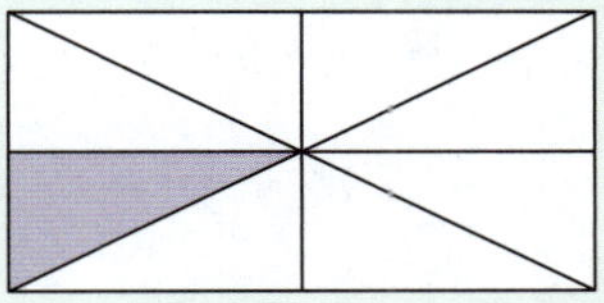

____

Revision   Time

1 What time is shown on the clock?

a 

____________ past ____________

b 

____

c 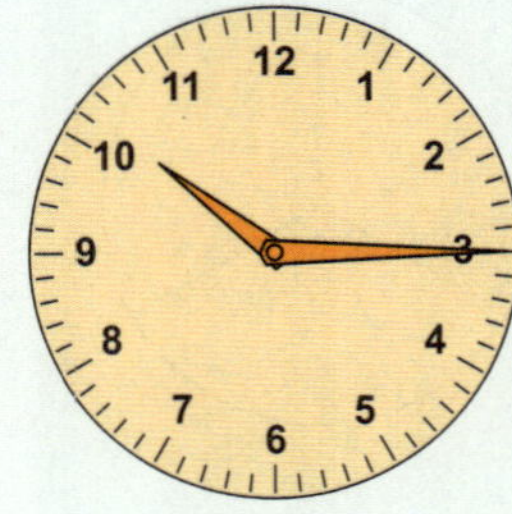

____________ past ____________

d 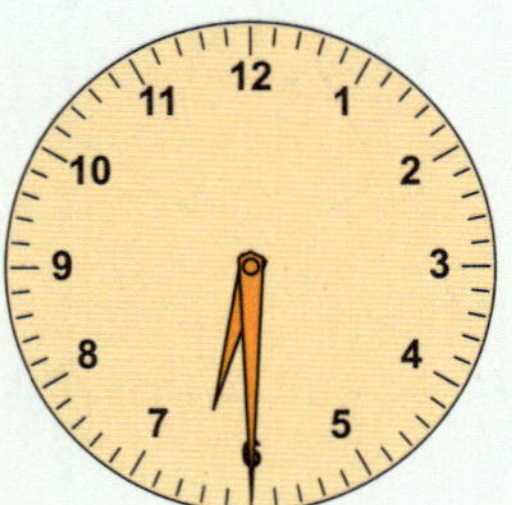

____

____

e 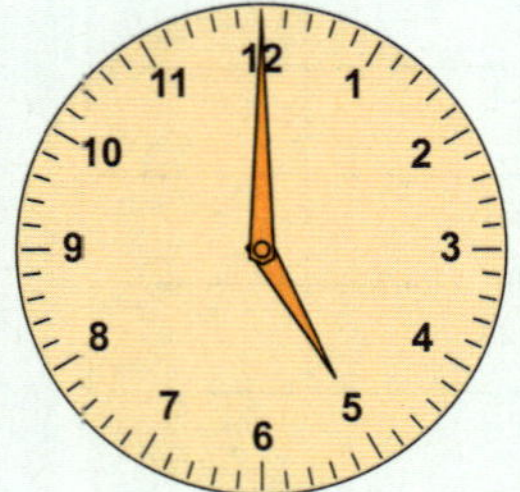

____

____

f 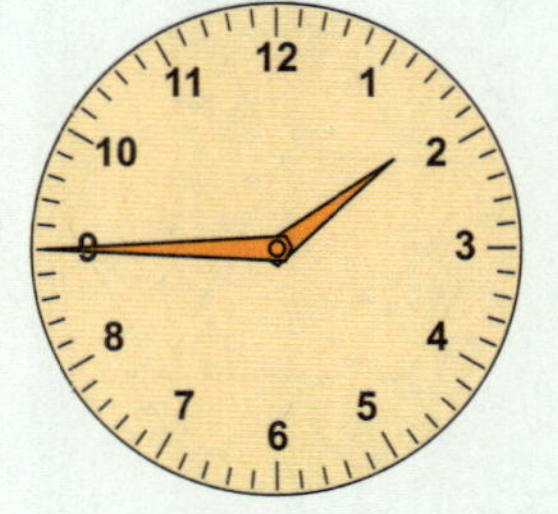

____

____

2 Draw hands on each clock face to show the time:

a 2 o'clock

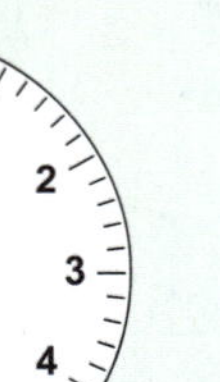

b half past 8

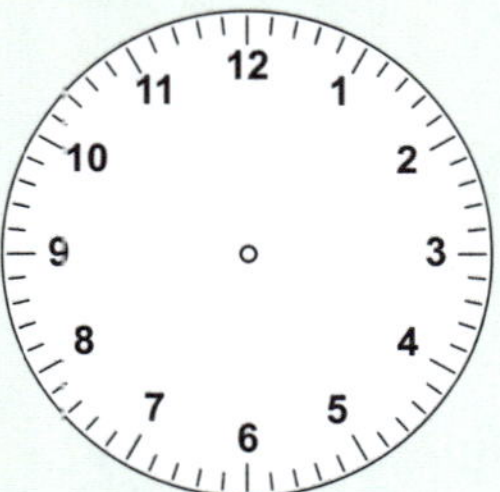

c 4 o'clock

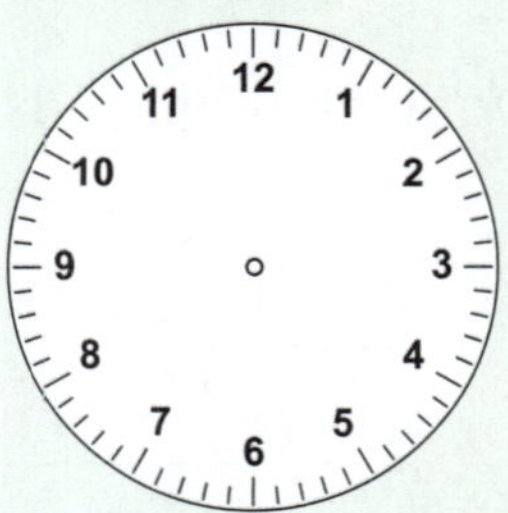

d quarter past 1

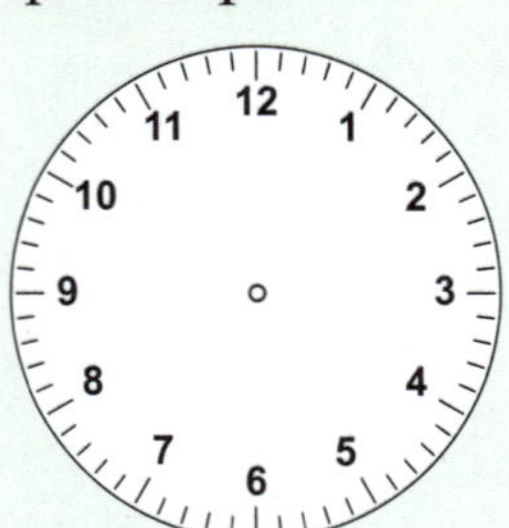

e half past 11

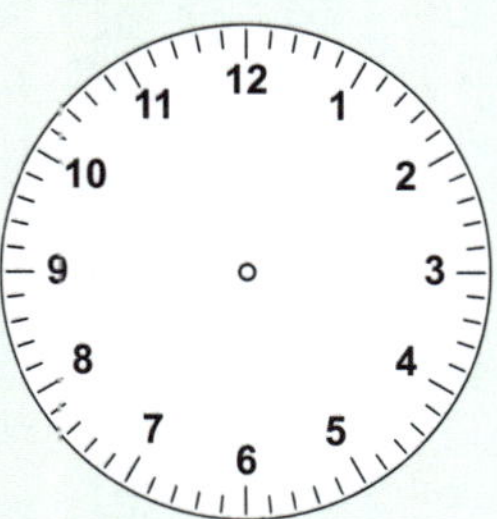

f quarter to 3

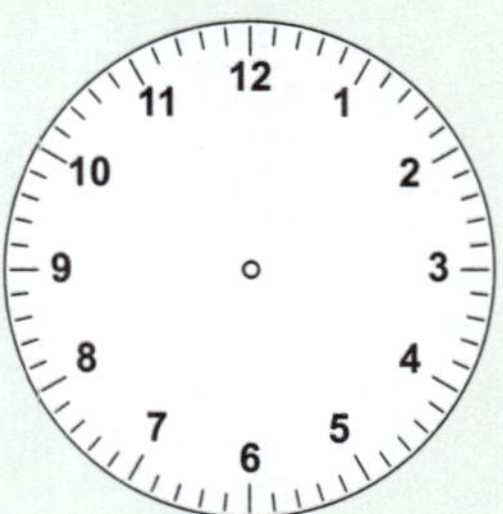

g quarter to 10

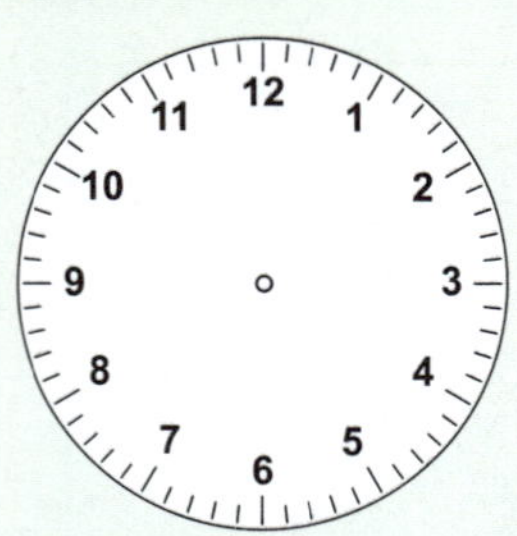

h quarter past 6

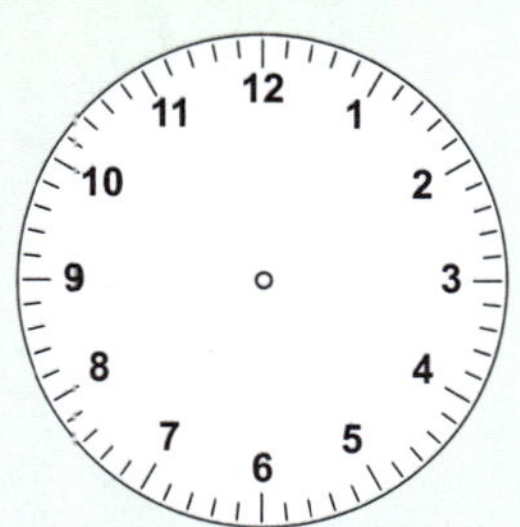

i quarter to 12

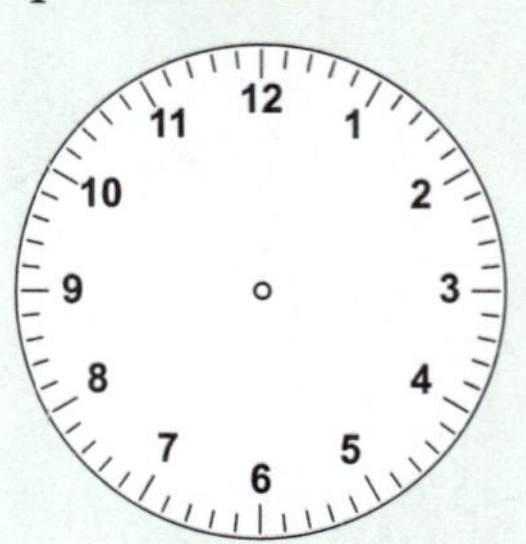

3 There are:

a ______ minutes in 1 hour   **b** ______ hours in 1 day

c ______ days in 1 week   **d** ______ months in 1 year.

4 How many months are there in 2 years?

In 2 years, there are ☐ months.

	T	U
+		

5 **a** The day after Thursday is ________________ .

b The day before Tuesday is ________________ .

6 **a** The month after May is ________________ .

b The month before November is ________________ .

c September has ______ days.

d March has ______ days.

7 **a** Use the calendar to fill in the day of the week:

________________ , 19th August

________________ , 10th August

________________ , 28th August

________________ , 2nd August

August 2025						
Su	M	T	W	Th	F	Sa
					1	2
3	4	5	6	7	8	9
10	11	12	13	14	15	16
17	18	19	20	21	22	23
24	25	26	27	28	29	30
31						

b The day before Monday, 25th August is

__________________________________ .

The day after Saturday, 16th August is

__________________________________ .

Revision   Chance

1 Simone is blindfolded. She has been asked to pick a
ball out of this bag.

Use *more* or *less* to complete each statement.

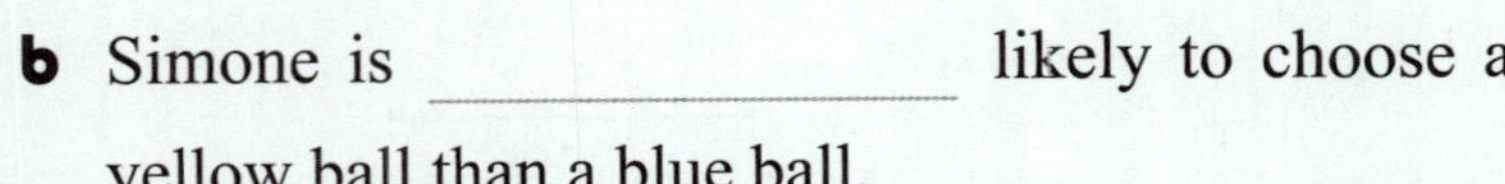

a Simone is ________________ likely to choose a
blue ball than a green ball.

b Simone is ________________ likely to choose a
yellow ball than a blue ball.

c Simone is ________________ likely to choose a green ball than a yellow ball.

2 Match each spinner to the chance it will stop on blue.

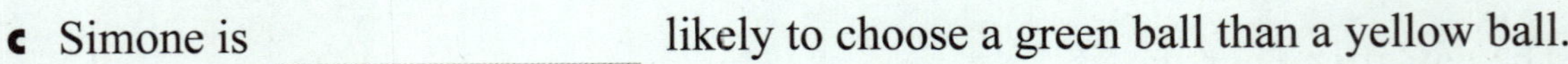

impossible   unlikely   50-50 chance   likely   certain

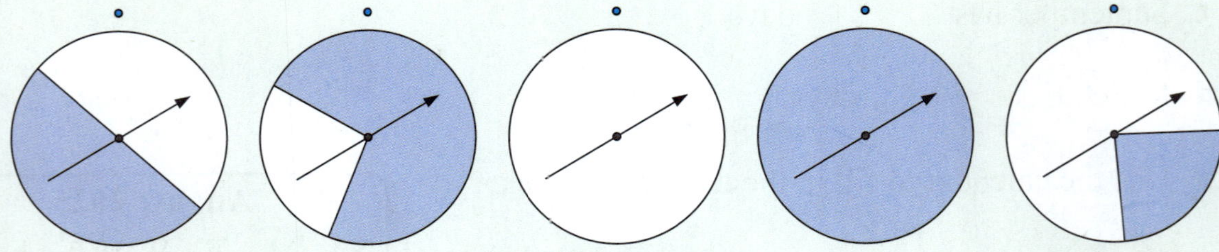

3 I am about to roll a die.

Use *certain*, *likely*, *50-50 chance*, *unlikely*, or *impossible* to describe the
chance that I will:

a roll an "8" ________________________________

b roll a number ________________________________

c roll a "5" or "6" ________________________________

d *not* roll a "3" ________________________________

e roll a "2", "4", or "6" ________________________________

Air Conditioning Service Guide

By Michael Prokup

ISBN: 978-1-930044-87-6

NOTE: This publication is general in nature and is intended for INSTRUCTIONAL PURPOSES ONLY.
It is not to be used for, nor supersedes, equipment manufacturer selection, application, installation or specific service procedures. Always follow manufacturer instructions. Neither Prokup Media Inc. nor ESCO Institute assumes no responsibility for equipment, property, or personal damage resulting from use of the information contained in this service guide.

In Memory of
Michael Prokup
1960-2021

Michael Prokup passed away unexpectedly, at the age of 60, on Tuesday, August 3, 2021. Well before his passing, Mike had been preparing this updated *Second Edition of the R-410A & R-22 Systems Air Conditioning Service Guide* and ESCO is honored to be able to continue Mike's work and share his talent for teaching complex subjects in an easy to understand format.

Mike was an admired and highly respected author, consultant, and business leader within the HVAC industry. Mike is best known for Prokup Media, Inc., his interactive training programs, and R-410A training seminars.

Mike tirelessly wrote and produced a library of training manuals, repair guides, and software titles over his 40-year career in an effort to move the industry forward and continue to be a benchmark in the industry. We bring this manual to you with a heavy heart, and we hope, in some way, you become a better technician and diagnostician for having read and studied this book.

Rest in Peace, Michael Prokup.

Dedication

The publication of this book is dedicated to the Prokup family. Kim, Kyle, Michael, Stephanie, Joe, Andrew, and Buster.

Service Procedure Quick-Lookup

Comfort Cooling System Basics

Basic Cooling System Overview

The illustration below depicts an overview of a typical refrigeration system found in comfort cooling air conditioning systems. The system consists of a metering device, evaporator coil, suction line, compressor, hot gas line, condenser coil, and liquid line. The system uses a refrigerant to transfer heat from the conditioned space (home) to the outdoor air.

The refrigeration circuit consists of a high pressure/high temperature side, and a low pressure/low temperature side. The compressor and the metering device are the pressure changing devices that separate the two sides of the system.

The high pressure/high temperature side of the system rejects heat, and the low pressure/low temperature side of the system absorbs heat. The high pressure side of the system consists of the compressor, discharge line, condenser coil, and liquid line. The low side of the system consists of the evaporator coil, the suction line, and the compressor inlet. Notice the evaporator coil connects the refrigeration system to the heat source. This connection is between the refrigerant inside of the evaporator coil and the air passing across the surface of the evaporator coil.

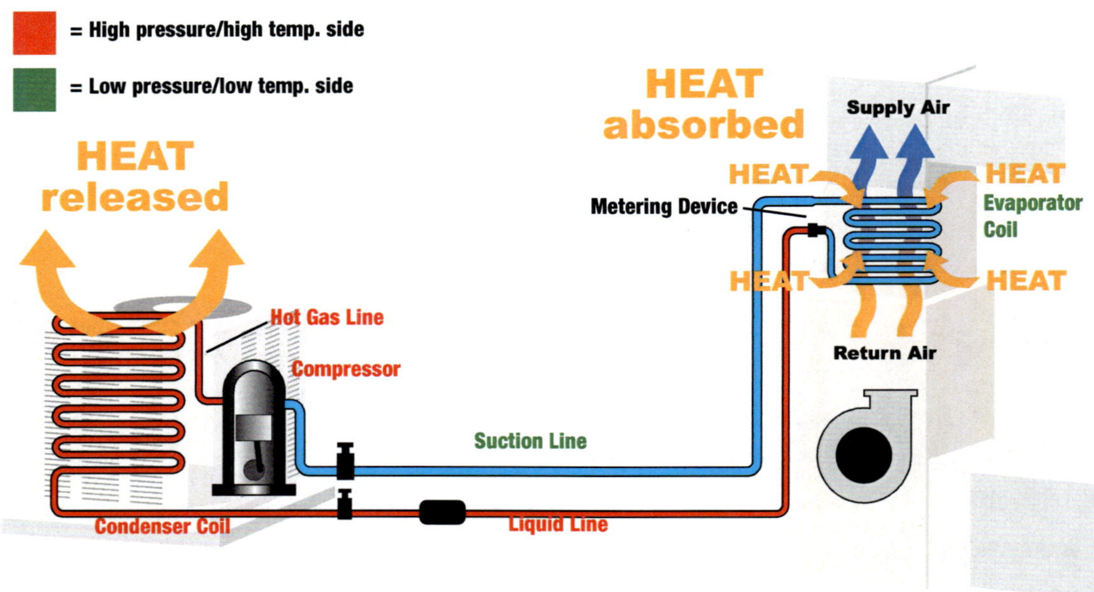

In an air conditioning system, heat from the home is absorbed into the cold refrigerant in the evaporator coil. The refrigerant carries the absorbed heat to the outdoor condenser coil, where the heat is released from the refrigerant by being absorbed by the cooler outdoor air passing through the condenser coil. The operation of the system works because heat always flows from a warmer object to a colder object. As the heat content of the air changes, it will have a profound affect upon the operation of the refrigeration system.

In this chapter, we will introduce the components of the system and how they work together to form the comfort cooling air conditioning system refrigeration cycle.

Refrigerant

Introduction to Refrigerants

Refrigerants are special chemicals developed as a heat transfer media for use in refrigeration systems. In nature, hot goes to cold. In other words, energy flows from high to low. Refrigerants, when warmer than air, will transfer their heat to the air. When the air is warmer than the refrigerant, the heat will flow from the air into the refrigerant.

There are two refrigerants found in modern residential air conditioning systems, R-22 and R-410A. R-22 is an ozone depleting refrigerant. R-410A is a non-ozone depleting refrigerant.

Refrigerant Saturation Temperature

Refrigerants have a special physical property called saturation temperature. Saturation temperature can be considered the temperature of the refrigerant where all the heat transfer power exists. A refrigerant that is at saturation temperature will be a mix of both liquid refrigerant and vapor refrigerant. Any addition or removal of heat from a saturated refrigerant will often cause a change of state. However, the refrigerant may also flash from a liquid to a vapor, or it may condense from a vapor to a liquid, depending on system conditions. These state changes occur when a heat exchange between the refrigerant and air takes place.

At saturation temperature the refrigerant exists as both a liquid and a vapor simultaneously.

A large amount of heat energy is required to make a refrigerant change its state. Therefore, when a refrigerant is at saturation temperature, and there is a source of heat that is warmer or cooler than the refrigerant, the refrigerant will either absorb or reject heat. Although the transfer of heat will cause the refrigerant to change from either a liquid to a vapor, or from a vapor to a liquid, <u>the heat transfer does not change the temperature of the refrigerant</u>!

Determining Refrigerant Saturation Temperature

When suction pressure is converted to saturation temperature, the saturation temperature of the refrigerant exiting the metering device and entering the evaporator circuit is found. When liquid pressure (discharge pressure) is converted to saturation temperature, the saturation temperature of the refrigerant that is condensing from a hot gas into a liquid in the condenser coil is determined. The use of these two saturation temperatures is critical to properly service an air conditioning system, because both saturation temperatures are used in the process of calculating superheat and subcooling levels.

<u>Reading the temperature of a refrigeration pipe will not indicate the pressure inside of the pipe</u>. The refrigerant in the pipe may be colder than saturation temperature (subcooled) or warmer than saturation temperature (superheated). (See Chapter C: Superheat and Chapter D: Subcooling.) The temperature of the pipe has no relationship to the operating pressure, as the refrigerants are not at saturation temperature.

Taking a pressure reading with a refrigeration gauge will enable the service technician to determine the saturation temperature of the refrigerant. Temperature/Pressure tables like that below are used to convert system pressure readings to saturation temperature.

The use of a temperature pressure table may not be necessary in the field, as the table values are typically printed on the face of the refrigeration gauge. Simply read the saturation temperature scale on the gauge to quickly find saturation temperatures.

R-410A PSIG	Saturation Temp. (F)	R-22 PSIG
12	-41	0
14	-36	2
16	-32	4
18	-28	6
22	-24	8
26	-20	10
30	-17	12
32	-14	14
36	-11	16
38	-8	18
42	-5	20
46	-2	22
48	0	24
50	2	26
54	5	28
58	7	30
60	9	32
64	11	34
68	13	36
70	15	38
74	17	40
78	19	42
80	21	44
84	23	46
86	24	48
90	26	50
98	30	55
106	34	60
114	38	65
122	41	70
128	44	75
138	48	80

R-410A PSIG	Saturation Temp. (F)	R-22 PSIG
146	51	85
154	54	90
160	56	95
168	59	100
178	62	105
184	64	110
192	67	115
198	69	120
208	72	125
216	74	130
222	76	135
230	78	140
240	81	145
248	83	150
256	85	155
262	87	160
270	89	165
280	91	170
288	93	175
292	94	180
300	96	185
308	98	190
318	100	195
322	101	200
330	103	205
340	105	210
346	106	215
356	108	220
364	110	225
370	111	230
380	113	235
386	114	240

R-410A PSIG	Saturation Temp. (F)	R-22 PSIG
396	116	245
400	117	250
412	119	255
416	120	260
422	121	265
434	123	270
440	124	275
450	126	280
456	127	285
462	128	290
474	130	295
482	131	300
494	133	310
512	136	320
526	138	330
544	141	340
560	143	350
572	145	360
596	148	370
608	150	380
624	152	390
640	154	400

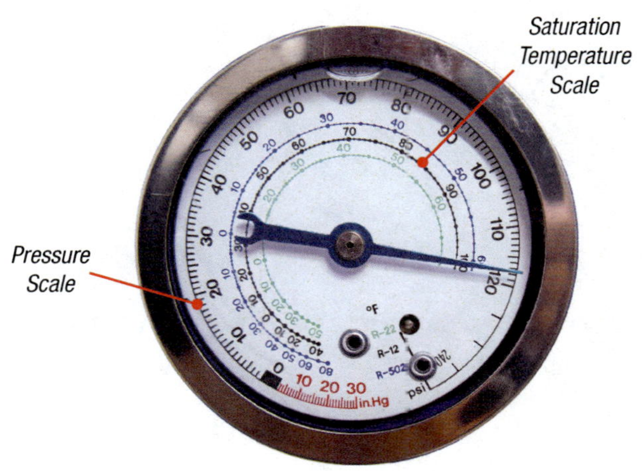

Refrigeration gauges display both the pressure and saturation temperature. This particular gauge shows the saturation temps for R-22, R-12, and R-502.

R-22 requires different refrigeration gauges than R-410A. R-410A refrigeration hoses are higher rated than R-22 gauge hoses. Make certain to use the proper refrigeration gauges and hoses for the two refrigerants.

Temperature/pressure tables are published by refrigerant chemical companies that plot gauge pressure and saturation temperatures. Each refrigerant will have its own unique table. This is an example of a combination R-22 and R-410A Temperature/Pressure Table.

Evaporator Coil

The evaporator coil is part of the low pressure side of the system. The coil is typically made from copper or aluminum tubing that has been expanded to form a bond with aluminum heat transfer fins. When the tubing is expanded at the manufacturing facility, the tubing becomes extremely thin. Briefly placing a brazing torch flame on the tubing is likely to melt the tubing if precautions are not taken. This makes repairing leaks extremely difficult for novices.

Configuration

The copper tubing is arranged together in what are called circuits. The circuits direct the refrigerant through the coil so there is near uniform flow of refrigerant through all parts of the coil. The refrigerant enters the evaporator circuits from distributor tubes that are on the outlet side of the metering device. If there is an inlet, then there has to be an outlet to the coil. At the outlet, the individual circuits are piped together. The outlet of the coil circuiting is called the Suction Line.

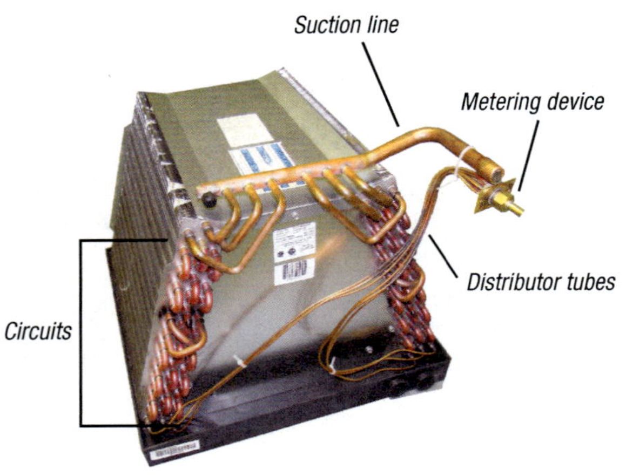

In a properly operating system, the majorirty of the coil has cold saturated refrigerant in its circuits. The cold refrigerant absorbs heat from air passing across the surface of the coil. The evaporator coil also removes moisture from the air just like a cold glass of ice water will do on a humid day.

The water that collects on the surface of the evaporator coil, called condensate, collects in the condensate drain pan, which is located directly under the evaporator coil. The drain pan has a fitting, which is typically threaded, where piping is connected. The piping carries the condensate to a convenient waste drain or pump.

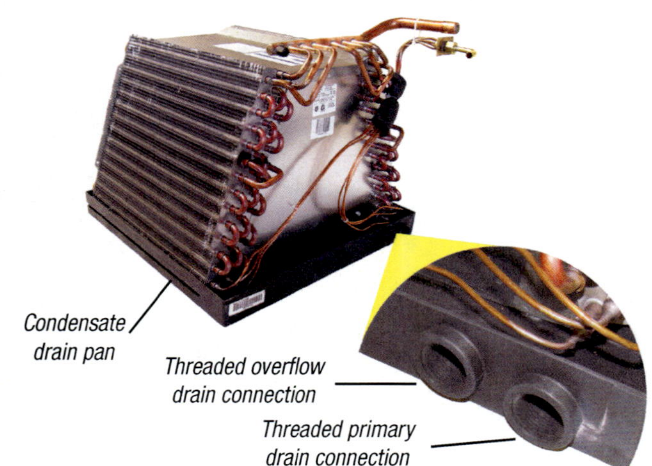

This drain must be connected to the evaporator coil for proper draining of condensate to occur. If the drain becomes clogged, water will overflow the condensate pan and may cause damage to the air handler, furnace, or structure. This is a common occurrence service technicians will deal with many times in their careers.

Evaporator coils used in residential homes can come in either an "A" type, or "slab" type coil configuration.

The "A" type coil is often installed on the supply air outlet side of a gas furnace. Air from the furnace blower assembly blows into the bottom of the coil and out the top. This type of coil should never be installed on the inlet return air side of the furnace. The cold air will cause condensation to drip on the furnace components, potentially damaging them.

Air handlers have coils installed internally within the air handler cabinet. These coils can be a single slab type or may be shaped like an "A" coil. In many cases, the coil is installed on the return air side of the blower assembly.

Horizontal duct coils are coils that are installed in duct systems where they are not directly attached to the furnace like an "A" coil is. These types of coils can be slab shaped or "A" shaped. They are installed on the supply air outlet side of the furnace.

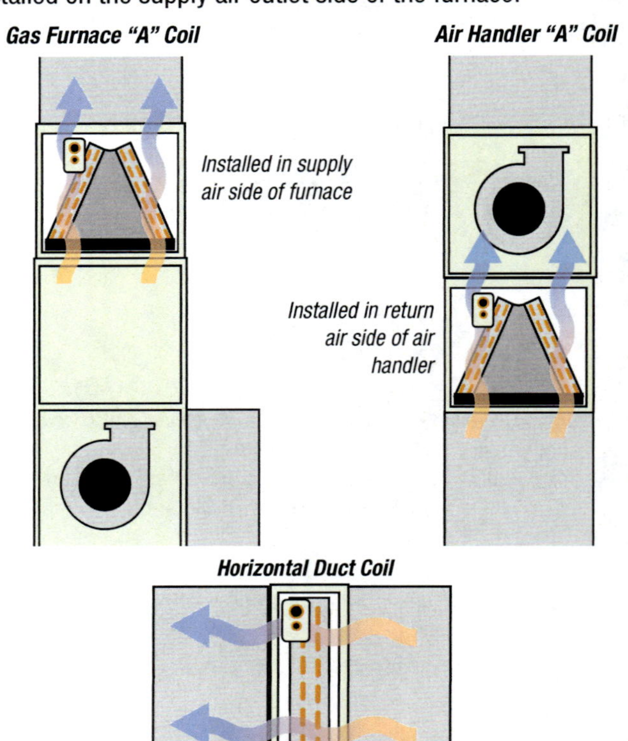

Comfort Cooling System Basics

Evaporator coils come in a variety of capacity sizes. They are selected to match the size of the outdoor condensing unit. For example, if the outdoor condensing unit is a 2 ton model, a matching 2 ton evaporator coil size is usually selected. To get higher efficiency, some systems have evaporator coils that have a larger capacity size than the outdoor condensing unit. These coils will run with higher suction pressure and warmer evaporator saturation temperatures. The elevated evaporator temperature may not be a good choice for comfort, but may be good for efficiency.

Function

The evaporator coil is placed in direct contact with the air from the space being cooled. An indoor blower assembly will circulate the air from the conditioned space through the evaporator coil.

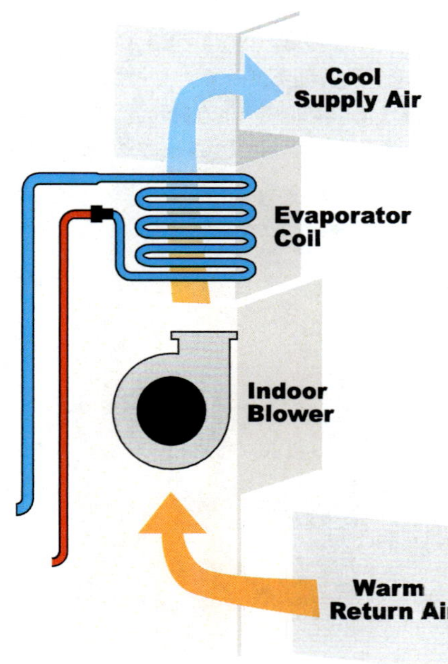

The refrigerant that enters the evaporator coil is in a saturated state, and any addition of heat at saturation temperature will cause the refrigerant liquid to vaporize as it travels through the evaporator coil's circuits. It is during this time in the evaporator coil that massive levels of heat are being transferred from the conditioned space's warmer air to the colder saturated refrigerant. The massive heat transfer taking place is due to the large amount of thermal energy required for the cold liquid refrigerant to change its state from a liquid into a vapor. This process is the same process that makes ice such a good refrigerant!

In this example, we have a suction pressure of 130 PSIG with R-410A refrigerant. The corresponding saturation temperature for R-410A at this pressure is 44°F. This is the temperature of the refrigerant as it flashes from a liquid to a vapor.

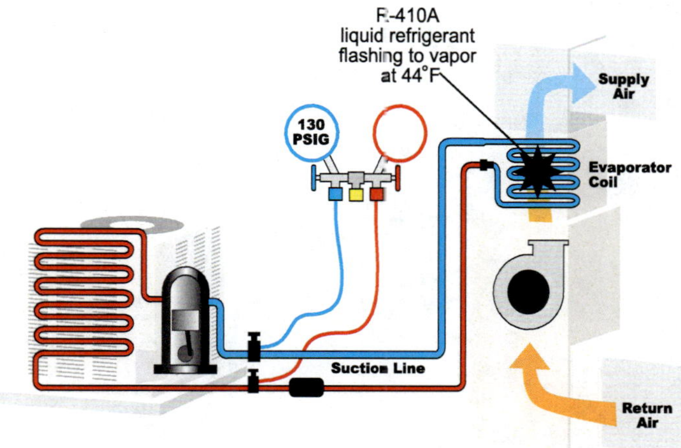

Assuming a properly working system, at some point in the evaporator coil circuit all of the cold saturated liquid refrigerant will be boiled off and only vapor will remain. At this point in the evaporator coil, the changing of state process ends. The vapor that remains is still at saturation temperature, which in the example is 44°F.

When all of the saturated refrigerant liquid has vaporized in the evaporator circuiting, only vapor will remain in the circuit. This condition occurs at the refrigerant's boiling point, which is the same as the saturation temperature. The remaining refrigerant vapor is still very cold. As the cold vapor travels through the remaining circuiting of the evaporator coil, it continues to pick up or absorb heat from the warm air. Because there is no liquid left, the refrigerant vapor temperature begins to rise to a temperature above saturation temperature.

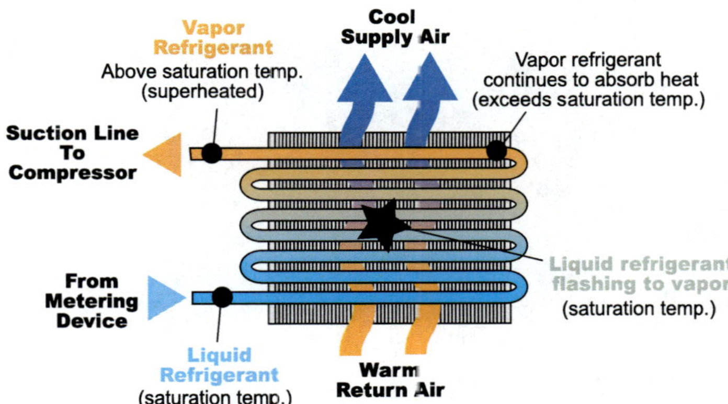

Once the temperature of the refrigerant vapor rises to a level that is higher than the saturation temperature, it is now called superheated. This heat added to the refrigerant vapor after all of the liquid has boiled off is called "Superheat". (Refer to Chapter C for more information on superheat.)

Suction Line and Suction Line Drier

The suction line is part of the low pressure side of the system, and connects the outlet of the evaporator coil to the suction line service valve on the outdoor condensing unit. The suction line is is cold, which will cause it to attract heat from the outdoor air and other sources. To prevent this from happening and to prevent condensate from forming on the line (sweating), the suction line is insulated.

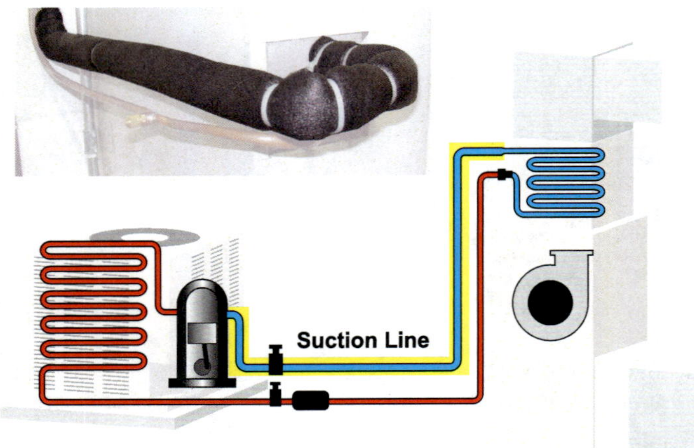

Suction lines come in different size diameters. Typical sizes are 5/8", 3/4", 7/8", and 1-1/8". The size of the line, measured by its outside diameter, or O.D., will depend upon the capacity of the system. Larger tonnage systems require larger suction lines than smaller tonnage systems. The size will also be determined by how far the line must run, and by how many elbows and other pipe fittings are installed in the piping run.

At times, a suction line filter drier is installed in the suction line. This drier is never installed by the manufacturer of the system, but rather by a servicing technician. Suction line driers are commonly used to clean up or remove contaminants from a system after a compressor motor burnout.

Suction Line Drier

Suction line filter driers typically have pressure ports on the shell so that the pressure drop across it can be measured. If the pressure drop across the filter is above allowable limits, the filter should be removed and replaced with a clean one. These filter driers come in different sizes based on the capacity of the refrigeration system they must clean.

Suction Line Pressure Loss

Suction lines drop the pressure of the suction gas due to friction that occurs as the refrigerant vapor travels in the copper tubing. This pressure drop is bad for the system as it causes the capacity of the compressor to fall. This is due to the density (or weight) of the refrigerant vapor reducing as pressure reduces. The resulting lighter vapor reduces the number of pounds of refrigerant that can be pumped by the compressor each minute, resulting in a loss of system capacity. It is best to keep suction lines sized to drop no more than 3 PSIG for R-22 systems and 5 PSIG for R-410A systems. Both of these pressure drops reduce the system capacity by 3%. Causes of excessive suction line pressure loss include kinks in the line, long line sets, or undersized refrigerant piping. (See Chapter B: Refrigerant & Refrigerant Piping)

Compressor

Configuration

Compressors found in newer residential split systems are typically scroll models, while older systems were commonly equipped with reciprocating compressors. Scroll compressors have orbiting scrolls that compress the refrigerant vapor. Reciprocating models have a piston and cylinder arrangement that closely resembles a car engine. Regardless of the type of compressor present in the system, both types are vapor pumps.

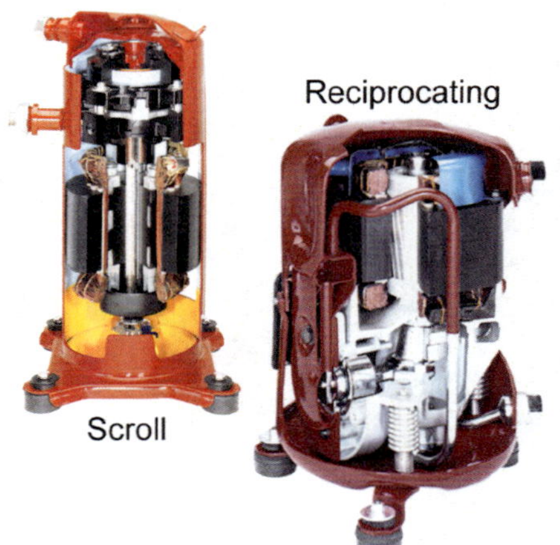

Scroll and Reciprocating Compressors. Both are sealed and unable to be serviced by a servicing technician.

The compressor has two refrigerant line connections. One connection is for the suction vapor and the other connection is the discharge line connection. These connections are typically brazed connections, but can be mechanical on some compressors.

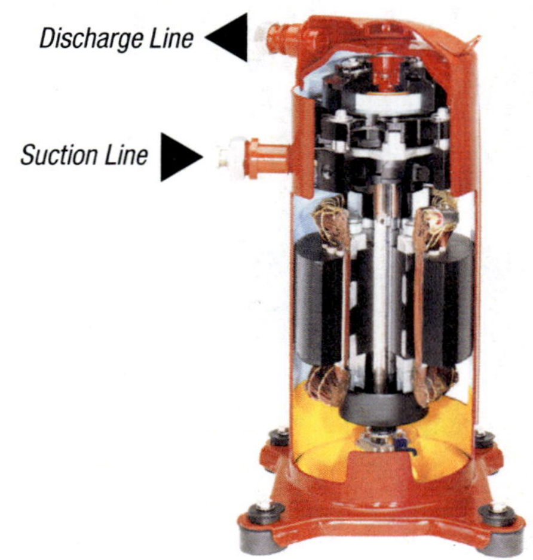

Comfort Cooling System Basics

Compressors that are used in residential air conditioners are called hermetic compressors. Hermetic compressors are sealed, so the internal components of the compressor cannot be accessed by service technicians.

Compressors connect to the electrical circuit of the system with a special plug called a Fusite Plug. This plug is located behind a protective cover that should never be removed if electrical power is supplied to the air conditioner. Once the power has been disconnected, the cover can be removed and the fusite plug's electrical terminals accessed.

Fusite Plug

Function

The compressor is a vapor pump. The vapor refrigerant enters the compressor where it is compressed to a high pressure/high temperature discharge gas into the discharge line, which connects to the condenser coil where the refrigerant gives up its heat to the cooler outdoor air.

When the suction vapor is drawn from the suction line into the compressor, the cold refrigerant vapor absorbs heat from the compressor motor windings. If there is an adequate amount of vapor present, the compressor will receive adequate cooling. If there is a lack of vapor due to a condition such as an undercharge, the compressor will run very hot.

After cooling the motor, the vapor is compressed to a high pressure before being discharged to the condenser..

Compressor Oil

Compressors have oil in them. The oil is a type that mixes with the refrigerant in the system. Many R-22 systems have mineral oil, while R-410A systems have POE (Polyolester) oil. The majority of the oil lays in the bottom of the compressor shell and is circulated by internal lubricating components to critical bearing areas.

During normal operation, a small amount of this oil will circulate throughout the refrigeration circuit, but the majority of the oil should remain in the compressor shell.

The temperature of the oil is critical to keeping compressor bearing wear to a minimum. If the system is running too hot, such as when the system is undercharged, the oil may break down and bearing damage may occur. This will not happen on systems that are operating at normal operating conditions.

Liquid Migration

There is a condition where liquid refrigerant may enter the compressor shell during periods where the air conditioning system is off. This condition is called liquid migration. When the liquid refrigerant enters the compressor, it mixes with the compressor oil. When the compressor is started, the heat from the compressor motor boils off the liquid. The boiling action of the refrigerant can cause the oil to foam and possibly mix with the refrigerant. If this occurs, it is possible that oil will be carried out of the compressor shell. This is called an oil pump out. With a reduced amount of oil in the shell, compressor bearing damage may occur.

To prevent liquid migration, a compressor accessory called a crankcase heater is used.

Crankcase Heater

This electric heater wraps around the shell of the compressor and keeps the compressor shell warm, which prevents liquid refrigerant from mixing with the refrigerant oil during periods of time where the compressor is off. Crankcase heaters are sometimes factory installed, but in most cases, they are field-installed.

Liquid can also enter compressors while the system is in operation. This is due to overcharging the system, or a reduced heat load on the indoor coil. The liquid will flood back to the compressor and cause an oil pump out. An accessory called an accumulator can be installed at the inlet of the compressor to catch liquid before it gets to the compressor suction line connection. Accumulators are usually installed by the manufacturer of the system. They can also be installed by service technicians.

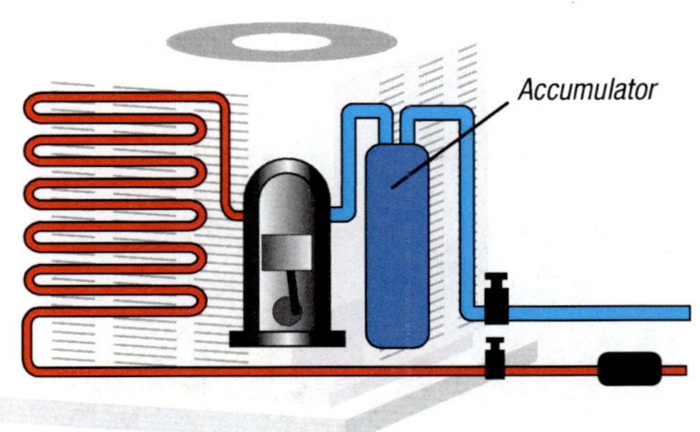

Discharge Line

The discharge line is part of the high side of the system. Discharge lines connect the outlet of the compressor to the inlet of the condenser coil. Some systems will have a pressure port installed on the discharge line for charging and servicing purposes. The discharge line is not insulated.

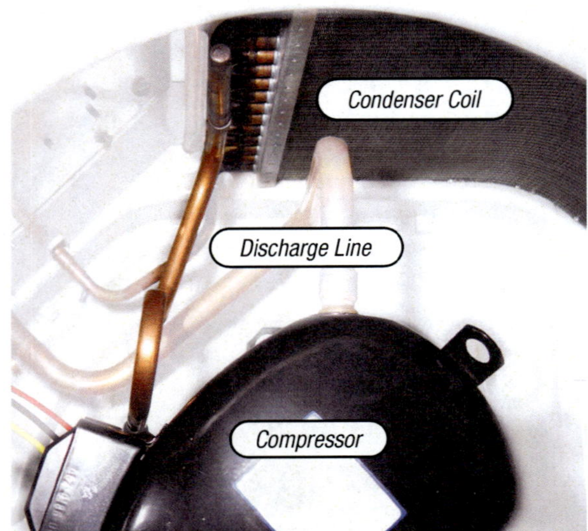

Condenser Coil

Configuration

The condenser coil is usually a copper tube/aluminum fin type coil similar in construction to evaporator coils. There are also high efficient heat transfer condenser coils that are configured as aluminum tubing wound with spines of aluminum. Other high efficiency condenser coils are of the microchannel type, which closely resemble an automobile's radiator.

Condenser coils have individual parallel refrigeration circuits that spread the hot gas from the compressor evenly through the coil. The hot gas gives up its heat and condenses into a hot saturated liquid. The liquid flows in the circuits where it is subcooled to a temperature that is lower than the temperature at which it condensed. (see chapter D: Subcooling). The subcooled liquid is directed to the liquid line.

Function

At the condenser coil, the heat from the evaporator and compressor will be removed from the refrigerant by transferring the heat to the cooler outdoor air. When the hot gas enters the condenser coil, it is highly superheated at a high pressure and high temperature. The purpose of compression was to raise the pressure of the refrigerant so that it's corresponding saturation temperature is above the temperature of the outdoor air.

In this example, we have a liquid pressure of 450 PSIG using R-410A refrigerant. The saturation temperature corresponding to 450 PSIG with R-410A is 126°F.

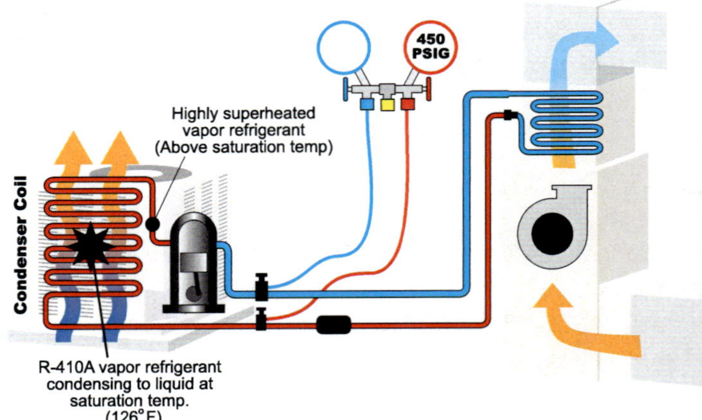

The hot gas leaving the compressor is highly superheated, so its actual temperature is going to be above the 126°F saturation temperature. The hot gas contains heat from the evaporator, suction line, compressor motor windings, and heat generated during the compression process. As the hot gas enters the condenser coil, it immediately begins to give up heat to the cooler outdoor air. The high pressure/high temperature gas travels through the condenser circuiting continually giving up heat until eventually enough heat is removed from the hot gas for it to reach its 126°F saturation temperature corresponding to the systems 450 PSIG liquid pressure.

At saturation temperature, any removal of heat will cause a change of state. In the case of the hot gas, it begins to condense into a liquid. The condensing liquid is still at 126°F saturation temperature, corresponding to the liquid pressure. As the hot gas condenses to a liquid, a massive amount of heat is transferred out of the refrigerant and into the cooler outdoor air being circulated by the condenser fan.

Once the refrigerant has completely condensed into a liquid, the condenser coil rejects additional heat from the liquid until the liquid temperature is well below the saturation temperature corresponding to the liquid pressure. The process of removing heat from the refrigerant so that its temperature is below saturation temperature is called "Subcooling." (See Chapter D: Subcooling)

Condenser Fan

The condenser fan pulls air through the surface of the condenser coil. The assembly consists of a fan motor mounted to the condensing unit cabinet, and a fan blade attached to the shaft of the motor. Each assembly is unique for the design of a specific condensing unit. These assemblies are typically not interchangeable as they are designed to produce a specific amount of air flow through the condenser coil.

The motors used to drive the condenser fan blade are usually PSC single or two speed motors. They will have an associated run capacitor. In some high efficiency models, a variable speed ECM type motor is used . This type of motor uses much less electricity than a PSC type motor.

Liquid Line and Liquid Line Drier

The liquid line which is part of the system's high side, connects the outlet of the condenser to the inlet of the metering device. The liquid line connection to the outlet of the condenser is often made at the liquid line service valve on the condensing unit.

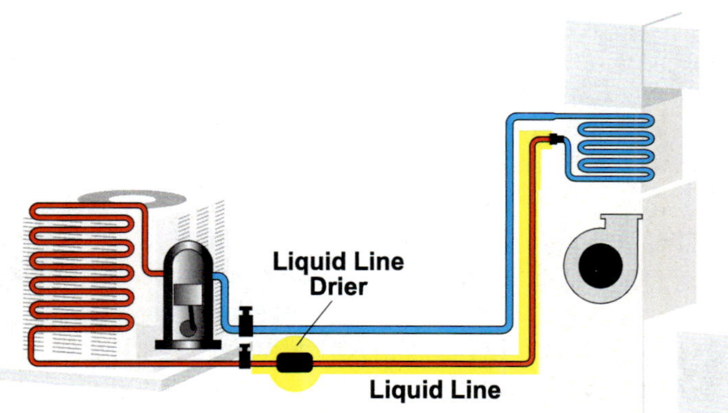

The liquid line is not insulated unless it runs through hot spaces where it can pick up heat. Liquid lines come in different size diameters. Typical sizes are 5/16", 3/8", and 1/2". The size of the line will depend upon the capacity of the system. Larger tonnage systems require larger liquid lines than smaller tonnage systems. The size will also be determined by how far the line must run, how much vertical lift is present, and by how many pipe fittings are installed in the line.

All liquid lines have a liquid line drier. The liquid line drier is either installed by the manufacturer of the system in the liquid circuit of the condenser coil, or it is field installed by the installing technician.

Liquid Line Drier

These driers collect debris and moisture that may be trapped in the system during manufacturing or installation. The size of the liquid line drier is determined by the capacity of the condensing unit. These driers may be brazed into the liquid line or may be connected using a flare fitting.

Liquid Line Pressure Loss

As the liquid refrigerant exits the condenser coil, friction between the refrigerant and piping causes the pressure of the liquid to fall. The further the liquid travels, the more pressure it loses. This pressure drop is bad for the system as it creates the potential for metering device problems.

If the pressure falls to a level that is equivalent to the liquid line temperature at saturation, the liquid will begin to flash off to a vapor before it gets to the metering device. Liquid line pressure drop can be controlled by proper line sizing. (See Chapter B: Refrigerant & Refrigerant Piping)

Metering Device Overview

The metering device is a pressure-dropping device that separates the high pressure side of the system from the low pressure side of the system. The metering device can be a simple piston that has a hole through its center, commonly called a "Fixed" or "Piston" metering device, or a more complicated thermostatic expansion valve commonly called a TXV metering device.

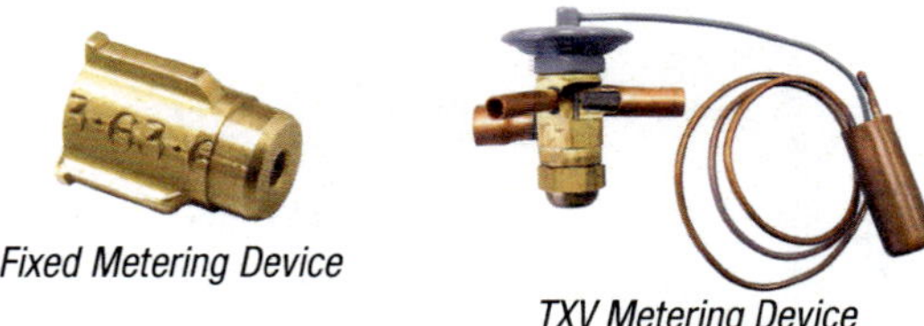

Fixed Metering Device

TXV Metering Device

Regardless of what type of metering device is being used, the purpose of the metering device is to act as a pressure drop that separates the high temperature area of the refrigeration system from the low temperature area. Each type of metering device

places a restriction in the liquid circuit between the liquid line and the evaporator coil circuiting.

When the high pressure warm liquid refrigerant from the liquid line enters the metering device, the warm liquid is converted into a low-pressure, cold saturated refrigerant mixture of liquid and vapor. Heat energy from the liquid is given up as some of the liquid turns into a vapor in the pressure drop area. This energy transfer is called the "Refrigerating Effect".

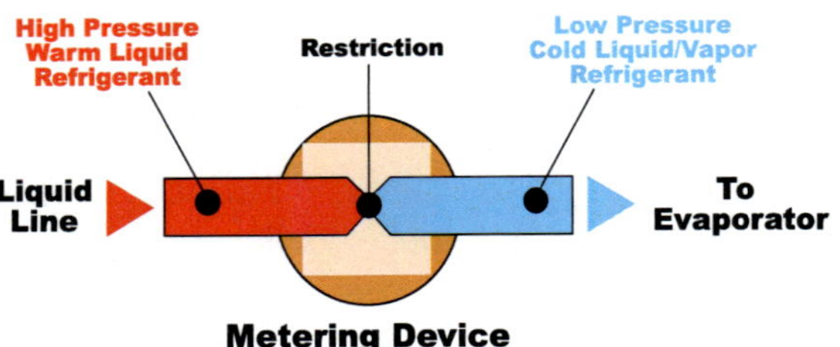

There are some differences in the operation of a mechanical refrigeration system using an expansion valve metering device in place of a fixed type device. Differences in cycle operation include how the system reacts to changes in evaporator heat load, system charge, and changes in the outdoor air temperature.

Both expansion valves and metering pistons are sized by the manufacturer of the system. It is unlikely to have the wrong size expansion valve, however, pistons are changed depending on the outdoor unit and indoor evaporator coil combinations. In some cases, the wrong size piston is installed in the evaporator coil.

Metering Device: Fixed Piston

Fixed metering devices are nothing more than a brass cylinder (piston) with a narrow hole bored through it. The piston resides in a fitting that allows the pressure of the liquid refrigerant to push the piston into a seat, where it creates a seal that prevents liquid refrigerant from bypassing the piston's bore hole. A screen in the fitting catches debris that may have bypassed the filter drier.

Fixed Metering Piston Assembly

Fixed bore metering devices have limited control over the flow of refrigerant into the evaporator coil circuiting. Refrigerant flow into the evaporator is determined by liquid line pressure and indoor heat levels.

Since indoor heat load and liquid pressure are constantly changing, the amount of refrigerant entering the evaporator circuiting is constantly changing. The changing flow of refrigerant into the evaporator causes superheat levels to fluctuate. In other words, the operating pressures and superheat levels are constantly fluctuating in response to changes in the outdoor air temperature, liquid line pressure, and indoor air temperature.

With a fixed type metering device, the indoor coil receives less refrigerant as the outdoor air temperature drops. This condition is due to the loss of liquid pressure at lower outdoor air temperatures. Fixed type systems operating in this state lose some performance of the evaporator circuit due to a lack of refrigerant during these periods of operation. With decreased refrigerant flow, the superheat level of fixed type metering devices can rise to significant levels during periods of cool outdoor air temperature operation.

Since the operational characteristics of these systems involve a constantly fluctuating flow of refrigerant into the evaporator circuiting, these systems are charged using the superheat charging method. See Chapter C: Superheat.

Metering Device: TXV

Typically factory installed, a TXV (Thermostatic Expansion Valve) is a metering device designed to maintain a constant level of evaporator superheat over a wide range of heat load and outdoor air temperature range combinations. The superheat level maintained by the TXV is typically between 10°F and 15°F. The TXV is able to maintain a constant superheat level by regulating the size of its internal orifice in response to changes in heat load and liquid pressure.

The size of the TXV orifice is controlled by an opening force and a closing force. The opening force consists of pressure from the TXV sensing bulb. This pressure acts on the top side of the TXV's diaphragm. The closing force is the combination of an internal spring tension and the evaporator pressure pushing on the bottom of the orifice, and evaporator outlet pressure placed upward on the underside of the TXV diaphragm. The amount of spring tension determines how much superheat will be maintained by the expansion valve. *Typically, this pressure is about 25 PSI on R-410A systems.*

A sensing bulb attached to the suction line at the outlet of the evaporator circuiting contains an internal charge of refrigerant. The sensing bulb capillary tube transmits pressure to the top of the TXV diaphragm assembly. When the temperature of the suction line changes, the temperature of the sensing bulb changes. This, in turn, causes the pressure inside the sensing bulb to change as well. If the sensing bulb senses a warm suction line, the pressure in the sensing bulb rises, pushing the valve open, feeding more refrigerant into the evaporator. If the suction line temperature falls, the pressure in the sensing bulb will also fall, pushing the valve closed, feeding less refrigerant into the evaporator.

The operation of the valve is very simple. If the sum of the spring tension and evaporator outlet pressure is greater than the sensing bulb pressure, the valve will move to a more closed position, thereby reducing refrigerant flow. If the sum of evaporator outlet pressure and spring tension is lower than the sensing bulb pressure, the valve will move to a more open position, thereby increasing refrigerant flow. The two opposing pressures eventually equalize, allowing the TXV to feed the proper amount of refrigerant into the evaporator coil.

Residential systems equipped with a TXV typically use a balanced port type expansion valve. Balance port type expansion valves cancel out the affect of liquid line pressure on the pressure drop across the valve. By using this type of valve, system efficiency can be enhanced by lowering the head pressure and corresponding liquid pressure without affecting the ability of the expansion valve to flow adequate refrigerant to the evaporator circuiting. The lower head pressure increases the volumetric efficiency of the compressor and overall system capacity.

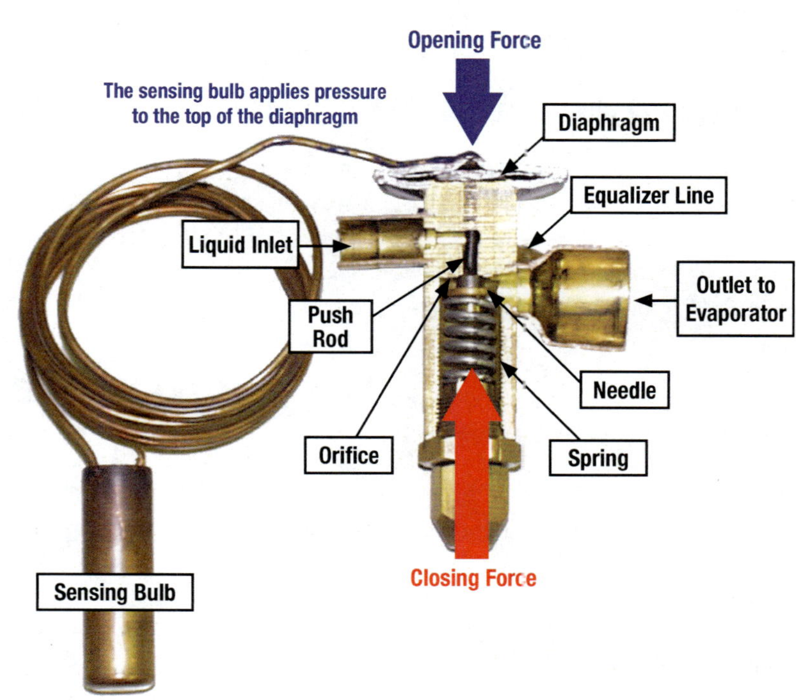

TXV and Evaporator Heat Load

A TXV will adjust to changes in heat load entering the refrigerant at the evaporator coil. If the heat load is very high, the suction pressure will rise and the saturated liquid entering the evaporator coil will quickly boil off in the evaporator circuit. The vapor will become highly superheated due to the amount of circuiting it must travel through before exiting the evaporator coil. Because of the high superheat, the suction line will be warm.

If the heat load is excessively high, the TXV may not be able to maintain its design superheat level. In this state, there is simply not enough refrigerant in the system to keep up with the heat load requirement. The TXV will completely open but may not be able to decrease the superheat. High suction pressure and evaporator superheat will result. In extreme cases, the suction pressure could rise high enough to overload the compressor motor. High heat load conditions can exist due to infiltration air, excessive indoor airflow, new system start-up, or an undersized system.

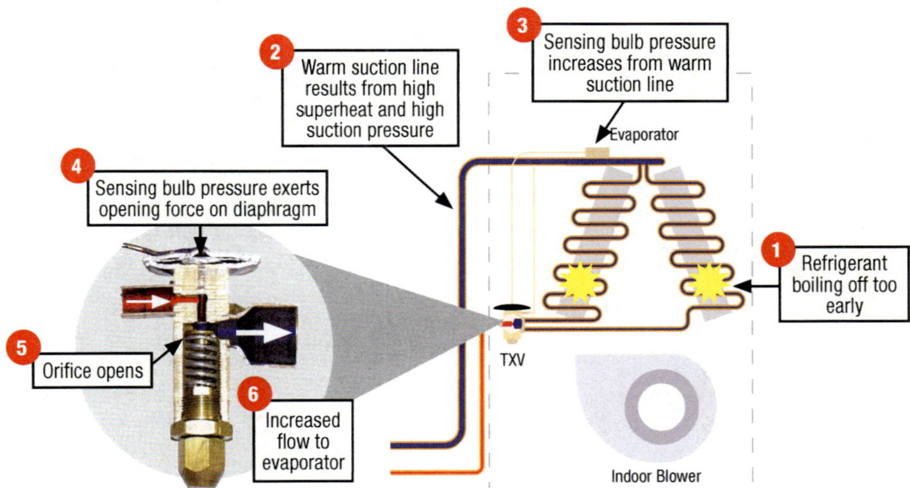

With the orifice in a more open position, the evaporator coil receives more refrigerant. With more refrigerant entering the evaporator coil, the suction pressure will rise and the superheat level will begin to drop.

Conversely, when the heat load is low, such as when the return air filter is dirty, the TXV will again try to maintain its design superheat. In this state, the suction pressure is low and the evaporator coil struggles to boil off all of the saturated liquid refrigerant.

The TXV sensing bulb (located at the suction line outlet of the evaporator coil) senses the cold suction line, and the sensing bulb internal pressure falls. The low sensing bulb pressure is directed to the top of the TXV diaphragm assembly. With very low sensing bulb pressure at the top of diaphragm, evaporator pressure and spring tension push the valve to a more closed position.

With less refrigerant entering the evaporator coil, the superheat level rises, and the suction pressure falls slightly.

With inadequate airflow, the TXV will struggle to reach a steady state of balance due to the lack of adequate suction pressure. The valve will continually open and close in a failed attempt to balance the refrigerant level with the heat load. This struggle is called "hunting" and will show up as a fluctuating suction pressure gauge and corresponding changes in the temperature of the suction line. The lack of suction pressure creates a very cold evaporator coil that may eventually develop a coating of ice.

TXV and Outdoor Air Temperature

Expansion valve systems offer greater efficiency because they can adjust their orifice size to allow additional refrigerant into the evaporator coil at cooler outdoor air temperatures. With more refrigerant in the evaporator coil, the heat transfer process is enhanced and higher overall efficiency ratings are obtained.

When the system has low liquid pressure due to low outdoor air temperature, the flow of liquid refrigerant into the TXV decreases. With less refrigerant flowing through the TXV, the evaporator coil starves for refrigerant. The suction pressure will fall and the liquid is quickly boiled off in the evaporator. The suction vapor superheat is high, and the suction vapor leaving the evaporator is warm.

The sensing bulb pressure rises in response to the warm suction line. The pressure at the top of the TXV diaphragm assembly increases to a point greater than spring tension and evaporator outlet pressure. The valve reacts to higher opening pressure by moving to a more open position and the evaporator coil receives more refrigerant.

With more refrigerant in the evaporator coil, the suction pressure rises and superheat falls. In this state, the TXV ensures that good use of the evaporator coil surface area is maintained despite the low liquid pressure.

The ability of the expansion valve to increase its orifice size to maintain a relatively low level of superheat has an effect upon the condenser coil. The condenser must give up some of its refrigerant charge to meet the refrigerant demand of the evaporator coil.

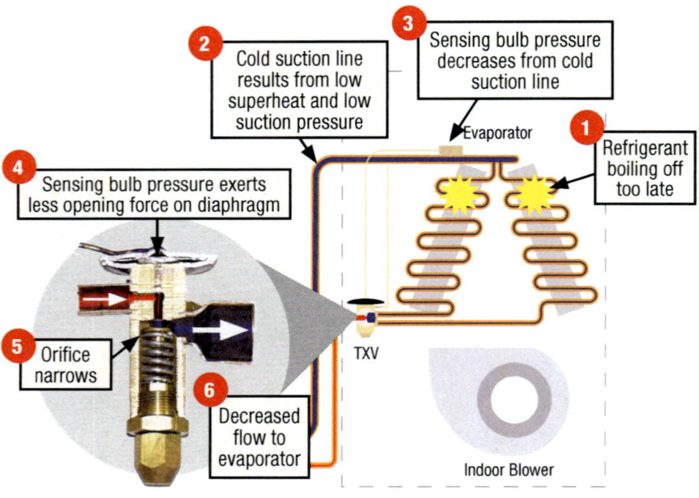

The low heat load drives the evaporator pressure and corresponding saturation temperature very low. As the cold saturated liquid travels through the evaporator circuiting, it will boil off very late in the circuit. Because of the low saturation temperature and late boil off point, there is little superheat and the suction vapor is very cold.

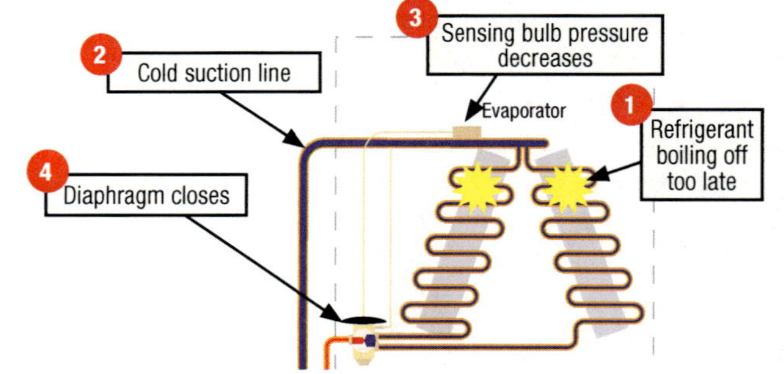

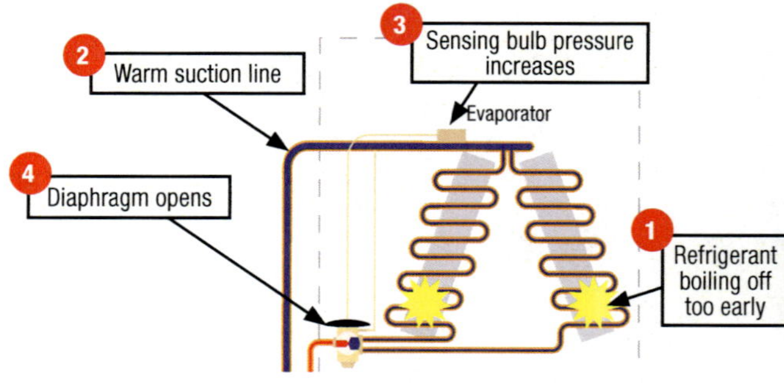

Indoor Air Blower

Indoor air blower assemblies consist of a housing, motor, and blower wheel. A mounting bracket holds the blower motor to the blower housing. Almost all residential furnaces made for use with air conditioning will have a direct drive motor. Direct drive means the blower wheel is attached directly to the motor shaft.

The blower motor type may be a PSC multi-speed motor with an associated run capacitor, or it may be a high performance ECM, Electronically Commutated Motor that does not use a run capacitor.

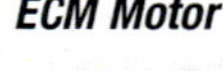
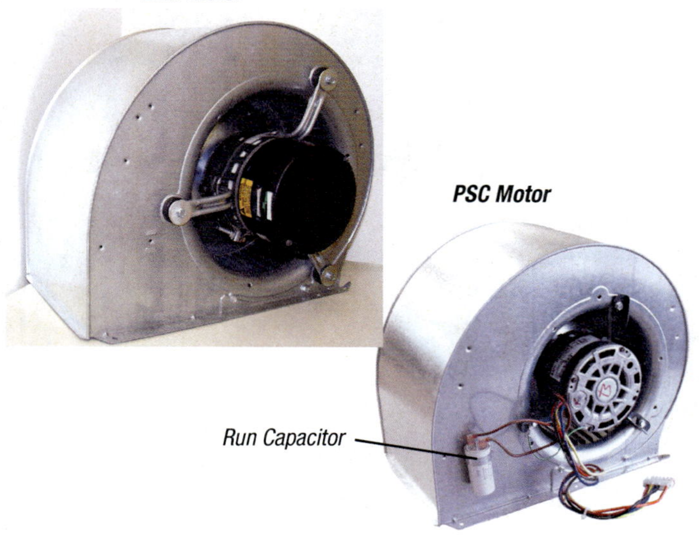

When a call for cooling operation occurs, the indoor blower assembly will energize and blow air across the surface of the evaporator coil. Blowers come in different air delivery capability ratings (CFM). The amount of air that blows will be determined by how restrictive the duct system in the home is. If the duct system is restrictive (sized too small), the air volume will be lower than it should be. Properly sized duct systems will allow the blower to deliver the proper amount of air to the evaporator coil.

Pressure Ports

There are different types of pressure access valves used in residential air conditioning systems. Some access valves are simple copper tubes with an access port brazed onto the end of the tube. These are called Service Tubes. Connecting to the tube will allow access to read system pressures.

Other types of pressure ports are part of a valve called a Service Valve. These types of valves can open or close the system to isolate the outdoor condensing unit from the refrigerant lines and the indoor

evaporator coil. These valves are opened and closed using a hex wrench. When this type of valve is closed, the pressure port on the valve can still be used to read system pressures.

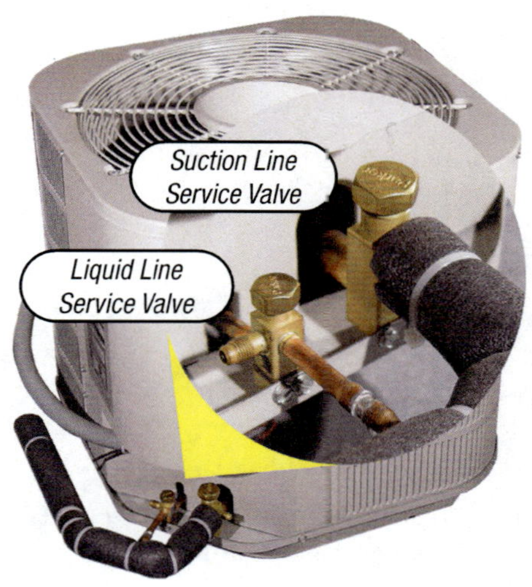

A special valve called a "King Valve" is also found on some air conditioning systems. This type of valve is a three way valve. There are three positions the valve can be in: back-seated, cracked, or front-seated.

In the back-seated position, the refrigerant can flow through the valve but the pressure port is shut off. When connecting or removing gauge hoses from the system, this is the position the valve should be in.

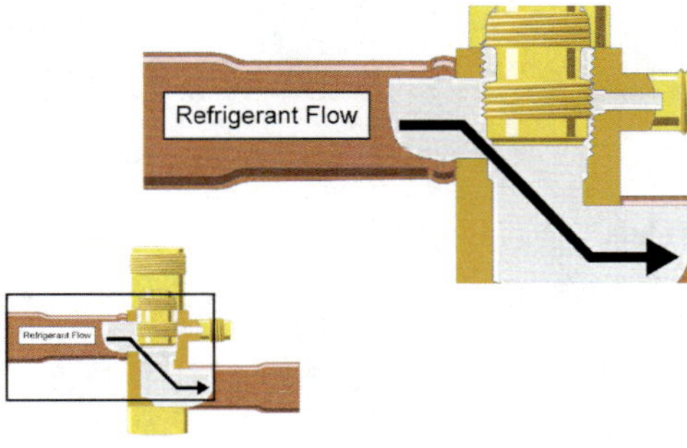

When measuring the pressure, turn or crack the valve open a few turns and the gauge port will now be open to the system and pressure can be read. The valve will continue to allow refrigerant flow.

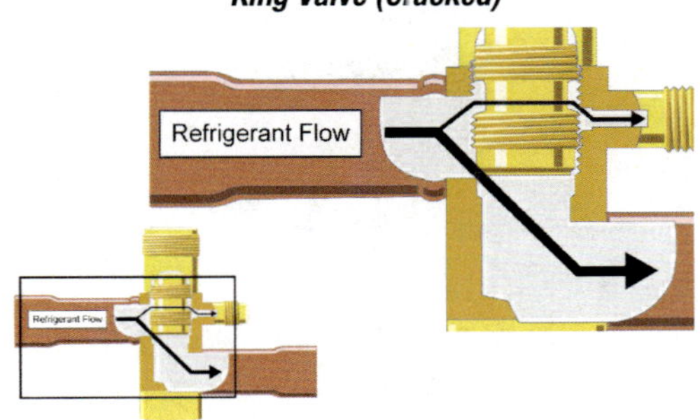

When the valve is front-seated, the refrigerant flow through the valve is blocked and the gauge access port is open to read system pressure.

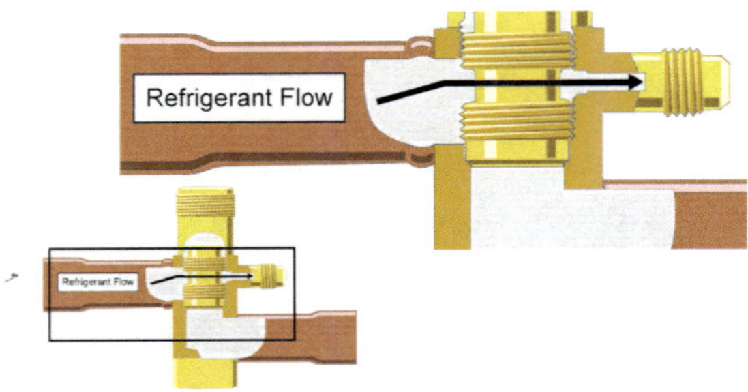

Pressure Port Locations

Residential air conditioning systems typically come with pressure access ports located at the outdoor unit service valves. These ports are located on the suction and liquid lines and used for charging and evacuation purposes. In some cases, the system may have a high pressure port located on the compressor discharge line. This port is commonly found on heat pump systems.

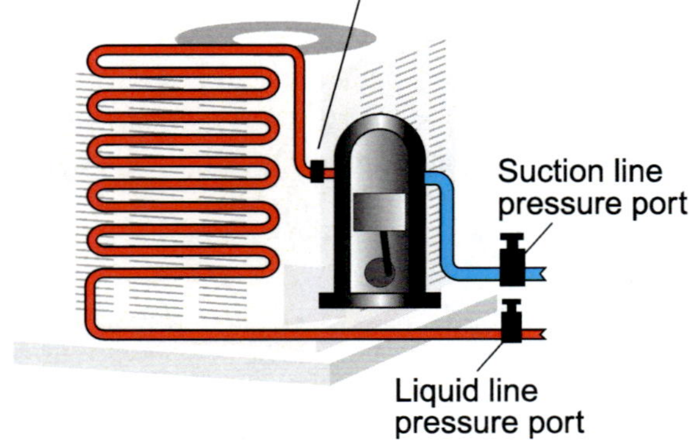

Installing Additional Pressure Ports

It is possible to add special pressure ports to systems that are already installed and have been evacuated. These ports are comprised of a saddle fitting that fits to the copper refrigerant lines. The port is brazed to the refrigerant line, and then a special piercing tool is screwed down to pierce a hole in the line. These ports may be added to a system where flash gas problems exist, or where excessive suction line pressure drop is suspected.

Exploded view of pressure port installation kit

If complete access to pressure at important diagnostic positions is desired, two additional pressure access ports could be added to the system. These ports should be installed in the liquid line about 1 foot in front of the metering device, and another pressure port on the suction line at the outlet of the evaporator coil. These two additional ports will allow measurement of the system pressure at the suction line outlet to the evaporator coil, and at the point where liquid is about to enter the metering device. Pressure at these ports would then be measured and compared against the pressure at the condensing unit service valves to detect excess pressure drop in the refrigerant lines.

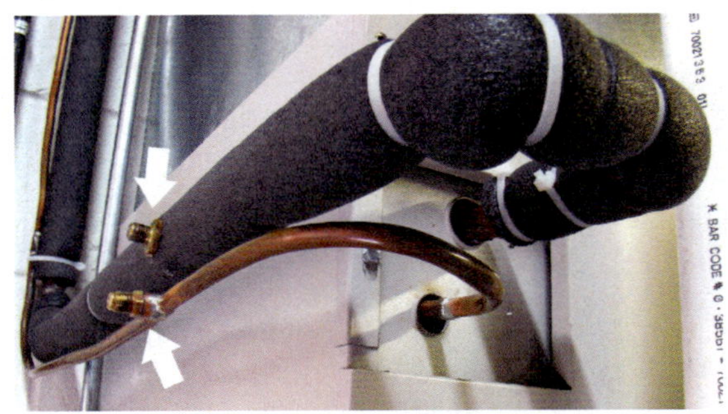

The indicated ports are about one foot in distance from the evaporator coil and metering device as noted above.

Replacing Leaking Pressure Port Cores

The valve stem cores used in pressure ports can be damaged by heat and leaked refrigerant. In the event a core is damaged, it must be replaced.

Valve Stem Core

Special valve core removal tools can be used to remove and replace a damaged core without losing system charge. The tool's ball valve is then shut to seal the refrigerant charge in the system. The damaged core is then replaced with a new core and screwed back into the system service valve.

Typical valve core removal tool

Comfort Cooling System Basics

Mechanical Refrigeration Cycle Overview

With a fundamental understanding of the function and characteristics of the individual components as explained in the previous sections in this chapter, the mechanical refrigeration cycle and how it transfers heat can now be summarized in terms of a complete system.

To move heat, a refrigerant is placed in the refrigeration system, and the refrigerant pressure is regulated from high to low. By changing the pressure in the system, the saturation temperature of the refrigerant can be changed. At high pressure, refrigerants have a high saturation temperature, and at low pressure, they have a low saturation temperature.

The high pressure liquid refrigerant travels from the condenser coil through the liquid line, and to the metering device located at the inlet of the evaporator coil. The metering device simply drops the refrigerant pressure from a warm, high pressure liquid to a cool, low-pressure saturated mix of refrigerant liquid and vapor.

The cool saturated refrigerant exits the metering device and enters the evaporator coil circuiting. The indoor blower moves air from the conditioned space across the surface of the evaporator coil.

Since "hot goes to cold", heat from the indoor air will be absorbed into the colder saturated refrigerant traveling through the evaporator coil. As the refrigerant circulates through the evaporator coil circuiting, it continues to pick up heat. The amount of heat the refrigerant picks up is greatest when the refrigerant is in a saturated liquid state and is vaporizing (evaporating) as it absorbs heat. As the liquid flashes to vapor, the refrigerant now contains much more heat, yet the temperature of the vapor refrigerant is the same temperature as the liquid refrigerant.

The air, after passing through the evaporator, now contains less heat than when it came in contact with the evaporator coil surface. The superheated refrigerant vapor carrying the absorbed heat exits the evaporator coil and travels through the suction line and into the compressor. The refrigerant absorbs additional heat from the compressor.

To reject heat from the vapor refrigerant, the compressor must raise the pressure of the refrigerant so its temperature is above the temperature of the outdoor air. The refrigerant is compressed to a high pressure and high temperature. The high pressure/high temperature refrigerant exits the compressor and travels through the discharge line and into the condenser coil circuiting.

With the saturation temperature of the refrigerant high, the heat absorbed by the refrigerant in the evaporator coil, suction line and compressor can now be absorbed by the cooler outdoor air circulated over the condenser coil surface, thus completing the heat transfer process.

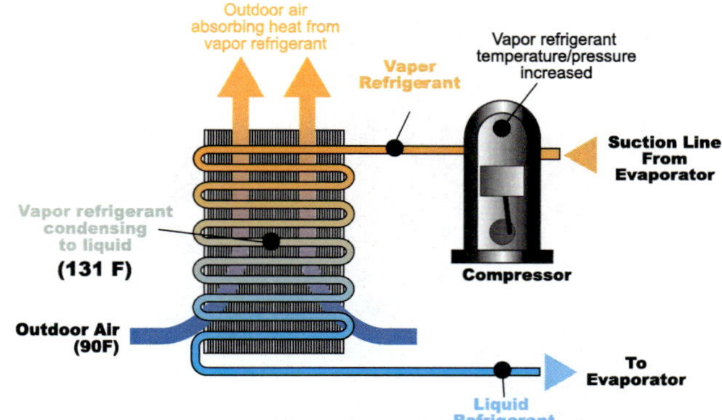

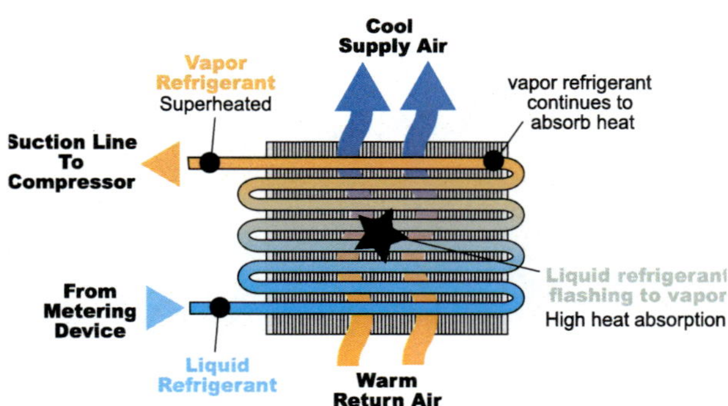

Eventually, all liquid refrigerant in the evaporator coil circuiting will boil off, and only vapor will remain. The vapor is still very cold, and it too absorbs additional heat from the air. The amount of heat transfer will begin to fall as all changing of state has ceased. This phase of heat transfer in the evaporator coil now involves heat going into the vapor and raising the temperature of the vapor above the saturation temperature. This phase is called superheating.

Basic Refrigeration Cycle Summary

The refrigerant cycle we have just covered is a sealed cycle, meaning that heat enters the refrigerant at the evaporator coil and exits the refrigerant at the condenser coil. For the system to operate properly, the following must be present:

- A properly sized metering device
- A clean condenser and evaporator coil
- The proper charge of refrigerant in the system
- Heat load within acceptable minimum and maximum ranges
- Properly sized refrigerant line set
- A mechanically sound compressor
- Proper airflow through the evaporator and condenser coils

In other sections of this book, diagnostic troubleshooting of system problems will be covered, along with the necessary test procedures needed to make a diagnostic decision.

Refrigerant & Refrigerant Piping

Continued....

R-22 & R-410A Refrigerant Characteristics

Simple Chemistry: R-22 vs. R-410A

R-22 is an HCFC refrigerant that will cause ozone depletion. R-410A is a non-ozone depleting HFC refrigerant. Both refrigerants may contribute to global warming and cannot be vented to the atmosphere.

R-22 is a single compound refrigerant. R-410A, on the other hand, is a blend of two refrigerants. The mixture is 50/50 HFC32 and HFC125. Although it is a mixture, the two refrigerants do not easily separate in a system, nor do they change composition when there is a leak. This behavior is called Azeotropic behavior and is highly desirable. A refrigerant blend that is made up of refrigerants that can easily separate is called a zeotropic blend. Because R-410A has a very stable composition, it is serviced in the same manner as R-22 systems.

Although R-22 and R-410A are handled in similar manners, <u>R-410A must be added to an air conditioning or refrigeration system as a liquid only.</u> If the refrigerant is added to a system as a vapor, some refrigerant separation, referred to as fractionation, may occur during charging.

Heat Transfer Ability: R-22 vs. R-410A

R-410A carries about 40% more heat per pound of refrigerant vapor than R-22. For example, at 50°F saturation temperature in the evaporator coil, R-410A has about 124 BTU/lb. versus 109 BTU/lb. for R-22.

Suction Vapor Weight Comparison: R-22 vs. R-410A

R-410A is heavier per cubic foot of vapor than R-22. At 50°F saturation temperature R-410A weighs about 2.3 lb./cubic foot versus 1.8 lb./cubic foot for R-22.

Refrigerant Oil Compatibility

Many R-22 refrigerant systems use mineral oil lubricant because the refrigerant and the oil mix together well under most conditions. The ability of two liquids to mix together well and form one liquid is called miscibility. R-410A refrigerant mixes well with polyolester oil (also called POE oil), but not with mineral oil. If mineral oil was added to an R-410A system, the refrigerant and oil would not mix, two separate liquids would be present and the compressor would experience oil loss.

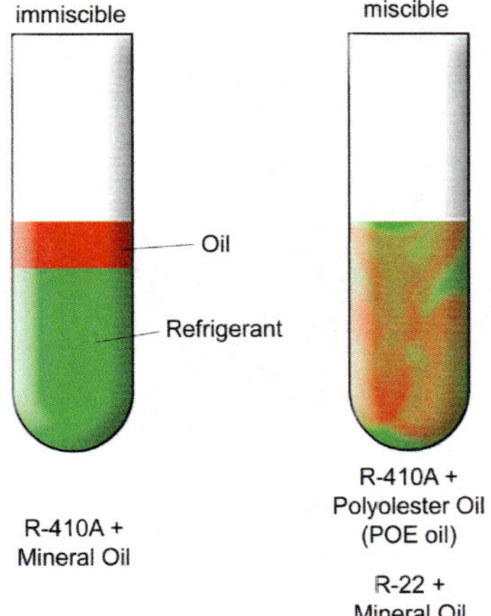

Because of the inability of mineral oil and R-410A to mix well, care must be taken when replacing an R-22 system with an R-410A system. If the existing refrigerant lines are to be reused, for example, they must be thoroughly cleaned before connecting them to the new equipment.

Removing Mineral Oil From Lines

Some manufacturers recommend purging mineral oil with nitrogen. Other manufacturers state that refrigerant lines can't be reused. If the lines are buried in the walls of a building, odds are the lines will have to be reused!

The existing refrigerant lines should be flushed out with a commercial cleaning solvent such as Nu-Calgon RX-11-Flush. This solvent will remove mineral oil from the line sets. If the mineral oil is not flushed out of the system, oil slugs will circulate and may cause compressor damage.

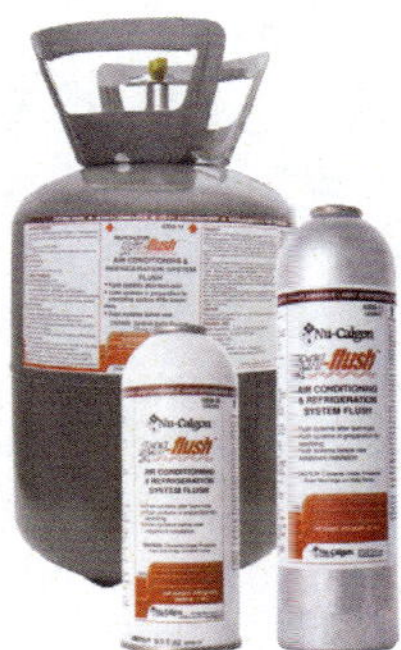

Carefully follow equipment manufacturer's recommendations and the flush kit manufacturer's instructions to ensure proper oil removal and prevent equipment damage.

<u>Not all manufacturers approve of the use of additives in their systems. Check equipment manufacturer recommendations prior to use of system additives.</u>

Mixed Refrigerants & Non-Condensable Gases

Mixed Refrigerants

R-22 and R-410A should never be mixed or substituted for one another in a system. Metering device size, oil compatibility, compressor displacement, and coil surface area will all be incorrect. The result will be an unpredictable system operating with no level of reliability whatsoever.

It is important to recognize that R-22 hoses fit on R-410A systems, and vice versa. Therefore, it is possible to add the wrong refrigerant into a system. Fortunately, there is a way to detect this problem if it were to occur.

If the system were to operate with a mixture of R-22 and R-410A, the system's operating pressures would be incorrect and compressor performance would be affected. This is because the two refrigerants are so dissimilar in their physical properties that neither the coils nor the compressor come close to matching operating design specifications.

If it is suspected the system has a mix of the two refrigerants, recover the charge and follow the procedure that follows. This procedure will find both mixed refrigerants and non-condensable gases. The test requires recovery of the total system charge into a clean, evacuated, recovery cylinder. The temperature and pressure of the cylinder are measured. If the refrigerant is in the correct state, the pressure and temperature will follow temperature/pressure chart values.

Non-Condensable Gases

Non-condensable gases found in residential systems are often the result of air infiltration into the refrigeration circuit during installation and service or when low pressure side leaks allow air to be pulled into an operating system. Non-condensable gases include oxygen, nitrogen, and carbon dioxide. In the condenser, these gases will not condense, causing a reduction in the effective area of the condenser. The condensing pressure will be high, heat transfer from the refrigerant to the outdoor air will be reduced, and the liquid line will be hot.

Along with these gases, systems that run hot can produce non-condensable gases as the insulation in the compressor deteriorates due to the high temperature.

Tools:

- Digital Temperature Probe
- Refrigerant Gauges

STEP 1

Start with an empty, evacuated, recovery cylinder. Recover the system charge into the cylinder.

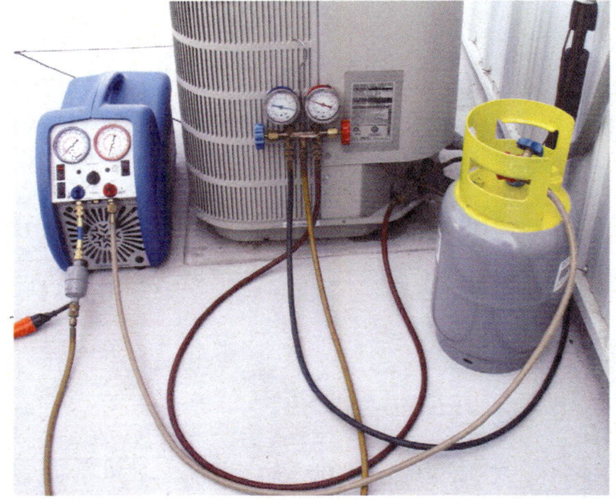

STEP 2

Allow the cylinder to sit (out of direct sunlight) until the cylinder temperature and the surrounding air temperature are equal.

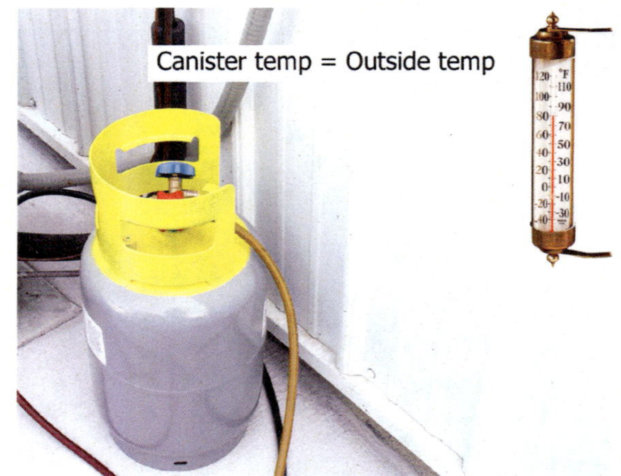

Step 3

Read the pressure in the refrigerant recovery cylinder.

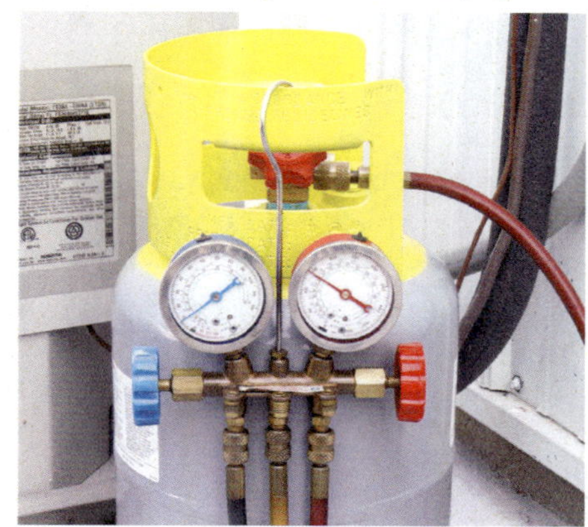

Step 4

Measure the air temperature surrounding the recovery cylinder.

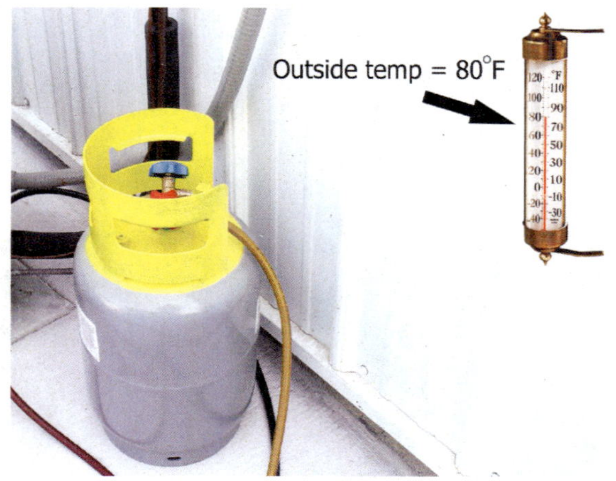

Step 5

The cylinder pressure should be equal to the saturation pressure for the air temperature. If the pressure and temperature are not correctly matched for the type of refrigerant that should be in the system, mixed or incorrect refrigerant, or non-condensables are present. In this example the recovery cylinder is at a pressure of 140 PSIG @ 80°F. These temperatures and pressures are correct for R-22. If the system were an R-410A machine, this would be an example of someone putting the wrong refrigerant into the system.

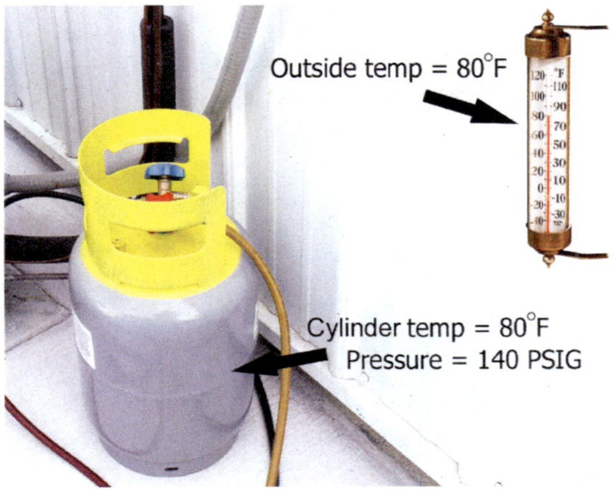

Moisture & Acid

Since air contains moisture, moisture can get into the refrigeration system when the system is installed or opened for service. Moisture, when mixed with refrigerants and refrigerant oils, can produce acid. Acid in a system can cause compressor failure due to corrosion, motor winding damage, and copper plating of compressor bearing areas.

When a system is operating in an acidic condition, the acid attacks copper surfaces in the refrigeration circuit. The copper is deposited on weight-bearing areas within the compressor. Acid will also attack the compressor motor windings and cause electrical failure of the compressor motor.

This compressor damage was caused by copper plating due to excess moisture and acid formation in the system.

Moisture and POE Oil

R-410A systems use POE (polyolester oil) that is produced from organic acid and alcohol. POE oil is hygroscopic and will attract moisture at a faster rate than mineral oil based systems. When exposed to moisture, POE oil can change back into organic acids, alcohol, and water through a process called hydrolysis. The organic acid is highly corrosive and will cause damage to compressor components and internal refrigeration system surfaces.

When opening a R-410A system for servicing, the potential for oil contamination must be minimized. One way to accomplish this is to bleed a small amount of nitrogen into the system during repairs. The nitrogen will keep the system slightly above atmospheric pressure and minimize the potential for oil contamination.

When servicing a R-410A system, do not allow the system to be open the atmosphere for more than 15 minutes. If a system has been left open to atmosphere (such as if there was a catastrophic leak and the total refrigerant charge was lost), it is best to replace the compressor oil.

Evacuation of the system will not separate moisture from the POE oil. The only method of removing moisture from POE oil is through

the use of a filter drier. Evacuation should always be performed when opening a system regardless of the fact that moisture cannot be separated from the oil.

Moisture Indicating Sight Glass

Color-changing moisture indicators can be used to monitor potential problems due to moisture contamination. When the system contains too much moisture, the color of the indicator will change to indicate there is excessive moisture in the system. An acid test kit should be used on the system to determine if excessive acid levels are present.

Moisture Meter and Sight Glass

Driers and Moisture Removal

Filter driers contain a special media, called desiccant, that can absorb water and acid from the circulating refrigerant. An internal filter is used to capture solid debris and foreign matter. Driers are made for use in liquid lines and suction lines. They are rated by the amount of moisture they can hold. Installation of a filter drier may not stop acid formation if the moisture content in the system exceeds the capacity of the filter drier.

Driers will operate with a 1-2 PSIG pressure drop across them when clean. If the drier is too small for the system, the pressure drop will be greater than 1-2 PSIG and will contribute to flash gas problems if installed in the liquid line. At some point during the life of an air conditioner, the filter drier may become plugged with debris. When the drier loads up with moisture or debris, the pressure drop will increase.

Excessive pressure drop can be detected across a drier with a digital thermometer. Measure the temperature of the refrigerant piping entering the drier, and then leaving the drier. If the temperature drop is in excess of 3°F, the pressure drop across the drier is at least 4 PSIG. If this occurs, the drier should be changed. The 3°F maximum temperature drop applies to both R-22 and R-410A driers.

Liquid Line Drier

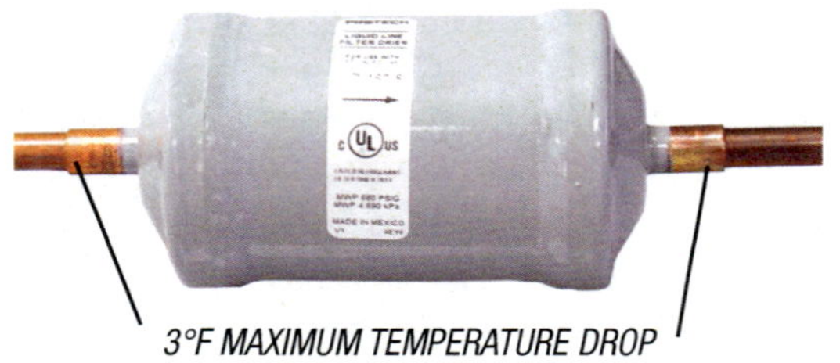

Generally, liquid line driers remain in the system until the system is opened. At that time the drier is replaced. Suction line driers should be removed from the system after about 48 hours of total run-time operation.

System Evacuation

When a refrigeration system has been opened to atmosphere, the moisture in the system must be removed. The best method of drying a system is to remove as much moisture as possible before installing a drier. To get moisture out of the system, the refrigeration system is pulled into a vacuum using a vacuum pump. In a vacuum, moisture will boil off into a vapor at low temperatures. The vacuum pump then pumps the vapor out of the system.

Vacuum pumps come in different sizes based upon how many cubic feet of air they can pump in a minute. This is referred to as the pump's CFM rating. The greater the CFM rating of the pump, the quicker the evacuation process can be completed. For residential systems, a vacuum pump rated between 5 and 6 CFM is sufficient.

Buy a higher CFM-rated vacuum pump for maximum performance. 5-6 CFM is a good range for residential systems.

When evacuating a system, the vacuum pump oil should be replaced frequently. If oil is reused, it may contain moisture that will make it very difficult to get the evacuation level down to 500 microns.

Remove the cores from the system pressure ports where the hoses are installed to increase pump capacity and reduce evacuation time.

Core removal tools allow a technician to remove and/or replace the pressure port cores without losing system refrigerant.

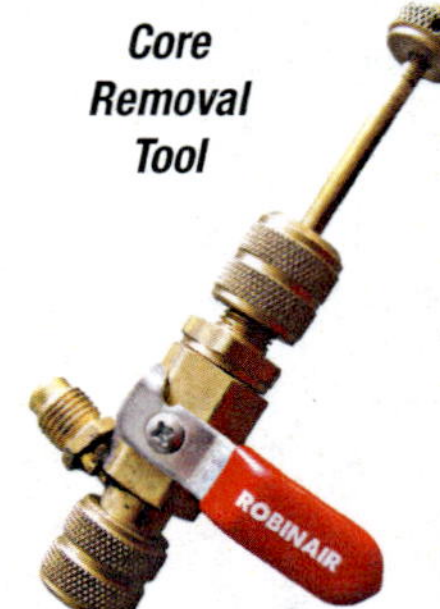

Micron Gauges

When the system is in a vacuum, below atmospheric pressure, a special instrument called a micron gauge is used to measure the pressure. As the vacuum pump pulls the refrigeration system pressure below atmospheric pressure, the micron level will fall. When the vacuum pump is shut off after isolating the pump from the system, the micron level will rise as water vapor boils off. The pump is started again and the process repeated until the vacuum level holds steady at 500 microns or lower.

Digital Micron Gauge by UEi

Sludge

Oil will break down and begin to form sludge when exposed to excessive heat. The hottest spot in the refrigeration circuit is the compressor discharge line. The discharge gas will gain temperature when suction vapor superheat is high. Breakdown of the oil begins with extended compressor operation when the discharge line temperature is excessively high.

Sludge can clog metering devices, refrigerant driers, and strainers. To prevent oil breakdown and limit the potential for sludge, system moisture must be limited and proper air flow and charge level established. Acid can be detected in the system by taking an oil sample and testing the oil with an acid test kit. If the kit detects acid, clean up measures will need to be taken.

Compressor Motor Burn-Out

If the compressor motor fails electrically and the motor insulation burns, sludge and acid may be produced. If the compressor failure occurred gradually, there may be a large amount of these materials circulated throughout the refrigeration circuit. The debris will find its way into driers, metering devices, coil circuits, and refrigerant line sets. If the compressor is going to be replaced and the system repaired, all signs of contamination must be removed.

- A flush kit made for this purpose can be used to clean the sludge out of the system.
- The old liquid line drier should be replaced with one that has a larger capacity rating than the original.
- A suction line drier should be added to the suction line. Suction line driers often have at least one pressure port on them, making it easier to measure the pressure drop across them. If there is no pressure port, install one on the suction line between the evaporator coil and the suction line drier.
- Clean any debris that may be at the metering device.
- Evacuate and charge the system.
- Run the system for 4-6 hours and monitor the pressure drop across the suction drier.
- If the pressure drop in the suction line drier exceeds 4 PSIG, replace the drier. Keep monitoring drier pressure loss until the pressure drop is stable.
- Remove the suction line drier after 48 hours of operation.
- Monitor the liquid line drier for excess pressure drop. Replace the drier if excess pressure drop is present.
- Repeat drier changes as needed until drier pressure drop is stable.
- Perform a final acid test on the new compressor oil to determine acid level of replacement compressor.

Refrigerant Recovery

R-22 is a hydrochlorofluorocarbon (HCFC) refrigerant that includes chlorine atoms. Because chlorine can damage the Earth's ozone layer, this refrigerant has been phased out of production. Venting this refrigerant to atmosphere is a violation of federal law. This refrigerant must be recovered.

R-410A is a hydrofluorocarbon (HFC) refrigerant that has no chlorine atoms, thus does not damage the ozone layer. However, it is considered a greenhouse gas and must also be recovered. It is illegal to vent R-410A to the atmosphere.

Suction Line Piping

Introduction

The suction line connects the evaporator outlet to the inlet of the compressor and carries suction vapor and refrigeration oil back to the compressor. Improperly sized suction lines can lead to low system capacity, improper charge, and poor oil return to the compressor. Technicians working on split system heat pumps and air conditioning systems need to know how to evaluate the suction line to ensure it is properly sized.

The importance of recognizing a potential pressure drop problem with a suction line prior to charging the system cannot be overstated. Since the suction pressure port is located at the outdoor unit, the servicing technician may be unknowingly charging downstream from a large pressure drop caused by an undersized or kinked line. If the problem is not identified, there is a tendency to overcharge the system in an attempt to achieve proper suction pressure.

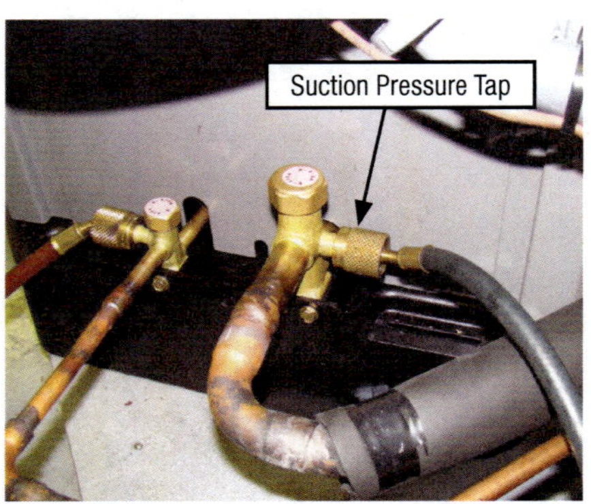

ASHRAE Recommended Limits for Suction Line Pressure Drop

ASHRAE standards for recommended maximum suction line pressure drop is 3 PSIG for R-22 systems, and a maximum of 5 PSIG for R-410A systems. These pressure drop allowances, if followed, will limit the capacity loss of either type system to a maximum of 3% of its rated cooling capacity (R-22 loses 1% capacity for every 1 PSIG of suction pressure drop. R-410A loses 0.6% capacity for every 1 PSIG of suction pressure drop).

Piping Connection Size

The pipe connection size on the outdoor unit does not always indicate the correct suction line size for the installation. The connection size is based on a standard line length of 25 feet used during the certification testing for the unit. If the suction line length is longer than 25 feet, and the standard connection size is used, the pressure loss in the line may be excessive. To determine what the proper line size should be when the distance between the indoor and outdoor units exceeds 25 feet, special pipe sizing tables are used.

Pipe Sizing Tables

The table below can be used to size R-22 and R-410A suction lines. The chart lists the system capacity, suction line size, and how much pressure the line will drop at 100 equivalent feet of pipe. Equivalent feet of piping is the sum of the linear length of the pipe plus the equivalent length, expressed in feet, of each fitting used in the line.

	Suction Line Sizing Table			
	R-22		R-410A	
Tons	Tube Size (in.)	Pressure Drop (PSIG per 100 ft. Equiv.)	Tube Size (in.)	Pressure Drop (PSIG per 100 ft. Equiv.)
1.5	5/8	4.7	1/2	10.8
	3/4	1.8	5/8	3.1
	-	-	3/4	1.2
2.0	5/8	8.1	5/8	5.4
	3/4	3.0	3/4	2.0
	7/8	1.3	7/8	0.9
2.5	5/8	12.7	5/8	8.2
	3/4	4.6	3/4	3.0
	7/8	2.0	7/8	1.3
3	3/4	6.5	5/8	11.7
	7/8	2.8	3/4	4.3
	-	-	7/8	1.9
3.5	3/4	8.8	3/4	5.8
	7/8	3.8	7/8	2.5
	1-1/8	1.0	-	-
4.0	7/8	4.9	3/4	7.4
	1-1/8	1.3	7/8	3.2
	-	-	1-1/8	0.9
5.0	7/8	7.5	3/4	11.5
	1-1/8	2.0	7/8	4.9
	1-3/8	0.7	1-1/8	1.3

The table below is used to determine how many feet of pipe each elbow equals.

Equivalent Length Brazed 90⁰ Elbows (Ft.)		
OD Tube (in.)	Short Radius	Long Radius
5/8	5.7	3.9
3/4	6.5	4.5
7/8	7.3	5.3
1-1/8	2.7	1.9
1-3/8	---	---

These tables are easy to use, and will be used to calculate the approximate pressure drop that will occur at peak system performance. <u>It is important to confirm there is not an excessive pressure drop in the suction line prior to charging the system.</u>

Service Procedure: Determining Suction Line Pressure Drop

Step 1

Identify the type of refrigerant used in the system and select the proper chart.

Example: Let's assume the system is an R-22 unit that has been installed for a few years. We would select the R-22 Suction Line Sizing Chart.

Step 2

Identify the line size and tonnage of the system. Use the table to determine the pressure drop that would occur at 100 equivalent feet of suction line. Note the pressure drop figure.

Example: The system is a 2 ton system, and the suction line size installed is 5/8. The chart indicates that a pressure drop of 8.1 PSIG would occur at 100 equivalent feet of piping.

Suction Line Sizing Table

Tons	R-22		R-410A	
	Tube Size (in.)	Pressure Drop (PSIG per 100 ft. Equiv.)	Tube Size (in.)	Pressure Drop (PSIG per 100 ft. Equiv.)
1.5	5/8	4.7	1/2	10.8
	3/4	1.8	5/8	3.1
	-	-	3/4	1.2
2.0	5/8	8.1	5/8	5.4
	3/4	3.0	3/4	2.0
	7/8	1.3	7/8	0.9
2.5	5/8	12.7	5/8	8.2
	3/4	4.6	3/4	3.0
	7/8	2.0	7/8	1.3
3	3/4	6.5	5/8	11.7
	7/8	2.8	3/4	4.3
	-	-	7/8	1.9
3.5	3/4	8.8	3/4	5.8
	7/8	3.8	7/8	2.5
	1-1/8	1.0	-	-
4.0	7/8	4.9	3/4	7.4
	1-1/8	1.3	7/8	3.2
	-	-	1-1/8	0.9
5.0	7/8	7.5	3/4	11.5
	1-1/8	2.0	7/8	4.9
	1-3/8	0.7	1-1/8	1.3

Step 3

Measure the linear length of the suction line.

Example: The suction line length is 40 ft. Note this length is not the equivalent length, it is the linear length.

Step 4

Count the number of elbows in the line and identify whether they are short radius or long radius elbows.

Example: The line has 6 short radius 90 degree elbows installed between the evaporator coil and the outdoor condensing unit.

Step 5

Reference the Equivalent Length Fitting Table to find the equivalent length of each fitting.

Example: Each 90 degree short radius elbow is equal to 5.7 ft. of pipe.

Equivalent Length Brazed 90° Elbows (Ft.)

OD Tube (in.)	Short Radius	Long Radius
5/8	5.7	3.9
3/4	6.5	4.5
7/8	7.3	5.3
1-1/8	2.7	1.9
1-3/8	---	---

Step 6

Determine the total combined equivalent length of all pipe fittings.

Example: There are 6 elbows in the line.
5.7 ft. x 6 elbows = 34.2 ft.

Step 7

Add the total linear length of pipe to the total equivalent fitting length. The sum is the total equivalent length of the suction line.

Example: 40 linear ft. of pipe + 34.2 fitting length = 74.2 ft.

Step 8

Divide the equivalent length of the line by 100 to determine the multiplier to use in Step 9.

Example: 74.2/100 = .742

Step 9

From Step 2, get the pressure drop per 100 feet of pipe for the system. Multiply this value by the multiplier calculated in Step 8. The result is the approximate pressure drop that will occur when the system is at maximum capacity.

Example: 8.1 x .742 = 6 PSIG of pressure loss.

Step 10

If the pressure drop is within ASHRAE standards, the capacity loss caused by the line is within acceptable limits.

Example: ASHRAE standards recommend a maximum 3 PSIG of pressure drop in the suction line for R-22 systems. The estimated pressure drop in this example is far above this limit and will cause a 6% loss of system capacity.

If operating costs are excessive, or there is a lack of cooling complaint, the line will need to be replaced.

R-410A Suction Line Performance Versus R-22 Systems

For comparison, let's use the example in the previous service procedure and see what would have happened if the system used R-410A refrigerant:

With R-410A systems, we have a maximum pressure drop allowance of 5 PSIG.

If the system is a 2 ton system, and the line size installed was 5/8, the chart indicates that a pressure drop of 5.4 PSIG would occur at 100 equivalent feet of piping.

	Suction Line Sizing Table				
	R-22		R-410A		
Tons	Tube Size (in.)	Pressure Drop (PSIG per 100 ft. Equiv.)	Tube Size (in.)	Pressure Drop (PSIG per 100 ft. Equiv.)	
1.5	5/8	4.7	1/2	10.8	
	3/4	1.8	5/8	3.1	
	-	-	3/4	1.2	
2.0	5/8	8.1	5/8	5.4	
	3/4	3.0	3/4	2.0	
	7/8	1.3	7/8	0.9	

The line length is 40 ft., Each 90 degree short radius elbow is equal to 5.7 ft. of pipe.

Equivalent Length Brazed 90⁰ Elbows (Ft.)		
OD Tube (in.)	Short Radius	Long Radius
5/8	5.7	3.9
3/4	6.5	4.5
7/8	7.3	5.3
1-1/8	2.7	1.9
1-3/8	---	---

There are 6 elbows in the line: 5.7 ft. x 6 elbows = 34.2 ft.

40 ft. + 34.2 ft. = 74.2 equivalent ft. of pipe.

74.2/100 = .742

.742 x 5.4 PSIG pressure drop per 100 feet of pipe =

4 PSIG of pressure loss.

ASHRAE standards recommend a maximum 5 PSIG of pressure drop in the suction line for R-410A systems. The estimated pressure drop in this example is within the limit and will cause around a 2% loss of system capacity.

Identifying Pressure Drop Caused by a Kink

An installer may accidently kink the suction line. Most attempts to remove the kink involve trying to re-round the pipe by crimping the tubing with channel lock pliers. The still results in a pressure drop in the pipe.

Since refrigerants drop temperature as they drop pressure, a touch type thermometer can be used to find a potential pressure drop by checking for a temperature drop across the suspected kink. A 2°F temperature drop across a kink indicates a pressure drop of 3 PSIG for R-22 and 5 PSIG for R-410A.

A kink with a 2°F temperature drop would, by itself, account for the total pressure drop allowance for the entire line, and will drop the system capacity by at least 3%. Thus, a temperature drop across a kink that is 2°F or greater is excessive and should be corrected.

Restricted suction line driers will also cause a suction line pressure drop.

Buried Suction Lines

Refrigerant migrates to the coldest point of the system during the off cycle. If the suction line is buried in the ground or concrete floor, refrigerant liquid will migrate to the buried portion of the line. When the compressor is started, the compressor will receive a slug of liquid refrigerant through the suction line.

Suction Line Accumulator

If this problem is pre-existing, compressor bearing wear is likely occurring due to liquid refrigerant floodback during start up. The compressor may also experience starting problems. The addition of a suction line accumulator at the outdoor unit will help prevent further damage to the system. The accumulator will catch the liquid as it floods back to the outdoor unit from the buried suction line.

Improper Suction Line Sizing: Beware of Charge

When servicing a system where the suction line is determined to have an excessive pressure drop, it is likely that refrigerant may have been added at some point in an attempt to raise the suction pressure to the proper level. The overcharged condition will need to be corrected. If the line is undersized and an excessive pressure drop is present, adding a line pressure tap on the suction line at the outlet of the evaporator coil may be necessary to check the evaporator superheat and charge to a normal evaporator superheat level.

Improper Suction Line Sizing: Beware of Compressor Liquid Migration

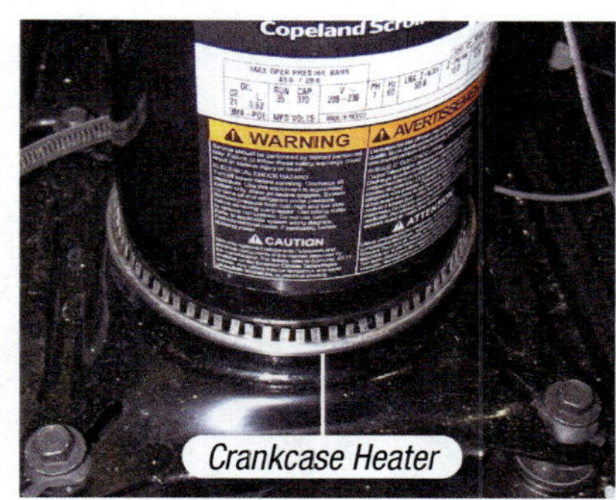
Crankcase Heater

A high charge level requires the addition of a crankcase heater at the compressor. Systems that operate with small suction lines tend to be overcharged in an attempt to correct system pressures. Excess charge may cause scroll compressors to flood during early morning starts due to liquid migration during off cycle periods. Adding a crankcase heater to the compressor will help prevent liquid refrigerant from migrating into the compressor shell during off cycle periods.

Using Existing Line Sets with R-410A Systems to Increase Capacity

R-410A systems have a higher suction line pressure drop tolerance than R-22 systems. In cases when it is impractical to replace the line sets due to excessive run length or lack of accessibility, it may be possible to correct system capacity problems by replacing an R-22 system with an R-410A system when the capacity drop is due to excessive suction line pressure drop. Using the existing line set by cleaning residual mineral oil within the piping can be accomplished with flush kits that are available at HVAC/R wholesalers. Always check with your equipment supplier to ensure the manufacturer of the air conditioning system approves the use of flush solvents in their systems prior to using.

Uninsulated Lines

Uninsulated suction line piping will allow the cold suction vapor to absorb additional heat as it travels through the piping. This heat, in most cases, will decrease the capacity of the system and contribute to excessive compressor heat.

Suction Line Evaporator Circuit Pressure Tap

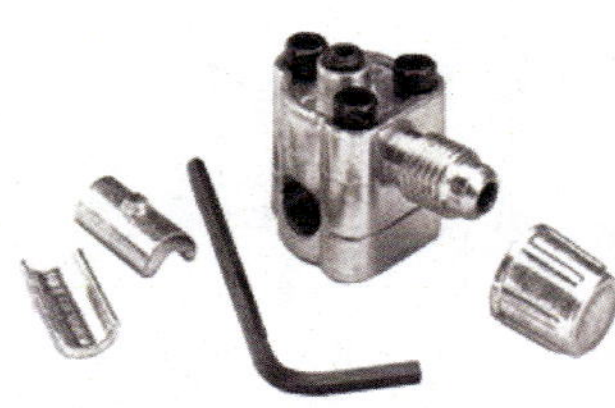
Pressure Tap Kit

When excessive pressure drop is suspected, a pressure tap should be installed at the outlet of the evaporator circuit. Measure the suction pressure at this port and then compare it to the suction pressure at the outdoor unit service valve. In situations where the suction line size problem cannot be corrected, it is best to charge the system at the indoor pressure tap port if using the superheat method. This will ensure the system is charged using the true evaporator superheat conditions. Line tap valves, such as the one shown in the figure must not be left on the system. Leaving a line tap valve on a system is a violation of EPA Section 608 guidelines. Once the pressure drop has been calculated, the line tap valve must be removed.

Suction Line Piping Summary

- Suction lines drop pressure. For R-22 systems, every 1 PSIG of pressure drop results in 1% capacity loss. R-410A loses .06% capacity for ever 1 PSIG of suction line pressure drop. 3% is normally the maximum capacity that a system should lose to suction line pressure drop.

- Some systems may only slightly exceed this value. It is then a decision of economy as to whether to leave the system as-is, or replace it.

- R-410A has better suction line performance when compared to R-22.

- Suction lines can be re-used when changing from one refrigerant to another. The lines should be cleaned to remove any residual oils and other contaminants.

- Kinks cause pressure drop. Check for excessive temperature drop across deformations in a suction line tube. A 2°F temperature drop indicates a pressure loss equal to the maximum allowed loss of the line itself!

Suction Line Evaluation Worksheet

System Specifications

Refrigerant Type: R-22 R-410A   System Tons: _______   Line Size: _______   Pressure Drop per 100 ft: _______ PSIG

Total Equivalent Line Length Calculation

Line Linear Length: _______ ft.

Number of Short Radius Elbows: _______ x Equivalent ft. each: _______ ft. = _______ ft.

Number of Long Radius Elbows: _______ x Equivalent ft. each: _______ ft. = _______ ft.

Total Equivalent ft. Elbows _______ + Plus Line Linear ft. _______ = Total Equivalent Line Length _______ ft.

Total Pressure Drop Calculation

Total Equivalent Line Length (_______ / 100 = _______) x Pressure Drop per 100 ft. _______ = _______ PSIG DROP

Capacity Loss Calculation

R-410A:
PSIG DROP _______ x .06 = _______% **Capacity Loss**   ASHRAE recommended maximum capacity loss = 3%

R-22:
PSIG DROP _______ x 1 = _______% **Capacity Loss**

(R-410A = .06% capacity loss for every 1 PSIG of suction pressure drop)
(R-22 = 1% capacity loss for every 1 PSIG of suction pressure drop)

Suction Line Sizing Table

Tons	R-22 Tube Size (in.)	R-22 Pressure Drop (PSIG per 100 ft. Equiv.)	R-410A Tube Size (in.)	R-410A Pressure Drop (PSIG per 100 ft. Equiv.)
1.5	5/8	4.7	1/2	10.8
	3/4	1.8	5/8	3.1
	-	-	3/4	1.2
2.0	5/8	8.1	5/8	5.4
	3/4	3.0	3/4	2.0
	7/8	1.3	7/8	0.9
2.5	5/8	12.7	5/8	8.2
	3/4	4.6	3/4	3.0
	7/8	2.0	7/8	1.3
3	3/4	6.5	5/8	11.7
	7/8	2.8	3/4	4.3
	-	-	7/8	1.9
3.5	3/4	8.8	3/4	5.8
	7/8	3.8	7/8	2.5
	1-1/8	1.0	-	-
4.0	7/8	4.9	3/4	7.4
	1-1/8	1.3	7/8	3.2
	-	-	1-1/8	0.9
5.0	7/8	7.5	3/4	11.5
	1-1/8	2.0	7/8	4.9
	1-3/8	0.7	1-1/8	1.3

Equivalent Length Brazed 90° Elbows (Ft.)

OD Tube (in.)	Short Radius	Long Radius
5/8	5.7	3.9
3/4	6.5	4.5
7/8	7.3	5.3
1-1/8	2.7	1.9
1-3/8	---	---

Liquid Line Piping

Introduction

The smaller liquid line connects the outlet of the condensing coil to the inlet of the metering device, and carries liquid refrigerant at high pressure. The small line size and high pressure result in a high pressure drop between the condensing coil and the metering device. If the pressure drop is excessive due to improper line sizing, the system may experience low capacity, compressor failure and potential damage to the metering device.

Technicians who are investigating poor system performance or compressor failure, should inspect and evaluate the liquid line circuit to rule out potential problems caused by the liquid line.

Subcooled Liquid and its Role in Proper System Operation

The role of the liquid line and subcooled liquid refrigerant is extremely important to overall system performance and reliability. When the liquid refrigerant temperature is lower than its saturation temperature at a given pressure, it is in a subcooled state. The liquid refrigerant should be subcooled late in the condenser coil circuiting, and should remain subcooled all the way from the condensing coil outlet to the inlet of the metering device. If the refrigerant at the inlet of the metering device has reached saturated or superheated temperatures, vapor refrigerant is present in the liquid line. This vapor is called flash gas. Flash gas will cause the metering device to feed an inadequate amount of refrigerant to the evaporator. The presence of flash gas will result in a system that is operating with a starved evaporator coil.

Symptoms of flash gas include noisy liquid lines, overheating compressors, high superheat, low system capacity, and intermittent comfort complaints.

How Flash Gas Forms in the Liquid Line

Referring to the figure below, the liquid line receives subcooled liquid, at 250 PSIG, from the condensing coil, Point 1. The pressure at this point in the line is 250 PSIG. At 250 PSIG, R-22 is at a condensing temperature of 117°F. In this example, the liquid line is 107°F which is 10°F cooler than the saturation temperature of 117°F. The liquid at this pressure port is therefore subcooled by 10°, and is in a pure liquid state.

At Point 2 in the figure, the liquid line pressure has dropped to 235 PSIG. At 235 PSIG, R-22 is at a saturation temperature of 113°F. The 107°F liquid reaches this port, and the subcooling has now dropped to 6°F. Since the liquid is still cooler than the corresponding saturation temperature at the second gauge port, the refrigerant is still in a subcooled state and is still pure liquid.

At Point 3, the 107°F liquid refrigerant is at a pressure of 220 PSIG. At 220 PSIG, R-22 is at a saturation temperature of 108°F. The refrigerant at this point is still subcooled by 1°F, and will remain pure liquid. However, If the pressure at Point 3 had dropped below 220 PSIG due to an undersized liquid line or excessive fittings, or if the refrigerant temperature rose above 107°F because the liquid line was run through a hot attic where the refrigerant was allowed to absorb heat, the refrigerant would return to a saturated state and flash gas vapor would form in the line.

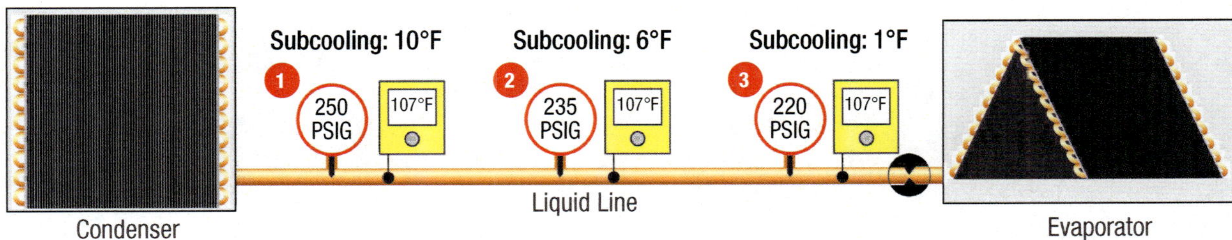

In the example above, the amount of subcooling that is measured at the beginning of the liquid line is not the same amount of subcooling that is present at the end of the liquid line.

A sight glass may be installed on the liquid line to visualy detect flash gas. In many cases, the sight glass is installed at the outdoor unit. Other than serving as a moisture indicator, a sight glass installed at that location provides no information as to the state of the liquid as it reaches the metering device. It is best to install the sight glass at the end of the liquid line to get a look at the condition of the refrigerant as it enters the metering device.

Causes of Flash Gas Formation

Flash gas occurs when there is inadequate subcooling. Inadequate subcooling means there is not enough subcooling to overcome the pressure drop of the liquid line. Even if the system is properly charged, but has an undersized liquid line or too many pressure dropping components, flash gas will form.

Flash gas can result when an uninsulated liquid line runs through hot spaces, or when the liquid line is located in direct sunlight where the refrigerant in the liquid line absorbs heat and loses subcooling. An undercharged system can also cause flash gas.

Flash Gas with Proper Charge

It should be clear that even when a system has been properly charged using the subcooling method, but the liquid line is undersized, refrigerant can still flash into a vapor. If the charge is too low, and the liquid line is properly sized, flash gas will occur. When flash gas is present, system capacity will be reduced and the compressor will run hot. Evaluating the size of the liquid line will confirm if the liquid line is capable of delivering subcooled liquid refrigerant to the metering device.

Pressure Loss in Liquid Lines

The pressure drop that takes place in the liquid line is due to friction, fittings, accessories, and line lift. Lift is the vertical distance of the piping when the condenser is located below the metering device and evaporator coil. When determining the pressure drop in the liquid line, all of these factors must be taken into consideration.

The pressure drop due to liquid lift is calculated by PSIG of drop per foot of lift. R-22 systems will drop 0.5 PSIG for every 1 ft. of liquid lift. R-410A systems will drop 0.43 PSIG for every 1 ft. of liquid lift.

The pressure drop due to friction can be calculated using special piping tables. By using piping tables, the equivalent length of the liquid line is calculated and plotted onto the table to determine the estimated pressure drop of the line. If the pressure loss is greater than the subcooling charge can overcome, the line must be re-sized.

Liquid Line Pipe Sizing Tables

The top of the R-22 Liquid Line Sizing Table (shown on oppoite page) shows that 10°F of subcooling with R-22 refrigerant can sustain adequate subcooling through a liquid line that drops up to 30 PSIG. This relationship is standard for R-22 liquid lines. Systems that can subcool by 12°F can handle a larger pressure drop of up to 35 PSIG. In summary, the more the condenser can subcool the liquid, the greater the amount of allowable liquid line pressure drop. The example table below is used for evaluating R-22 liquid lines used with residential systems. The table gives the maximum allowable pressure drop allowance based on the subcooling capability of the condensing unit, and then plots the liquid line equivalent length against line size and capacity to determine the estimated total pressure loss of the line. If the loss exceeds the maximum allowable drop, the line is either too small, has too many fittings, or the lift is too high.

Refrigerant & Refrigerant Piping

Liquid Line Pipe Sizing Example

R-22 Liquid Line Sizing Table

Maximum Allowable Pressure Drop
- 10°F Subcooling = 30 PSIG
- 12°F Subcooling = 35 PSIG
- Subtract From Maximum Allowable Any Lift Pressure Drop   0.50 PSIG X _______Ft. of lift = _________LOSS
- Total Available PSIG After any Lift Loss = _________PSIG Maximum piping pressure drop

Tube OD	Rated BTUH	20'	40'	60'	80'	100'	120'	140'	160'	180'	200'	220'	240'
1/4"	18000	6	12	18.1	24.1	30.1	-	-	-	-	-	-	-
	24000	10.2	20.3	30.5	-	-	-	-	-	-	-	-	-
5/16"	18000	1.6	3.2	4.7	6.3	7.9	9.5	11.1	12.6	14.2	15.8	17.4	19
	24000	2.7	5.3	8	10.6	13.3	16	18.6	21.3	23.9	26.6	29.3	31.9
	30000	4	8	11.9	15.9	19.9	23.9	27.9	31.8	-	-	-	-
	36000	5.5	11.1	16.6	22.2	27.7	33.2	-	-	-	-	-	-
	42000	7.3	14.6	22	29.3	-	-	-	-	-	-	-	-
3/8"	18000	0.6	1.1	1.7	2.2	2.8	3.4	3.9	4.5	5	5.6	6.2	6.7
	24000	0.9	1.9	2.8	3.8	4.7	5.6	6.6	7.5	8.5	9.4	10.3	11.3
	30000	1.4	2.8	4.2	5.6	7	8.4	9.8	11.2	12.6	14	15.4	16.8
	36000	1.9	3.9	5.8	7.8	9.7	11.6	13.6	15.5	17.5	19.4	21.3	23.3
	42000	2.6	5.1	7.7	10.2	12.8	15.4	17.9	20.5	23	25.6	28.2	30.7
	48000	3.3	6.5	9.8	13	16.3	19.6	2.8	26.1	29.3	32.6	-	-
	60000	4.9	9.8	14.7	19.6	24.5	29.4	34.3	-	-	-	-	-
1/2"	36000	0.4	0.8	1.2	1.6	2	2.4	2.8	3.2	3.6	4	4.4	4.8
	42000	0.5	1	1.6	2.1	2.6	3.1	3.6	4.2	4.7	5.2	5.7	6.2
	48000	0.7	1.3	2	2.6	3.3	4	4.6	5.3	5.9	6.6	7.3	7.9
	60000	1	2	3	4	5	6	7	8	9	10	11.3	12

Courtesy of The Trane Company

Service Procedure: Evaluating Liquid Line Size

Step 1

Determine which type of refrigerant is being used, and select the proper liquid line sizing table. Measure the operating subcooling of the system.

Example: An R-22 system operating with 10°F subcooling. (See the table on the next page to follow this example.)

Step 2

Measure the length of the liquid line. Take note of the pipe size and capacity of the system.

Example: The liquid line size is 5/16", and the capacity of the system is 42,000 BTUH. The line is 61 ft. long.

Step 3

Count the number of elbows or other fittings on the line.

Example: The line has 6 long radius elbows.

Step 4

Determine the equivalent length for all fittings in the line using the equivalent length table for fittings. (Note the table is for brazed fittings. Err on the safe side and also use these lengths for bends.)

Example: There are 6 long radius elbows (3.1 ft. of pipe each). 3.1 ft. x 6 = 18.6 ft. of piping. Let's round this to 19 equivelant ft.

Step 5

Add the linear pipe length to the equivalent length of fittings to determine the total equivalent length of the liquid line.

> Example: 61 ft. (linear line) + 19 ft. (fittings) = 80 total equivalent ft. of piping.

Step 6

Determine if there is liquid line lift. Multiply the lift length by the multiplier on the piping table. This will determine how much pressure is lost in the lift process. This pressure loss must be subtracted from the allowable maximum liquid line loss. The pressure that remains after subtracting the lift pressure loss is the maximum the line can lose due to friction drop and any additional liquid line components, such as liquid line filter driers.

> Example: There is 20 feet of lift. From our chart, each foot of lift accounts for 0.5 PSIG of our allowance. 20 x 0.5 = 10 PSIG
>
> In our example, we have 10°F of subcooling so we can overcome up to 30 PSIG of liquid line pressure drop. Of the 30 PSIG, we will lose 10 in the lift. 30 PSIG – 10 = 20 PSIG of loss allowance for the liquid line friction.

Step 7

Using the liquid line pipe sizing table, plot the total equivalent length of line onto the data on the table. If the pressure drop is exceeds the allowance, and the condenser is not able to take any additional refrigerant without raising head pressure above the desired level, the liquid line size is too small. If the pressure drop is below the maximum allowance, the liquid line is properly sized.

> Example: On the table, plot the 5/16 line size @ 42,000 BTUH and cross over to 80 ft. At 80 ft., the line will drop 29.3 PSIG which is well above the allowance of 20 PSIG. The line is too small for an R-22 system that has a subcooling capability of 10°F. The line will need to be replaced with a larger line. If the line cannot be removed or changed, the system may need to be replaced with one that will work with the line. To see if the line can perform with an R-410A system, plot the line information onto the piping table for R-410A.

R-22 Liquid Line Sizing Table

Maximum Allowable Pressure Drop

Step 1 10°F Subcooling = 30 PSIG
12°F Subcooling = 35 PSIG
Subtract From Maximum Allowable Any Lift Pressure Drop 0.5 PSIG x __20__ ft. of lift = __10__ LOSS
Total Available PSIG After any Lift Loss = __20__ PSIG Maximum piping pressure drop   **Step 6**

Tube OD	Rated BTUH	20'	40'	60'	80'	100'	120'	140'	160'	180'	200'	220'	240'
1/4"	18000	6	12	18.1	24.1	30.1	-	-	-	-	-	-	-
	24000	10.2	20.3	30.5	-	-	-	-	-	-	-	-	-
5/16"	18000	1.6	3.2	4.7	6.3	7.9	9.5	11.1	12.6	14.2	15.8	17.4	19
	24000	2.7	5.3	8	10.6	13.3	16	18.6	21.3	23.9	26.6	29.3	31.9
	30000	4	8	11.9	15.9	19.9	23.9	27.9	31.8	-	-	-	-
	36000	5.5	11.1	16.6	22.2	27.7	33.2	-	-	-	-	-	-
	42000	7.3	14.6	22	29.3	-	-	-	-	-	-	-	-
3/8"	18000	0.6	1.1	1.7	2.2	2.8	3.4	3.9	4.5	5	5.6	6.2	6.7
	24000	0.9	1.9	2.8	3.8	4.7	5.6	6.6	7.5	8.5	9.4	10.3	11.3
	30000	1.4	2.8	4.2	5.6	7	8.4	9.8	11.2	12.6	14	15.4	16.8
	36000	1.9	3.9	5.8	7.8	9.7	11.6	13.6	15.5	17.5	19.4	21.3	23.3
	42000	2.6	5.1	7.7	10.2	12.8	15.4	17.9	20.5	23	25.6	28.2	30.7
	48000	3.3	6.5	9.8	13	16.3	19.6	2.8	26.1	29.3	32.6	-	-
	60000	4.9	9.8	14.7	19.6	24.5	29.4	34.3	-	-	-	-	-
1/2"	36000	0.4	0.8	1.2	1.6	2	2.4	2.8	3.2	3.6	4	4.4	4.8
	42000	0.5	1	1.6	2.1	2.6	3.1	3.6	4.2	4.7	5.2	5.7	6.2
	48000	0.7	1.3	2	2.6	3.3	4	4.6	5.3	5.9	6.6	7.3	7.9
	60000	1	2	3	4	5	6	7	8	9	10	11.3	12

Step 2 points to the 5/16" row. **Step 7** points to the 5/16" 42000 row at 80' = 29.3

Liquid Line Brazed Fitting Equivalent Lengths

OD Tube (in.)	45° ELL	90 Short Radius ELL	90 Long Radius ELL	Sight Glass
1/4"	2	4.6	3.1	1.2
5/16"	2.1	**Step 3**	3.1	1.4
3/8"	2.2	4.7	3.2	1.6
1/2"	2.4	4.7	3.2	1.7

Step 4 3.1 (Fitting eq. length) x 6 (# of fittings) = 19 ft.

Step 5 65 (Linear length) x 19 (total fitting eq. length) = 84 ft.

Refrigerant & Refrigerant Piping

R-410A versus R-22 Liquid Line Sizing

For comparison, let's re-evaluate the line set used as the example in the previous service procedure and see what would have happened if the system used R-410A refrigerant:

To begin, 10°F subcooling can overcome up to 50 PSIG of liquid line pressure drop. The job lift is 20 feet of pipe. R-410A will lose 0.43 PSIG per foot of lift.   20 x 0.43 = 8.6 PSIG of loss. Let's round that off to 9 PSIG of pressure drop due to liquid line lift.

Subtract 9 PSIG from the available 50 PSIG Maximum allowance, which leaves 41 PSIG for friction loss in the piping.

Plot the 5/16 inch piping @ 42,000 BTUH onto the piping chart and move over to 80 total equivalent feet of pipe column. The table indicates a loss of 30.7 PSIG. We are well within tolerance.

R-410A Liquid Line Sizing Table													
Maximum Allowable Pressure Drop													
10°F Subcooling = 50 PSIG													
Subtract From Maximum Allowable Any Lift Pressure Drop 0.43 PSIG x __20__ ft. of lift = __9__ LOSS													
Total Available PSIG After any Lift Loss = __41__ PSIG Maximum piping pressure drop													
Tube OD	Rated BTUH	20'	40'	60'	80'	100'	120'	140'	160'	180'	200'	220'	240'
1/4"	18000	6.3	12.6	18.8	25.1	31.4	37.7	44	-	-	-	-	-
5/16"	18000	1.6	3.3	4.9	6.6	8.2	9.8	11.5	13.1	14.8	16.4	18	19.7
	24000	2.8	5.5	8.3	11	13.8	16.6	19.3	22.1	24.8	27.6	30.4	33.1
	30000	4.1	8.3	12.4	16.6	20.7	24.8	29	33.1	37.3	41.4	45.5	49.7
	36000	5.8	11.6	17.3	23.1	28.9	34.7	40.5	46.2				
	42000	7.7	15.4	23	30.7	38.4	16.1						
3/8"	24000	1	1.9	2.9	3.8	4.8	5.8	6.7	7.7	8.6	9.6	10.6	11.5
	30000	0.4	2.9	4.3	5.8	7.2	8.6	10.1	11.5	13	14.4	15.8	17.3
	36000	2	4	6.1	8.1	10.1	12.1	14.1	16.2	18.2	20.2	22.2	24.2
	42000	2.7	5.3	8	10.6	13.3	16	18.6	21.3	23.9	26.6	29.3	31.9
	48000	3.4	6.8	10.2	13.6	17	20.4	23.8	27.2	30.6	34	37.4	40.8
	60000	5.1	10.3	15.4	20.6	25.7	30.8	36	41.1	46.3			

In summary, a R-410A system would perform properly with the example liquid line, yet an R-22 system would not. When an instance like this is discovered with an R-22 system, and it is impractical to replace the line sets due to excessive run length, lift, or lack of accessibility, replacing it with an R-410A system and reusing the existing piping (if allowed by the equipment manufacturer) will compensate for the liquid line pressure drop.

Tapping the End of the Liquid Line to Confirm Subcooling

A service port, such as a bullet valve, can be added to the liquid line to allow for pressure measurements to be taken at the end of the liquid line. A pressure reading at this point can be used to make a subcooling calculation at the end of the liquid line to confirm proper liquid line performance. Bullet valves are an inexpensive and efficient way of gaining access to the refrigerant circuit for pressure tapping without having to recover and recharge the system.

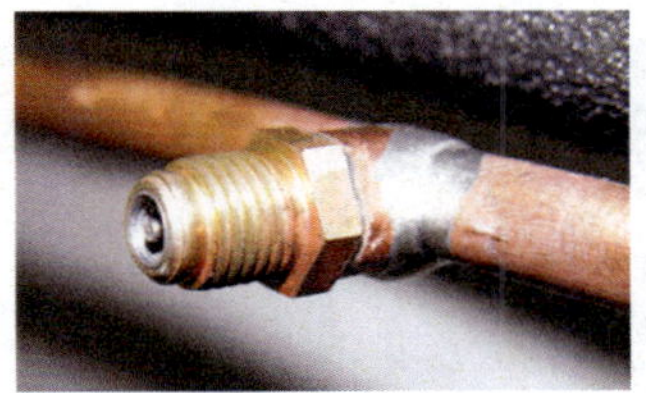

Systems Operating with Flash Gas Present

When flash gas is present in the liquid line, the liquid line may sound like there is air inside of the line. The evaporator coil may be noisy as the saturated liquid/vapor refrigerant mix leaves the metering device. The suction superheat will be high and the compressor may operate hotter than normal.

Liquid Line Piping Summary

- Liquid lines drop pressure. This pressure drop, if excessive, can overcome the subcooling level and cause liquid refrigerant to flash in the liquid line.
- R-410A has more pressure drop tolerance than R-22.
- Liquid line lift affects available liquid line pressure.
- Proper charge with incorrect line sizing can cause flash gas formation.
- Flash gas can lead to overheating compressors.
- Evaluate liquid lines to confirm they are capable of working with the available subcooling.

Liquid Line Sizing Tables

R-22 Liquid Line Sizing Table

Maximum Allowable Pressure Drop
  10°F Subcooling = 30 PSIG
  12°F Subcooling = 35 PSIG
  Subtract From Maximum Allowable Any Lift Pressure Drop   .50 PSIG x _______ft. of lift = _________LOSS
  Total Available PSIG After any Lift Loss = _________PSIG Maximum piping pressure drop

Tube OD	Rated BTUH	20'	40'	60'	80'	100'	120'	140'	160'	180'	200'	220'	240'
1/4"	18000	6	12	18.1	24.1	30.1	-	-	-	-	-	-	-
	24000	10.2	20.3	30.5	-	-	-	-	-	-	-	-	-
5/16"	18000	1.6	3.2	4.7	6.3	7.9	9.5	11.1	12.6	14.2	15.8	17.4	19
	24000	2.7	5.3	8	10.6	13.3	16	18.6	21.3	23.9	26.6	29.3	31.9
	30000	4	8	11.9	15.9	19.9	23.9	27.9	31.8	-	-	-	-
	36000	5.5	11.1	16.6	22.2	27.7	33.2	-	-	-	-	-	-
	42000	7.3	14.6	22	29.3	-	-	-	-	-	-	-	-
3/8"	18000	0.6	1.1	1.7	2.2	2.8	3.4	3.9	4.5	5	5.6	6.2	6.7
	24000	0.9	1.9	2.8	3.8	4.7	5.6	6.6	7.5	8.5	9.4	10.3	11.3
	30000	1.4	2.8	4.2	5.6	7	8.4	9.8	11.2	12.6	14	15.4	16.8
	36000	1.9	3.9	5.8	7.8	9.7	11.6	13.6	15.5	17.5	19.4	21.3	23.3
	42000	2.6	5.1	7.7	10.2	12.8	15.4	17.9	20.5	23	25.6	28.2	30.7
	48000	3.3	6.5	9.8	13	16.3	19.6	2.8	26.1	29.3	32.6	-	-
	60000	4.9	9.8	14.7	19.6	24.5	29.4	34.3	-	-	-	-	-
1/2"	36000	0.4	0.8	1.2	1.6	2	2.4	2.8	3.2	3.6	4	4.4	4.8
	42000	0.5	1	1.6	2.1	2.6	3.1	3.6	4.2	4.7	5.2	5.7	6.2
	48000	0.7	1.3	2	2.6	3.3	4	4.6	5.3	5.9	6.6	7.3	7.9
	60000	1	2	3	4	5	6	7	8	9	10	11.3	12

R-410A Liquid Line Sizing Table

Maximum Allowable Pressure Drop
  10°F Subcooling = 50 PSIG
  Subtract From Maximum Allowable Any Lift Pressure Drop   0.43 PSIG x _______ft. of lift = _________LOSS
  Total Available PSIG After any Lift Loss = _________PSIG Maximum piping pressure drop

Tube OD	Rated BTUH	20'	40'	60'	80'	100'	120'	140'	160'	180'	200'	220'	240'
1/4"	18000	6.3	12.6	18.8	25.1	31.4	37.7	44	-	-	-	-	-
5/16"	18000	1.6	3.3	4.9	6.6	8.2	9.8	11.5	13.1	14.8	16.4	18	19.7
	24000	2.8	5.5	8.3	11	13.8	16.6	19.3	22.1	24.8	27.6	30.4	33.1
	30000	4.1	8.3	12.4	16.6	20.7	24.8	29	33.1	37.3	41.4	45.5	49.7
	36000	5.8	11.6	17.3	23.1	28.9	34.7	40.5	46.2				
	42000	7.7	15.4	23	30.7	38.4	16.1						
3/8"	24000	1	1.9	2.9	3.8	4.8	5.8	6.7	7.7	8.6	9.6	10.6	11.5
	30000	0.4	2.9	4.3	5.8	7.2	8.6	10.1	11.5	13	14.4	15.8	17.3
	36000	2	4	6.1	8.1	10.1	12.1	14.1	16.2	18.2	20.2	22.2	24.2
	42000	2.7	5.3	8	10.6	13.3	16	18.6	21.3	23.9	26.6	29.3	31.9
	48000	3.4	6.8	10.2	13.6	17	20.4	23.8	27.2	30.6	34	37.4	40.8
	60000	5.1	10.3	15.4	20.6	25.7	30.8	36	41.1	46.3			
1/2"	42000	0.5	1.1	1.6	2.2	2.7	3.2	3.8	4.3	4.9	5.4	5.9	6.5
	48000	0.7	1.4	2	2.7	3.4	4.1	4.8	5.4	6.1	6.8	7.5	8.2
	60000	1	2.1	3.1	4.2	5.2	6.2	7.3	8.3	9.4	10.4	11.4	12.5

Liquid Line Brazed Fitting Equivalent Lengths

OD Tube (in.)	45° ELL	90 Short Radius ELL	90 Long Radius ELL	Sight Glass
1/4"	2	4.6	3.1	1.2
5/16"	2.1	4.6	3.1	1.4
3/8"	2.2	4.7	3.2	1.6
1/2"	2.4	4.7	3.2	1.7

_______ *(Fitting eq. length)* x _______ *(# of fittings)* = _______*ft.*

_______ *(Linear length)* x _______*(total fitting eq. length)* = _______*ft.*

Refrigerant & Refrigerant Piping

Condenser Circuit Piping

In the figure below, the refrigerant circuiting of a typical condenser coil can be followed. Starting at the compressor discharge line (1), the hot vapor refrigerant enters a hot gas header (2) that directs the hot vapor to the circuits in the condenser coil (3). The condenser fan pulls outdoor air through the condenser coil, where the cooler outdoor air will absorb heat from the hot vapor refrigerant (remember, heat travels from a warmer substance to a cooler one). Within the condenser circuits, this transfer of heat condenses the hot vapor to liquid. The liquid refrigerant exits the condenser circuits at the collection header (4), where it is directed to a final subcooling loop (5) at the bottom of the coil. This subcooling loop further cools the liquid refrigerant. A drier (6) is located in the liquid line circuit to catch debris and absorb moisture.

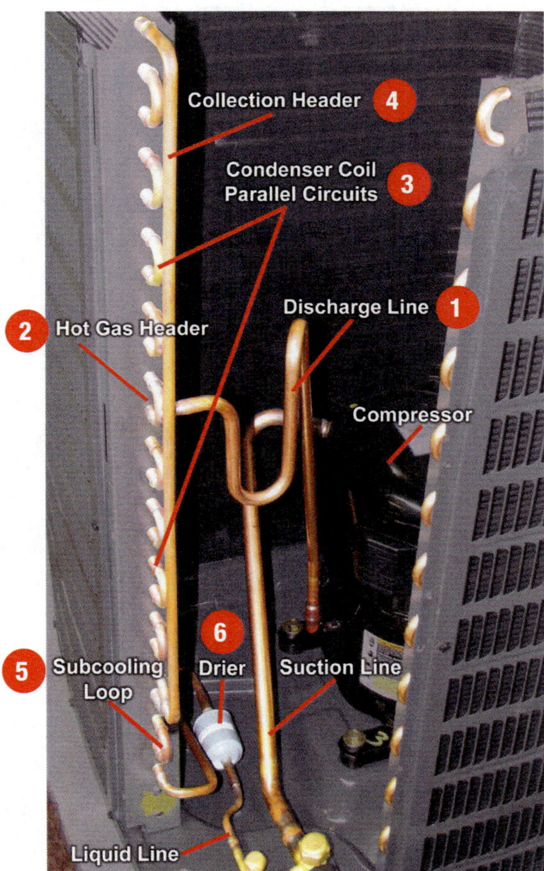

Condenser Circuit Restrictions

It is rare to have a refrigerant restriction in a condenser circuit. If a restriction is present, the compressor's Internal Pressure Relief (IPR) valve would likely open, or if a high pressure switch were on the unit, the switch would shut the system off. For example, if one of the two circuits had a restriction in it, the effective area of the condenser coil would be reduced and the system would likely cycle off on high head pressure.

If liquid refrigerant is leaving the condenser coil at a lower temperature than that of the outdoor air, there is a restriction causing a pressure drop in the coil. To find the restriction, use a digital thermometer to measure the inlet and outlet circuits of the coil. The temperature of these circuits will have some deviation, but if one circuit is found to be disproportionately cooler than the others, there is a pressure drop in that circuit.

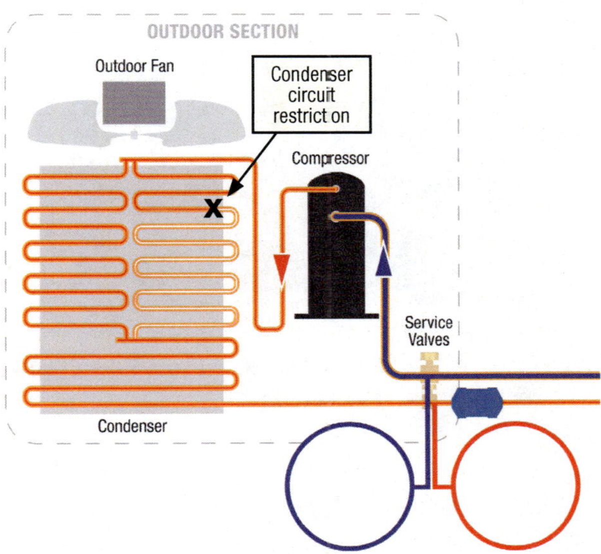

The restriction in the compressor coil will reduce surface area of the condenser. The restriction can be found with a digital thermometer.

Filter driers will always have a pressure drop across them. Typically, filter driers will drop 1 PSIG of pressure when new. As they load up with moisture and debris, this pressure drop will increase. Temperature measurements taken at the inlet and outlet of the drier can detect some levels of pressure drop. Line kinks can also be detected by measuring for a temperature drop.

A 3°F temperature drop equals a 3 PSIG pressure drop (R-22). Replace a drier or repair any kink with a temperature drop of 3°F or more.

To find a restriction in a condenser coil, the system must be operating. However, it may not be possible to run the system if the restriction is great enough to cause excessive discharge pressure.

Service Procedure: Checking for Restrictions in Condenser Circuits

Tools:

- Digital Temperature Probe

STEP 1

Run the system.

STEP 2

If there is a liquid line drier at the outlet of the subcooling loop, check the temperature of the liquid at the inlet and outlet of the drier.

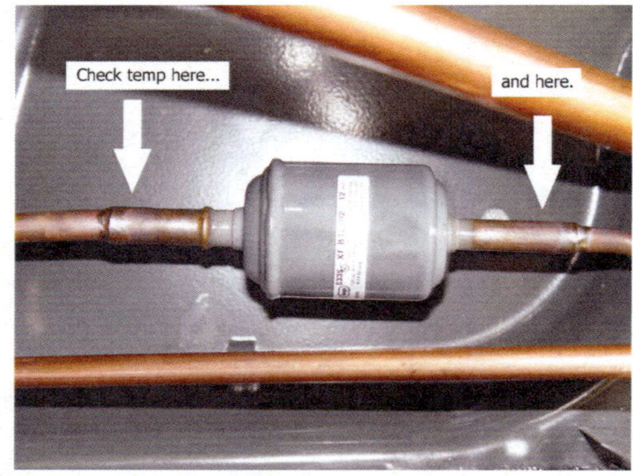

There should be no more than a 3°F drop in temperature across the drier. Replace the drier if needed. If there is no temperature drop detected, proceed to step 3.

STEP 3

Check the temperature of the liquid as it enters and leaves the outdoor liquid line service valve.

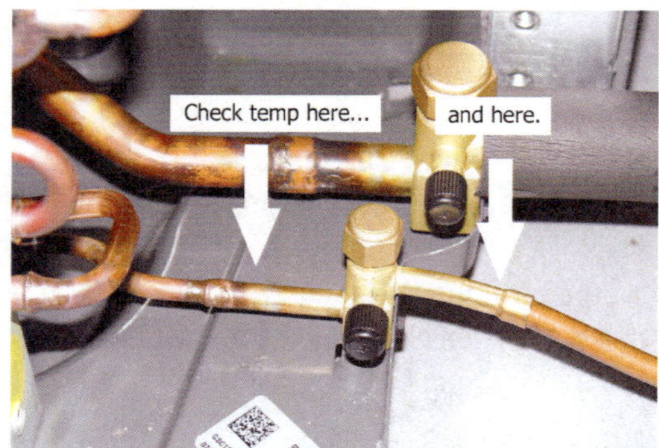

There should be no more than a 3°F temperature change, if any. If excess temperature drop is detected, make sure the valve is completely open. If there is no excessive temperature drop, proceed to step 4.

STEP 4

Measure the temperature of the refrigerant lines at the individual outlets of the condenser coil parallel circuits.

The temperatures will vary based upon the air flow pattern from the condenser fan, but no outlet temperature should be cooler than the outdoor air.

If a low temperature is detected, there is a pressure drop in the circuit where the temperature drop is detected. Repair or replace the coil as needed. If no noticeable temperature drop is found, proceed to step 5.

STEP 5

If the condenser coil has a subcooling loop, check the temperature at the exit of the subcooling loop.

The temperature of the liquid should be no cooler than the outside air temperature. If it is, there is a pressure drop in the subcooling loop circuit. If no problem is found, the condenser coil has no flow restrictions in it.

Dirty Condenser Coil

The most common problem encountered with condenser coils is dirt and other airborne particles that become trapped on its surface. The dirt insulates the refrigerant from the outdoor air, and reduces the amount of heat being rejected from the refrigerant to the outdoor air. The pressure in the condenser will be elevated and the liquid line will be hot istead of warm.

The combination of high condensing pressure and elevated liquid temperature will reduce the system's capacity. The excess pressure reduces the volumetric efficiency of the compressor and reduces condenser subcooling. The lack of subcooling may cause liquid refrigerant to flash in the liquid line.

Specially formulated coil cleaners are used to clean off the debris from the coil. Some cleaners are acid based, and others are environmentally friendly.

Superheat

Suction Vapor Superheat

Introduction

At the point where all of the saturated liquid refrigerant in the evaporator circuiting has been boiled off into vapor, the vapor is still very cold. As the cold vapor travels through the remaining circuiting of the evaporator coil, it picks up additional heat from the warm air passing over the evaporator coil. Because there is no liquid remaining in this part of the coil, the temperature of the vapor refrigerant begins to rise to a temperature above saturation temperature. <u>Once the refrigerant vapor temperature rises above saturation temperature, it is called superheated vapor.</u> The heat added to the refrigerant vapor after all of the liquid has boiled off is called suction vapor superheat, or simply, evaporator superheat.

In the example below, saturated R-410A refrigerant at 44°F enters the evaporator circuiting. We know this temperature because we converted our 130 PSIG suction pressure to saturation temperature. A digital temperature probe is placed at the suction line service gauge port and we measure 54°F. The temperature of the suction line indicates that the refrigerant vapor has been heated 10 degrees, therefore the superheat level is 10°F.

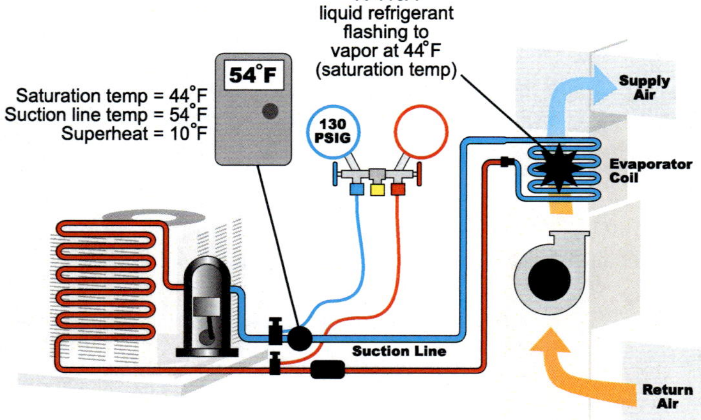

Since no change of state is taking place when the vapor is being superheated, the actual amount of cooling that takes place declines in the evaporator circuiting where superheating occurs. At this point, the vapor temperature will rise very rapidly. Consider the analogy of melted ice. The remaining cold water from the melted ice will produce very little refrigeration effect even though its temperature is colder than the product being refrigerated.

Although superheating of the refrigerant vapor in the evaporator coil provides only minimal cooling benefits, it is required. Superheating in the evaporator, and the suction line, ensures that no liquid refrigerant leaves the evaporator circuit and ultimately enters the compressor. It is necessary to prevent saturated refrigerant from leaving the evaporator because at saturation temperature, both liquid and vapor refrigerant may be present, and compressors cannot pump liquid. If liquid is introduced into the compressor, damage to the compressor could occur.

What Affects Superheat Level

The position within the evaporator coil where all of the liquid has been boiled off is determined by the amount of refrigerant entering the evaporator from the metering device, the liquid/vapor makeup of the refrigerant entering the evaporator coil, and by the amount of heat energy passing across the evaporator surface.

Fixed-bore metering devices allow the flow of refrigerant into the evaporator to be controlled by the high side, or liquid pressure, present at the end of the liquid line. The liquid pressure will vary with changes in outdoor air temperature and heat load. If the liquid pressure is high, the refrigerant flow through the metering device will be high, and the evaporator will flood with refrigerant. If the pressure at the liquid line inlet is low, the refrigerant flow through the metering device will be low, and the evaporator coil will be starved for refrigerant.

Since refrigerant flow through a fixed-bore metering device is inconsistent, due to fluctuations in high side pressure, the actual efficiency of the evaporator coil is affected.

On a cool day, the liquid pressure falls, causing the flow of refrigerant into the evaporator coil to decrease. The lack of adequate refrigerant flow in the evaporator coil will cause the liquid to boil off (vaporize) early in the evaporator circuit. The remaining vapor travels through the evaporator circuiting and becomes highly superheated. This results in a reduction of efficiency since no change of state is taking place in a large portion of the evaporator coil.

On hot days, when the liquid pressure is high, the evaporator coil floods with cold refrigerant. The liquid boils off very late in the evaporator circuit, which leaves very little space in the evaporator for superheating to occur. In this state, the capacity of the evaporator is high since most of the coil circuit is being used to boil off cold liquid to a vapor.

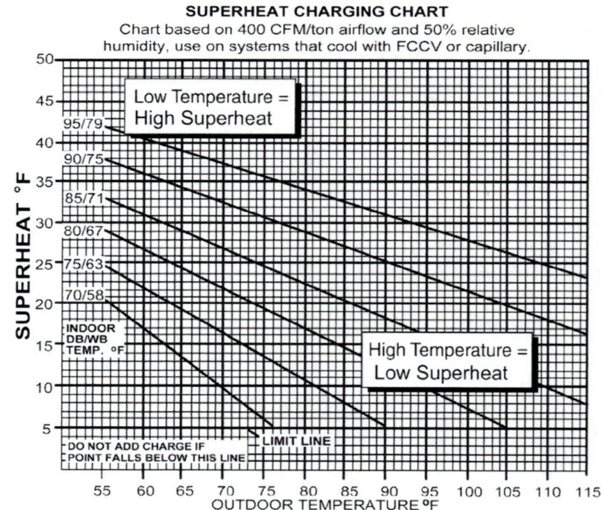

This chart is for instructional uses ONLY. Not for field use.
Provided courtesy of The Trane Company

By studying charging charts, good examples of how superheat changes with changes in the outdoor air temperature will be found. On cold days, the superheat level will be high, and on hot days it will fall. At no time should it be allowed to fall below 5°F.

The other factor affecting the point at which the last droplet of cold liquid refrigerant is boiled off in the evaporator circuit is the amount of heat energy passing through the evaporator coil from the conditioned space. The amount of heat energy from the conditioned space is determined by the volume of air, the sensible heat of the air, and the latent heat of the air.

If the air volume or heat contained in the air is high, the liquid will boil off early in the evaporator circuit and the vapor will be highly superheated.

If the air volume or heat contained in the air is low, the liquid will travel further into the evaporator coil before being boiled off to a vapor. In this case, there is very little circuiting left for superheating. In some cases, the refrigerant may leave the evaporator in a saturated state, where both liquid and vapor are present. If saturated refrigerant leaves the evaporator, there is no superheat.

It is typical for the indoor blower to deliver air to the evaporator coil at a rate between 350 and 450 Cubic Feet Per Minute (CFM) for every ton of cooling capacity present. For example, a 2-ton unit would typically need to move between 700 and 900 CFM. If the system is to operate properly, the amount of indoor airflow must be correct.

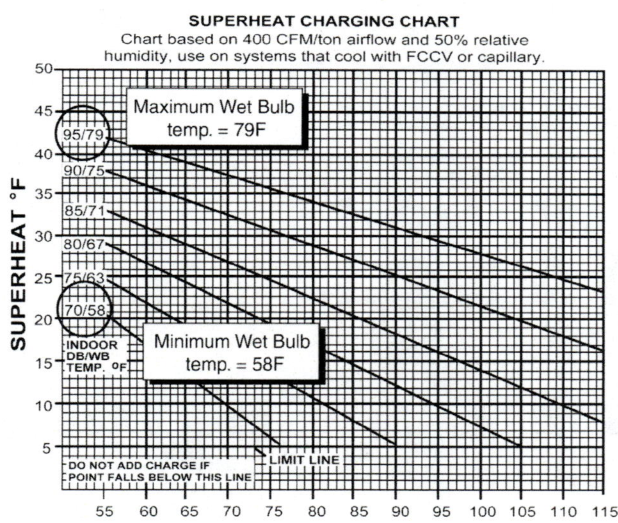

This chart is for instructional uses ONLY. Not for field use.
Provided courtesy of The Trane Company

The heat contained in the air is plotted on the charging chart. The indoor heat load can be anywhere from 58° degree Wet Bulb (Total Heat Content) to 79° Wet Bulb. At levels below 58° or above 79° Wet Bulb, the heat is either too low or too high for proper system operation. The ranges for Dry Bulb temperature are used in the superheat chart. Refer to the chart for the minimum and maximum limits of temperature.

How Superheat is Used by Service Technicians

Calculating superheat takes only a few moments to accomplish, and is used by service technicians as a means to determine if the evaporator coil has the correct amount of refrigerant in its circuits.

When the evaporator coil contains excessive liquid refrigerant, the liquid travels deep into the evaporator circuits. Little area is left for absorbing the additional heat required to ensure all the liquid refrigerant has vaporized. This may result in some liquid refrigerant escaping the evaporator coil and potentially damaging the compressor.

When the evaporator circuit lacks refrigerant, the liquid refrigerant vaporizes early in the evaporator circuit and travels longer through the coil as a vapor. The greater the distance the vapor travels, the more sensible heat it picks up. This additional heat may cause overheating and damage to the compressor.

By calculating the suction vapor superheat level, technicians can compare the measured level against charts that indicate what the superheat should be. When superheat is above the required level, the evaporator coil is either starving for refrigerant, or has excessive heat load. When the superheat is below the required level, the evaporator coil is either flooded with too much refrigerant, or the coil lacks heat load.

Suction Vapor Superheat with a Flooded Evaporator Coil

In a coil flooded with cold liquid refrigerant at saturation temperature, every inch of the circuiting is boiling off cold liquid to cold vapor. The refrigerant in this example is R-22. The suction gauge indicates a pressure of 85 PSIG, which corresponds to a saturation temperature of 51°F. When the liquid vaporizes, the vapor contains the latent heat that was used to change the liquid to the vapor. The vapor temperature has not risen above 51°F, yet it contains large amounts of heat.

A digital temperature probe is placed at the suction line service gauge port. The suction line temperature is 51°F. Since the entering saturation temperature and leaving refrigerant temperature are the same, the refrigerant is still saturated has not had a chance to completely vaporize and pick up superheat. This condition is producing maximum heat transfer in the evaporator, but it will likely result in liquid floodback to the compressor.

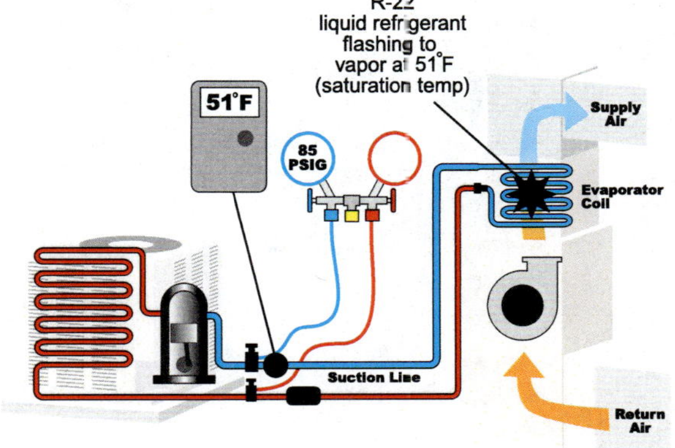

Suction Vapor Superheat with a Starved Evaporator Coil

If the refrigerant flow through the metering device is low, suction pressure will result. In this example, the refrigerant is R-22 at a gauge pressure of only 49 PSIG. R-22 at this pressure has a corresponding saturation temperature of 25°F. The temperature of the vapor leaving this coil is 70°F, which is 45°F warmer than the refrigerant saturation temperature at the outlet of the metering device. The high superheat level is a result of the liquid refrigerant completely boiling off into a vapor early in the coil circuiting, causing the vapor to travel an excessive distance in the evaporator coil circuiting.

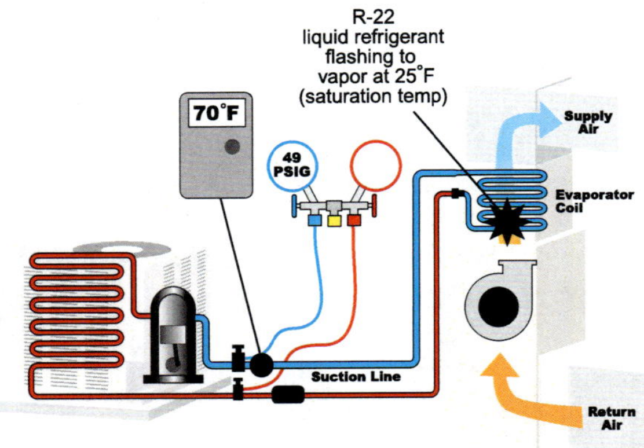

Suction Vapor Superheat with a Low Heat Load on the Evaporator Coil

When the evaporator coil lacks heat to absorb, the refrigerant pressure will be below the charging chart requirement and the evaporator coil will struggle to vaporize the cold liquid refrigerant. The liquid will travel at a low pressure deeper into the evaporator circuiting. For example, let's say we had a suction pressure of 60 PSIG. The corresponding saturation temperature is 34°F. The temperature probe at the suction line outlet of the coil indicates a suction vapor temperature of only 36°F. There is only 2°F of superheat, and the suction pressure is too low.

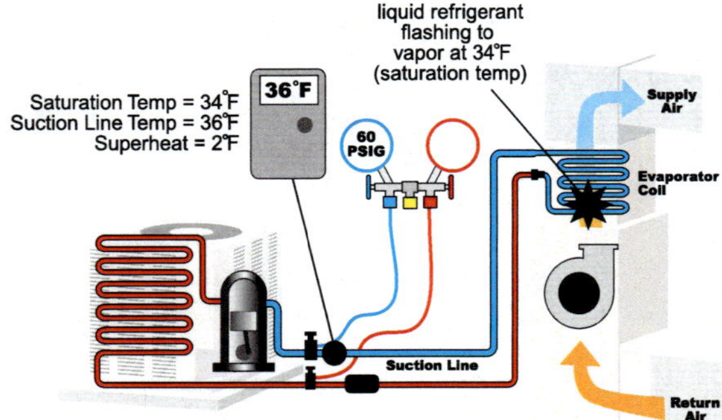

With a low heat load, the refrigerant will run at a lower pressure into the evaporator and will have low superheat.

Suction Vapor Superheat with a High Heat Load on the Evaporator Coil

When the refrigerant in the evaporator is absorbing too much heat, pressures will be high and the liquid refrigerant in the evaporator will completely vaporize early in the circuiting. In this example, the system suction pressure is up to 88 PSIG. At 88 PSIG the saturation temperature of the refrigerant is 53°F. The suction line temperature at the outlet of the evaporator coil is at a temperature of 83°F, which is 30°F warmer than saturation temperature. If the coil had too much refrigerant, the superheat would be low, but in this case, it is too high.

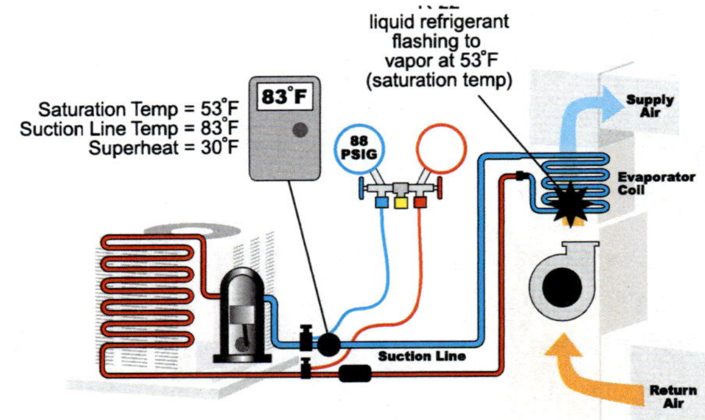

With a high heat load, the refrigerant will run at a higher pressure into the evaporator and will have a much higher superheat temperature.

Service Procedure: Calculating Suction Vapor Superheat

Tools:
- Digital Temperature Probe
- Refrigerant Gauges

STEP 1
Run the air conditioner and allow system pressures and temperatures to stabilize.

Superheat

STEP 2

Measure suction pressure and convert to saturation temperature.

STEP 3

Measure suction line temperature near suction gauge pressure port. (Use a digital thermometer. DO NOT USE INFRARED TYPE THERMOMETERS.)

STEP 4

Subtract saturation temperature from suction line temperature. The difference is the amount of superheat.

STEP 5

For fixed metering systems, compare the measured superheat to the charging chart requirement. TXV systems should measure around 10°F (with some exceptions as specified by equipment manufacturer).

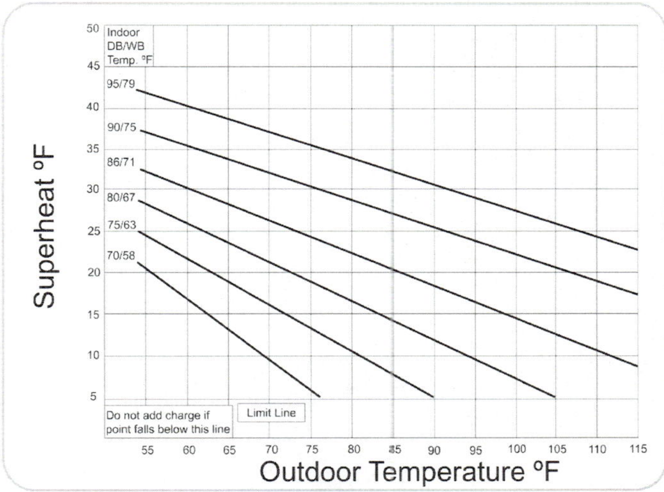

NOTE:

Changing superheat level (fixed orifice metering only):

- Adding refrigerant lowers superheat and raises pressure.
- Recovering refrigerant raises superheat and lowers pressure.
- Increased evaporator air flow raises superheat and raises pressure.
- Decreased evaporator air flow lowers superheat and lowers pressure.

Fixed Metering System Charging Using Superheat

Fixed Metering Systems and Superheat

Fixed metering devices are dependent on liquid line pressure. When the high pressure liquid travels through the piston hole, the liquid pressure at the outlet of the piston is much lower than the liquid pressure at the inlet of the piston. The amount of refrigerant that flows through the piston is affected by the pressure difference between the liquid line pressure at the inlet of the piston and the pressure at the outlet of the piston. The larger the pressure difference, the greater the rate of refrigerant flow through the piston.

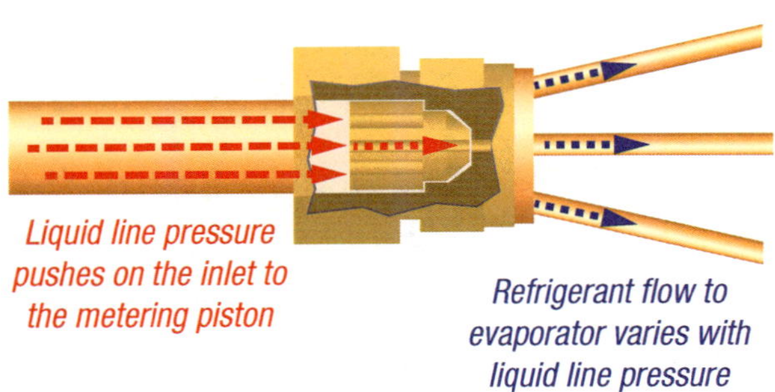

The operation of a fixed metering piston is very similar to the operation of a garden hose with an attached nozzle. When the pressure on the hose is high, the flow out of the nozzle is also high. When the pressure is reduced, the flow is reduced. Fixed metering pistons operate in a similar manner with changes in liquid line pressure. At high liquid line pressure levels, the flow is high. When the liquid line pressure is low, the flow out of the piston is reduced.

The resulting change in flow out of the piston results in fluctuations of refrigerant level in the evaporator coil circuiting. In other words, the evaporator coil will flood and starve based on the pressure in the liquid line.

Since a fixed-type metering device allows an inconsistent flow of refrigerant into the evaporator coil when the liquid pressure changes, the actual efficiency of the evaporator coil is affected both positively and adversely.

On a cool day, the liquid pressure falls, causing the flow of refrigerant into the evaporator coil to decrease. The lack of adequate refrigerant flow in the evaporator coil will cause the liquid to boil off (vaporize) early in the evaporator circuit. The remaining vapor travels through the evaporator circuiting and becomes highly superheated. This is poor for efficiency since no change of state is taking place in a large area of the evaporator coil.

On hot days with high liquid pressure, the evaporator coil is flooded with cold refrigerant. The liquid boils off very late in the evaporator circuit, which leaves very little circuiting for superheating. In this state, the capacity of the evaporator is high since most of the coil circuit is being used to boil off cold liquid to a vapor. Both evaporator coil conditions are illustrated below:

Figure 1

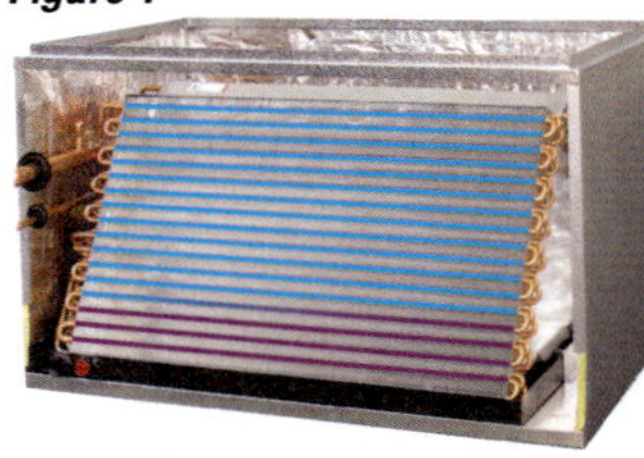

STARVED STATE
Cool outdoor air temperature drops the head pressure and corresponding liquid pressure.

Figure 2

FLOODED STATE
Hot outdoor air temperature increases the condensing pressure and corresponding liquid pressure.

The major cause of liquid pressure change is a change in outdoor air temperature. Superheat charging tables exemplify how superheat changes with changes in the outdoor air temperature. These tables are used to determine how much superheat should be present when operating with a correct system charge.

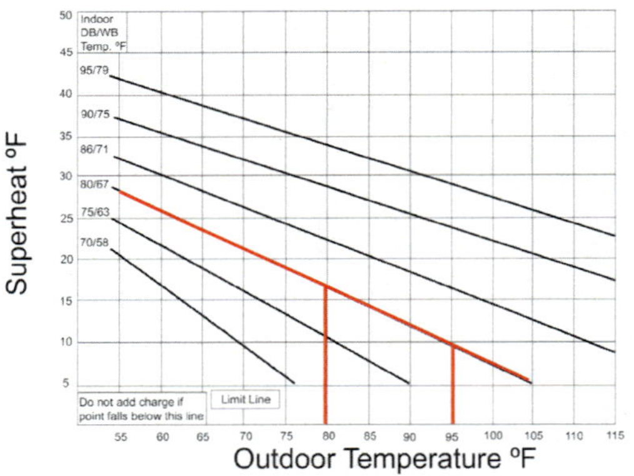

Typical Superheat Charging Table

Table is courtesy of The Trane Company

In this superheat chart example, the 80°/67° indoor air temperature line has been highlighted. This line will be used to illustrate how outdoor air temperature affects the superheat level requirement on the chart.

The first example shows that at an outdoor air temperature of 95°F, the suction vapor superheat requirement is 10°F.

At 80°F outdoor air temperature, the superheat requirement is 17°F.

The superheat requirement is rising as the outdoor air temperature is dropping.

Based on the above examples, it should be clear that the outdoor air temperature must be measured and plotted onto a superheat charging chart to determine how much superheat should be present when the fixed metering system is operating with the correct charge.

Suction Vapor Superheat and Heat Load

The amount of heat that comes in contact with the surface of the evaporator coil has an impact on the superheat level. At high heat load levels, the superheat will be high. At low heat load levels the superheat will be low.

Heat load is a combination of two factors: how much heat the air contains, and the amount of air being delivered by the indoor blower. The combination of these two factors is defined as the heat load. Both factors must be known before attempting to adjust the refrigerant charge.

The system is engineered to operate at a specific air volume level. Air volume is measured in cubic feet per minute (CFM) of air delivered across the surface of the evaporator coil. Capacity and efficiency tables published for the system are generated at these air volume levels. If the system is to operate as specified, the air volume (CFM) must be set to the level that was specified by the manufacturer. In the event the information is not available on either the charging charts or installation literature, adjust the air volume to 400 CFM of air per 1 ton of cooling capacity. (See Chapter J: Air Volume.)

The heat contained in the air can be referenced to the wet bulb temperature of the air returning to the evaporator coil. The relationship between return air heat content and wet bulb temperature is plotted on charging charts by the system manufacturer.

Think of heat load as 1 cubic foot blocks of air passing over the surface of the evaporator coil. Each block contains heat. If the number of blocks are reduced, the amount of heat available drops. If the amount of heat contained in each block is reduced, the heat available also drops.

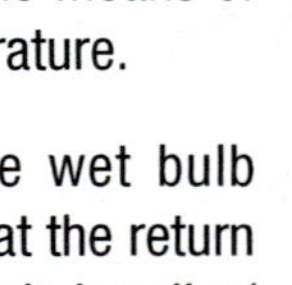

Sling Psychrometer

A tool called a psychrometer is used to measure the wet bulb temperature of the air. Sling psychrometers come in both mechanical and digital models. Both types of meters provide an acceptable means of measuring the wet bulb temperature.

When possible, measure the wet bulb temperature of the air directly at the return air filter rack in or near the air handler/ furnace. When the reading is taken at this spot, any infiltration air that may have entered the return air ducting is now measured as part of the total heat load. If the wet bulb temperature was measured in the conditioned space, and heat was infiltrating into the return air ducting between the conditioned space and the filter rack, the true wet bulb temperature of the air would be unknown. The result would be an inaccurate wet bulb temperature being plotted onto the charging chart. Incorrect charging would result.

Short-cutting the process by not measuring indoor air volume will also result in an improper charge. For example, if the air volume is too low, a technician will likely add refrigerant to the system in an attempt to raise pressure, when in fact the system needs a higher heat load.

Heat Load and the Superheat Charging Chart

In this superheat charging chart example below, the heat load is represented on the diagonal dry bulb/wet bulb temperature lines. Notice the increase in required superheat as the wet bulb temperature of the air increases. For example, at an outdoor air temperature of 90°F, and a return air temperature of 75°F dry bulb/63° wet bulb, the superheat with a correct charge should be about 5 °F. When the temperature of the air in the home increases to 80°F dry bulb/67°F wet bulb, the required superheat will rise to around 12°F.

Superheat Charging Table

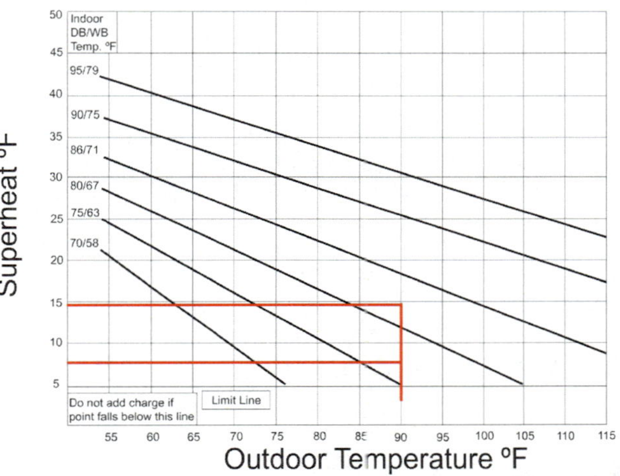

Note that these charts assume the air volume being delivered by the indoor blower motor is correct and constant. If the wet bulb temperature is measured with an incorrect air volume, the actual heat coming in contact with the evaporator coil will be incorrect. As a result, the required superheat, as read from the chart, will not be correct. The refrigerant charge added to the system will not be correct either, as the refrigerant level is adjusted to compensate for the wrong amount of heat coming in contact with the surface of the evaporator coil.

The example shows that as the temperature of the air in the home falls, the suction vapor superheat falls. When the air temperature in the home rises, the suction vapor superheat also rises. Therefore, when the thermostat first calls for cooling operation, the superheat will be higher than when the thermostat is satisfied and the system shuts off.

An important observation can be made by studying the superheat charging table. Notice in the example below that at an outdoor air temperature of 90°F, and a return air temperature of 80°F dry bulb and 67°F wet bulb (1), the evaporator coil with the correct charge should have a superheat level of about 12°F. When the return air temperature reaches 75°F dry bulb/63°F wet bulb (2), the required amount of superheat drops down to 5°F.

Superheat Charging Table

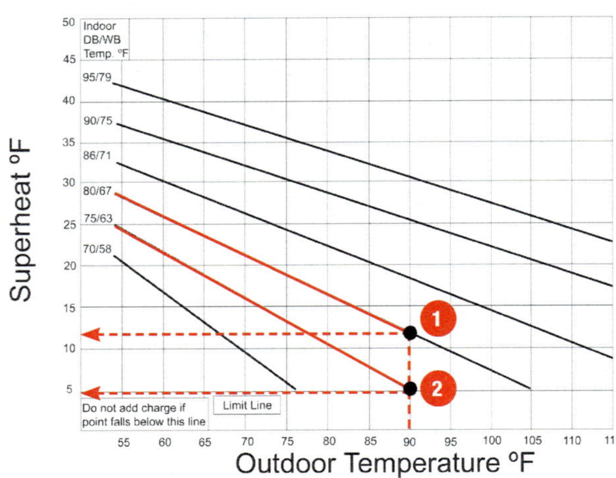

This is a good illustration of how critical it is to properly charge a fixed piston system. At 5°F degrees of superheat, the system is very close to the potential for saturated refrigerant entering the suction line back to the compressor. At saturation temperature, refrigerant is a mix of liquid and vapor. Compressors can be damaged by liquid floodback.

Suction Vapor Superheat Levels at Correct Charge Level

When the evaporator coil is operating with the correct charge, the refrigerant flow into the evaporator coil will fluctuate from a starved state to a flooded state. During periods where the air conditioning system is operated at cool outdoor air temperatures, OAT, the evaporator coil will starve and high suction vapor superheat will be present. During periods where the air conditioning system is operating at hot outdoor air temperatures, the evaporator coil will flood and low superheat levels will be present.

Indoor air heat load fluctuations will also affect the suction vapor superheat level. When the heat load is low, the evaporator coil will flood and the superheat level will drop. When the heat load is high, the evaporator coil will starve and the superheat level will rise.

With the system charge set properly, during flooded state operation (like on a hot day), a minimum level of suction vapor superheat will be maintained. The minimum level of suction vapor superheat will ensure that liquid floodback to the compressor does not occur.

When the evaporator circuit is operating in a starved state due to low outdoor air temperature, and charge is set properly, the compressor will operate within acceptable temperature ranges. Compressors will overheat when excessive suction vapor superheat levels are present. The overheating may open the compressor internal temperature overload or cause oil breakdown and potential bearing damage. Correct charging eliminates the potential for these problems.

Important System Conditions:

Evaporator Coil

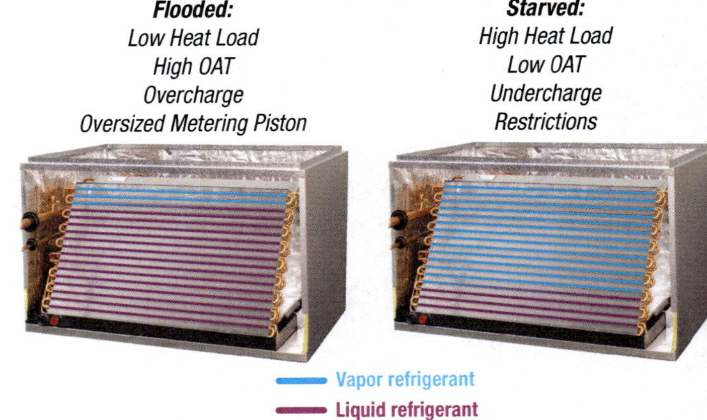

Heat Load And Superheat

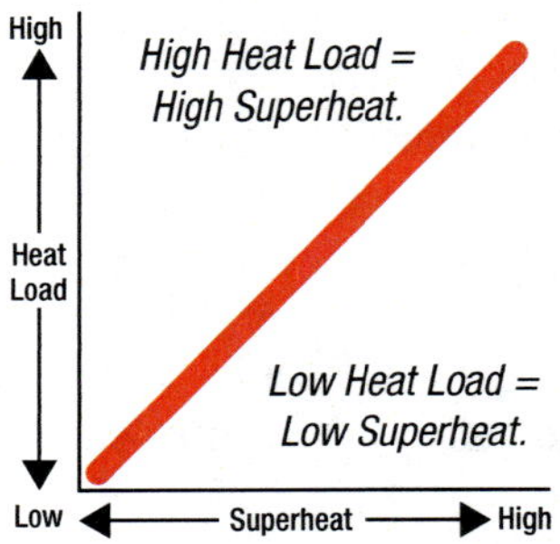

Compressor Vs. Superheat

High superheat will cause the compressor to run hot.

Floodback of liquid may cause compressor damage.

Determining Suction Vapor Superheat Requirement using a Superheat Charging Chart

There are different styles of superheat charging charts. Some charts are plotted on slide rules and others reference suction line temperature requirements at given suction pressures. Regardless of the type of chart used, they all plot the same information to determine the required superheat. The outdoor air temperature and the indoor air wet bulb temperature must be known. The two required temperatures are plotted and the required suction vapor superheat level is determined.

The next step is to measure the actual suction vapor superheat level of the system being charged. The superheat level is then compared to the charging chart required superheat. If they are the same, then charge adjustment is not required. If the required superheat level and actual superheat level are different, an adjustment in refrigerant charge is made.

If the superheat level is too high, ADD REFRIGERANT. If the superheat level is too low, RECOVER REFRIGERANT. Remember that adjustment of refrigerant charge should only be done after confirming the correct air volume, suction line size and condition, and if necessary, piston bore size.

Example of determining suction vapor superheat requirements for charging:

An R-22 air conditioning system is operating at a suction pressure of 68 PSIG. The suction line temperature is 55°F.

The outdoor air temperature is 90°F and the indoor air wet bulb temperature is 63°F. The air volume has been properly adjusted.

The first step is to determine how much superheat the system requires with correct charge. This is done by plotting the outdoor air temperature and indoor wet bulb temperature onto the superheat charging table. In our example, we require 5°F of superheat.

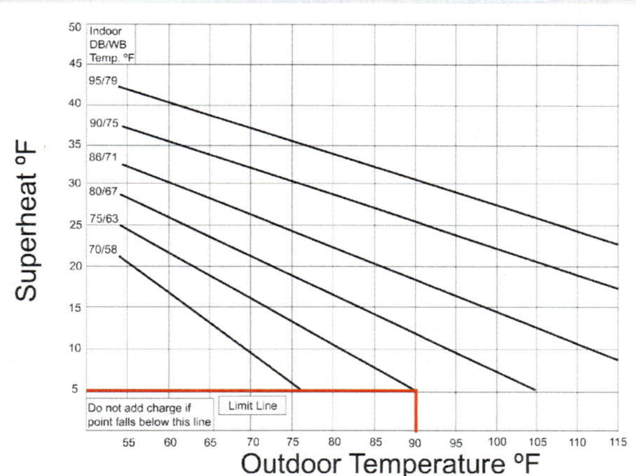

Superheat Charging Table

Next, determine how much superheat the system is currently operating with. In this example, there is a suction pressure of 68 PSIG. By referring to a pressure/temperature chart for R-22, we find that this pressure equals a saturation temperature of 40°F at the evaporator circuiting.

The suction line temperature is 55°F, which is 15°F warmer than the saturation temperature. Therefore, the superheat level is 15°F. The superheat measured is higher than it should be, so we must add refrigerant to the system.

This superheat calculation process should be repeated until the suction vapor superheat level falls to 5°F.

System Suction Pressure

The system suction pressure will fluctuate with changes in the flow rate of the liquid refrigerant into the evaporator coil circuiting, and by the heat load that is being absorbed by the refrigerant. The suction pressure will rise at high flow rates, and fall at low flow rates. When the heat load passing across the evaporator surface rises, the suction pressure will rise. When the heat load across the evaporator coil falls, the suction pressure will fall.

Suction Pressure Charts

Some manufacturers provide charts that are used to plot the required suction pressure at combinations of outdoor air temperature and indoor heat load. When charged properly to the suction pressure requirement indicated on the chart, the system will fluctuate its pressure properly as outdoor air and indoor air temperatures change.

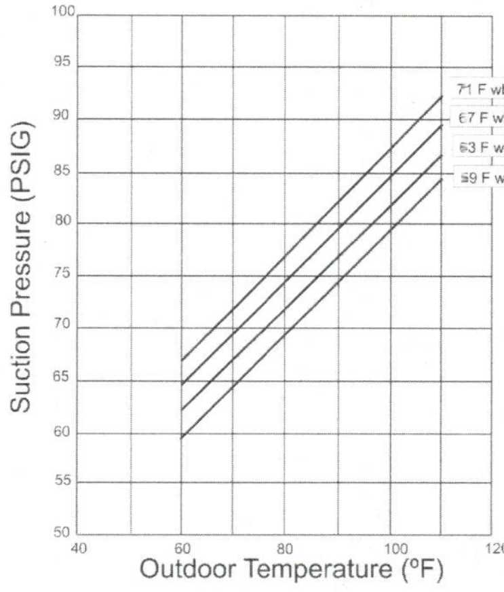

Suction Pressure Table

Indoor air temperature is plotted on 4 diagonal lines labeled with wet bulb temperature from a low of 59°F to a high of 71°F

Outdoor air temperature is plotted across the bottom of the chart from 60°F to 115°F.

It is very important for a technician charging the air conditioning system to get the system operating at the suction pressure requirement indicated by these charts. If a suction pressure chart is not available, remember that the suction pressure should stabilize at the correct charge level if the metering piston is properly sized and indoor air volume is at the correct level.

Suction Pressure and Capacity

With R-22 systems, the capacity of the system will fluctuate 1% for every 1 pound of suction pressure change. For example, if a suction pressure chart indicates the system should have 85 PSIG suction pressure and it was actually operating at 82 PSIG, the system capacity will be reduced by 3%.

R-410A has better capacity performance than R-22. Every 1 PSIG of pressure loss will reduce the capacity of the system by 0.6%. Therefore, a 5 PSIG loss of suction pressure will result in a 3% drop in cooling capacity.

The capacity changes that occur with changes in suction pressure are due to the vapor density of the refrigerant vapor changing with changes in pressure. When the suction pressure is high, the vapor entering the compressor is heavy. When the suction pressure is low, the refrigerant vapor entering the compressor is light. In other words, the compressor pumps more pounds of vapor refrigerant at higher suction pressures.

If the suction line has an excessive pressure drop due to a kink or undersizing, the suction pressure measured at the condensing unit will be lower than at the outlet of the evaporator coil. If this problem is suspected, place a pressure port in the suction line at the outlet of the evaporator coil and compare it to the pressure at the condensing unit pressure port. If excessive pressure drop is detected, confirm line size and check for a restriction in the line.

Fixed Metering Evaporator Coil Pressure at Correct Charge Level

Most R-22 and R-410A air conditioning systems are designed to operate at a full load evaporator coil temperature of 46°F to 55°F. This temperature occurs when the suction pressure of the system is at the correct level and with air temperature conditions of 95°F outdoor air and 80°F Dry Bulb/67°F Wet Bulb indoor air. At other combinations of outdoor air temperature and indoor air wet bulb temperature, the evaporator saturation temperature will vary from this range.

The evaporator saturation temperature range fluctuation is caused by changes in suction pressure. R-22 systems will operate at suction pressures within a pressure range of 56 PSIG up to 93 PSIG. At pressures below 56 PSIG, the evaporator coil may freeze. At pressures above 93 PSIG, most residential compressors are outside of their maximum operating pressure range.

R-410A systems will operate at similar temperatures as those on R-22 systems. The suction pressure range of the R-410A system will be higher than R-22 systems. For example, at 84 PSIG, R-22 systems will operate at a 50°F evaporator saturation temperature, while, an R-410A system will operate at a pressure of 143 PSIG. Although the pressures are different, the operating temperatures of the evaporator coils are the same.

The actual operating suction pressure and corresponding evaporator saturation temperature will depend on the outdoor air temperature and indoor heat load. The pressure change will occur with no change in system charge level. For example, at low outdoor air temperature the suction pressure is lower than at high temperatures.

Sample R-22 Suction Pressure Table

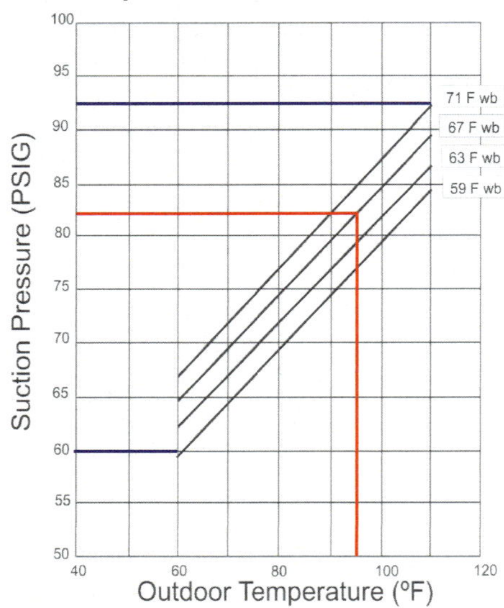

In this example suction pressure charging table, the lowest plotted suction pressure is under 60 PSIG and the maximum pressure is 93 PSIG.. Highlighted in blue.

Note operating suction pressure at 95°F outdoor air/67°F Indoor air wet bulb. Highlighted in red.

PRESSURES ARE R-22 REFRIGERANT. CONVERTING TO SATURATION TEMPERATURE SHOWS A RANGE OF TEMPERATURE FROM 32°F - 55°F.

THE SWING IN EVAPORATOR TEMPERATURE IS DUE TO CHANGES IN THE OUTDOOR AIR TEMPERATURE AND THE INDOOR AIR HEAT LOAD. (INDICATED BY WET BULB TEMPERATURE)

THE SWING IN PRESSURE OVER A WIDE RANGE OF OPERATING TEMPERATURES IS A GOOD CASE FOR NOT CHARGING A SYSTEM UNTIL IT "LOOKS RIGHT." SINCE IN A SHORT TIME, WHAT LOOKS RIGHT IS GOING TO CHANGE.

Superheat and Suction Pressure Changes as the Charge is Increased

When adding refrigerant to the system, the suction pressure will rise and the suction vapor superheat level will fall. The following series of figures show the changes to suction line temperature and suction pressure as charge is added to a piston equipped system.

Suction pressure and suction line temperature before charge is added:

Suction pressure and suction line temperature after addition of 1/2 Pound of R-22:

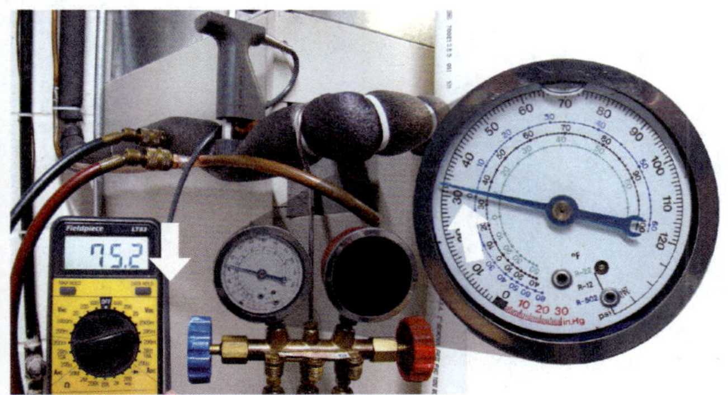

Suction pressure and suction line temperature after addition of 1 Pound of R-22:

Liquid Pressure

The correct amount of liquid pressure must be present for the fixed metering device to flow the correct amount of refrigerant into the evaporator coil. The pressure level is affected by cleanliness of the condenser coil, the outdoor air temperature, and charge level.

Some systems include liquid pressure charts that plot the required liquid pressure at a given outdoor air temperature and indoor heat load level. Other systems may not include these charts.

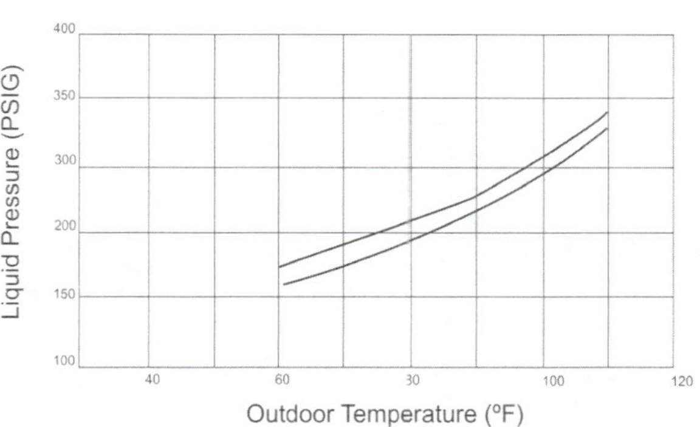

If the system comes with a liquid pressure chart, charge the system to the required superheat level. Then compare the operating liquid pressure to the value indicated by the liquid pressure chart. If the indicated chart value pressure and the measured pressure are far apart, the metering piston may be the wrong size, the air volume incorrect, or suction line pressure loss is out of the acceptable range.

Notice the outdoor air temperature range on the chart only goes as low as 60°F. At temperatures below 60°F, liquid pressure is too low to properly charge the system.

If system operating pressure charts are unavailable, make certain that suction vapor superheat is operating at the correct level and that suction pressure falls within ranges that equal typical evaporator coil saturation temperatures as indicated elsewhere in this chapter.

If the correct suction vapor superheat range cannot be reached, and the suction pressure appears to be off, check for correct piston sizing, air volume, and suction line sizing.

Liquid Pressure versus Piston Bore Size

Make certain the metering piston is the correct size when charging a fixed metering system. Piston sizes are matched to the operating head pressure of the outdoor condensing unit. Low SEER units with high head pressure will have smaller bore requirements than high SEER units operating with lower head pressure.

Incorrect piston size is predominantly found in high SEER equipment where the installer failed to replace the lower SEER piston shipped with the evaporator coil.

<u>Correct piston size can be found in the installation literature shipped with the condensing unit.</u>

Bore size is printed on pistons

Conditions With Improper Piston Size and Charging Attempt

Metering Piston Too Small:

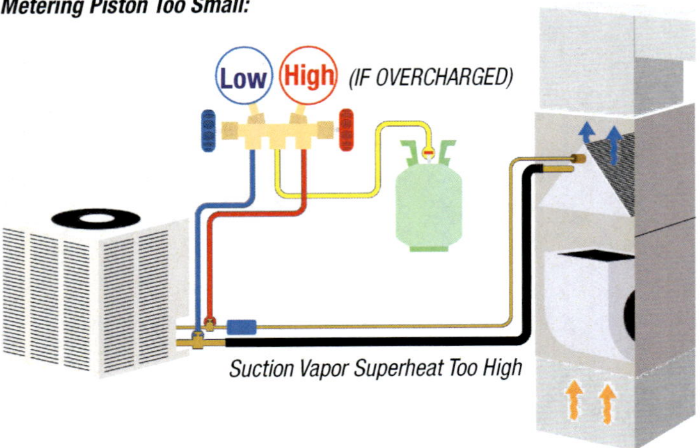

Metering Piston Too Large:

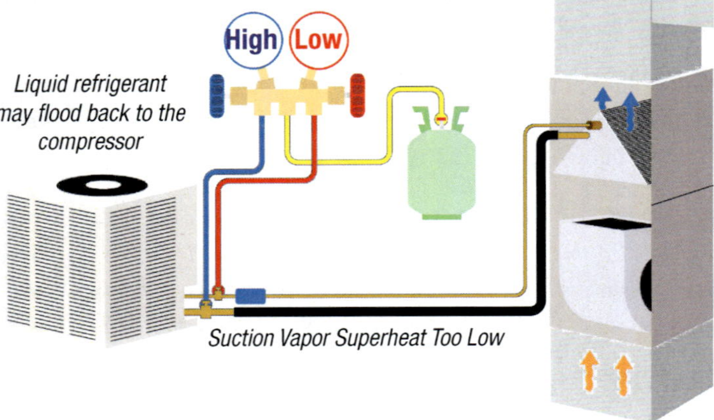

Liquid Subcooling at Correct Charge Levels

The liquid subcooling level in a fixed metering system will fluctuate with changes in outdoor air temperature and indoor heat load. Typical ranges of liquid subcooling under normal system operation for a piston metering system can be 8°F - 20°F.

At correct charge levels with the system operating at the correct discharge or liquid line pressures, the amount of subcooling produced in the condenser circuiting should be at the level required by the system manufacturer to overcome liquid line pressure loss.

Compressor Operation at Correct Charge Levels

Compressor performance changes with fluctuations in suction pressure, discharge pressure, and suction vapor superheat levels. Peak performance is achieved when the system is operating at high suction pressure, low discharge pressure, and low superheat levels. Performance drops when suction pressure drops, superheat rises, or discharge pressure rises. All of these changes occur naturally throughout normal operation of a fixed metering system.

Compressors will run cooler at lower superheat ranges and hotter as superheat rises. The temperature of the compressor oil will be kept at optimum temperatures when operating within correct charge limits.

Even when operating at the lowest superheat levels under correct charge, liquid refrigerant will not flood back to the compressor during run operation.

Compressors should be protected from liquid migration during off cycle periods. The amount of refrigerant charge affects the amount of liquid refrigerant that may migrate during the off cycle. As charge rises, the amount of liquid that migrates into the compressor oil rises. A good preventative measure is to always add a crankcase heater to systems where there is a high amount of charge required, for example systems with long line sets.

Summary: Fixed Metering System & Correct Charge Level

- Fixed metering systems fluctuate suction vapor superheat based on changes in outdoor air temperature and indoor heat load.
- Superheat charts are used to plot the required suction vapor superheat level.
- Suction pressure levels for R-22 systems will vary from about 56 PSIG up to 93 PSIG.
- Indoor air volume must be set to the correct level before attempting charge adjustment.
- The metering piston size must be correct before attempting to adjust the refrigerant charge.
- Improper piston size can be detected with abnormal superheat and pressure readings.
- System capacity will be reduced when systems operate at low suction pressure levels.
- System capacity will be reduced when systems operate at high liquid/head pressure levels.
- Do not charge at outdoor air temperatures below 60°F. (Liquid pressure is too low.)

Undercharged Fixed Metering Systems

Suction Pressure

When systems do not have enough refrigerant, the suction pressure will be lower than specified. The low suction pressure will cause the compressor capacity to fall. The low suction pressure will also cause the evaporator coil temperature to be below the design level established by the system manufacturer. If the charge level and the suction pressure are low enough, the evaporator coil may develop a coating of ice due to freezing condensate on the surface of the coil.

Suction Vapor Superheat

The suction vapor superheat level will be higher than specified. The high superheat level is caused by the fact that there is not enough liquid refrigerant in the evaporator circuiting. The lack of liquid will cause the liquid refrigerant to be boiled off to all vapor too early in the evaporator circuiting.

Compressor

The compressor will run hotter than design due to the high suction vapor superheat level and the lack of enough vapor to properly cool the compressor motor. The hot temperatures may trip the internal overload protection embedded in the compressor motor windings. The temperature of the compressor oil will also be too high, which can cause bearing damage over time.

The Condenser and Liquid Line Pressure

The lack of charge will cause the condenser to starve for liquid refrigerant. The condensing pressure will be lower than design and the liquid subcooling will be low. The lack of condensing pressure will cause low liquid line pressure. The low liquid line pressure and inadequate subcooling may cause flash gas to form in the liquid line.

Overall, the performance and long term reliability of the system will be degraded.

Overcharged Fixed Metering Systems

Suction Pressure

When systems have too much refrigerant, the suction pressure will be higher than specified. The high suction pressure will cause the evaporator coil temperature to be above the design level established by the system manufacturer. The high evaporator temperature will reduce the evaporator coil's ability to remove moisture from the air. The home may experience high humidity levels.

Suction Vapor Superheat

The suction vapor superheat level will be lower than specified. The low superheat level is caused by liquid refrigerant circulating too far into the evaporator circuiting. The system may experience liquid refrigerant floodback to the compressor during run operation.

Compressor

The compressor may experience liquid floodback during run operation. The floodback may cause the compressor to experience oil pump out. In extreme cases, the liquid refrigerant may cause internal damage to compressor components. In either case, the reliability of the compressor will be degraded.

The excessive refrigerant will also cause an increase in the amount of liquid refrigerant migration to the compressor oil that occurs during periods where the compressor is off. This liquid migration will cause the oil to foam during compressor start up and oil pump out will occur.

The Condenser and Liquid Line Pressure

The condenser coil will be flooded with excess liquid refrigerant. The excess refrigerant will cause an abnormally high condensing pressure and corresponding liquid line pressure. The liquid subcooling level will be above normal levels.

Summary: Fixed Metering System & Undercharge

- Suction pressure will be lower than required. The surface of the evaporator coil may freeze if the temperature of the evaporator is cold enough.
- The suction vapor superheat level will be too high.
- The compressor will overheat.
- The condenser will lack adequate liquid line pressure and liquid subcooling.
- System capacity will be low.
- The compressor may overheat and trip the internal temperature overload protection.

Summary: Fixed Metering System & Overcharge

- High suction pressure and abnormally high evaporator coil temperature.
- Superheat will be low compared to chart requirement.
- The compressor may have liquid flooding back during operation. During off cycle periods, liquid migration may cause starting problems.
- The condensing/liquid line pressure will be higher than it should be. The high pressure will reduce efficiency and capacity.
- Liquid subcooling will be high.

Service Procedure: Fixed Metering Charging Using Superheat

Tools:

- Digital Temperature Probe
- Psychrometer
- Refrigerant Gauges

STEP 1

Make certain the evaporator coil and condenser coil are clean. If the coils are not clean, the pressures will be adversely affected because the effective coil surface is altered from design conditions.

- *Confirm the correct metering piston size.*
- *Confirm the suction line is properly sized and not kinked.*
- *Outdoor air temperature must be at least 60°F to charge the system.*

STEP 2

Measure the outdoor air temperature near the condenser coil. Make certain to measure the air temperature entering the coil, not the temperature of the air leaving the coil.

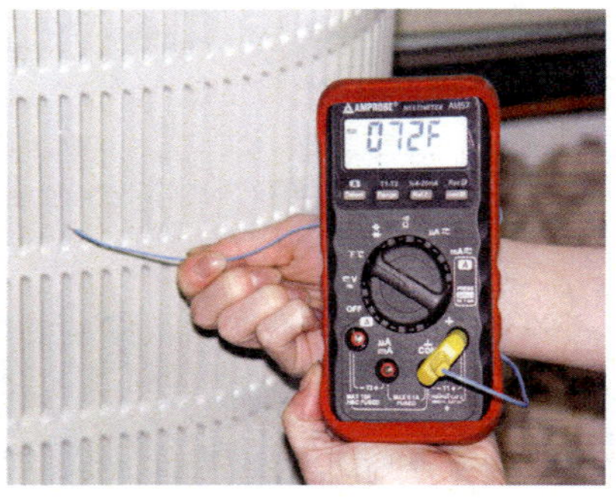

STEP 3

Use a psychrometer to measure the indoor air dry bulb and wet bulb temperatures.

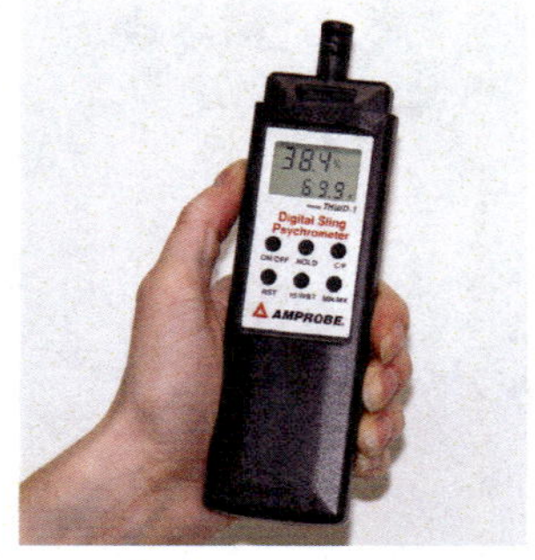

STEP 4

Measure the indoor air volume to determine if it is within specifications set by the manufacturer charging chart. Make adjustment to air volume if necessary. Failure to get the air volume to the correct levels will invalidate the superheat charge process.

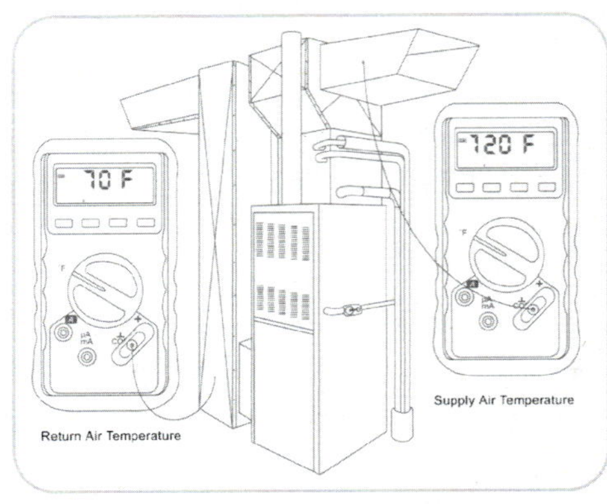

STEP 5

Using the charging charts, plot the required system pressures and superheat level. If one is not available at the jobsite, see the generic chart provided in this chapter.

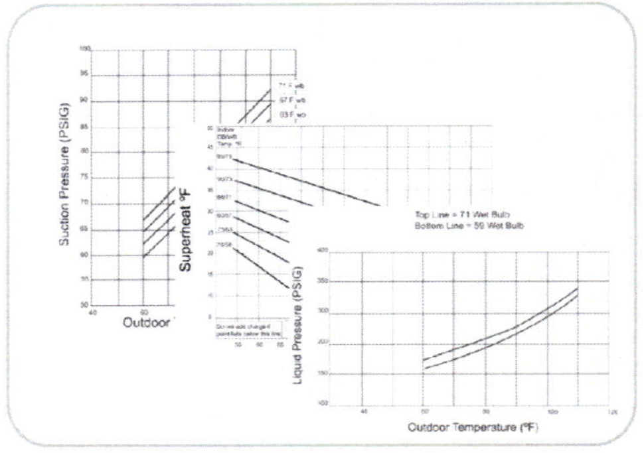

STEP 6

Connect the center hose of the refrigerant gauge manifold to the valve on the refrigerant cylinder. The blue suction hose should be connected to the suction service valve. The red high pressure hose should be connected to either the liquid line service valve or the discharge line pressure port. The charging chart will indicate which port the hose should be connected to. Turn the system on.

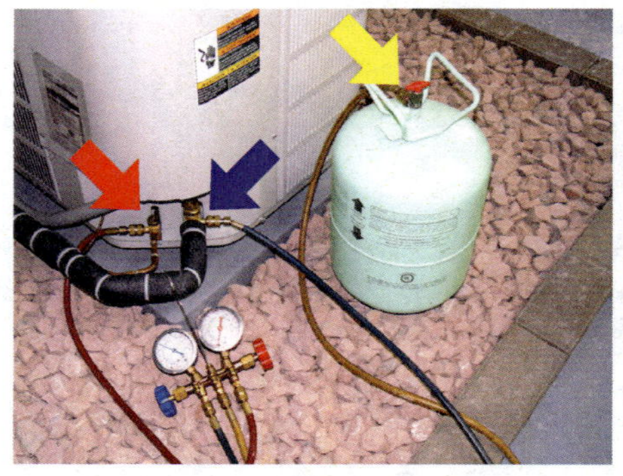

STEP 7

Open the refrigerant cylinder and purge the air from the hoses. Open the blue suction side dial on the refrigerant gauge. The refrigerant will now enter the system via the suction service valve. Close the gauge valve and wait for pressures to stabilize. Repeat as needed.

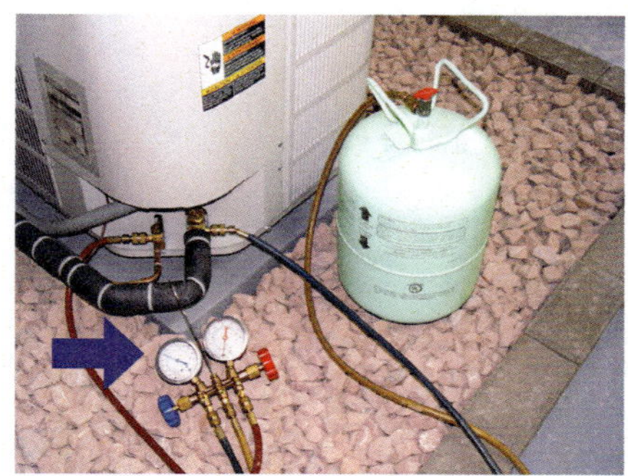

Note If the refrigerant used is R-410A, make certain to charge the system with the refrigerant drum in the correct orientation as indicated on the refrigerant cylinder. Add or recover refrigerant until pressures match the charging chart requirements.

STEP 8

Once system pressures are set to factory required levels, measure the temperature of the suction line near the service valve on the outdoor condensing unit. Calculate the superheat. (See Measuring Superheat.)

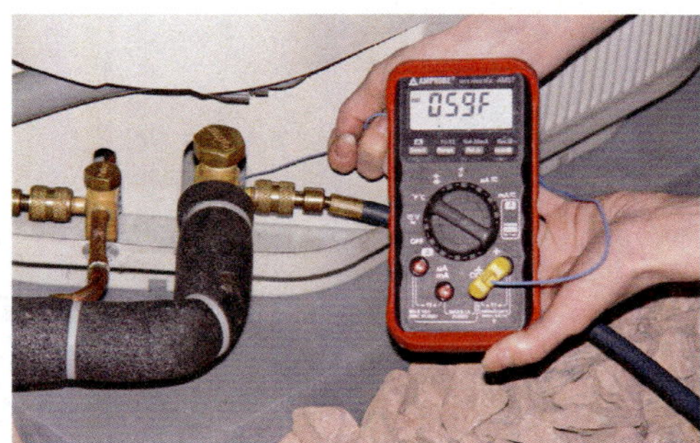

STEP 9

- *If the superheat is too low, recover refrigerant.*
- *If the superheat is too high, add refrigerant.*
- *Allow time for system temperatures and pressures to become stable and then re-measure.*
- *Continue process until the pressures and superheat are within chart allowances.*

Note: It is good practice to compare the temperature of the suction line at the outlet of the evaporator to its temperature at the inlet to the condensing unit. The temperatures should be close. If there is a significant drop in temperature from the evaporator outlet to the condensing unit, check for a kink or restriction in the suction line. If there is a significant rise in temperature, check for a lack of insulation on the suction line.

Generic Fixed Metering Superheat Charging Chart

<u>THIS CHART SHOULD ONLY BE USED WHEN THE ORIGINAL FACTORY CHART CANNOT BE FOUND</u>

OAT	INDOOR AIR DRY BULB / WET BULB TEMPERATURE			
	70/58	75/63	80/67	85/71
55°F	21°F	25°F	29°F	33°F
60°F	17°F	22°F	27°F	31°F
65°F	14°F	19°F	24°F	29°F
70°F	10°F	16°F	22°F	27°F
75°F	6°F	13°F	19°F	25°F
80°F	5°F	11°F	17°F	23°F
85°F	5°F	8°F	15°F	20°F
90°F	5°F	5°F	12°F	18°F
95°F	5°F	5°F	10°F	16°F
100°F	5°F	5°F	7°F	14°F

This table is provided as a general charging guide to use when the original chart cannot be found. In all cases, try to get a copy of the original chart. Use of this chart does not guarantee that the charge obtained by its use will deliver certified rated capacities and efficiencies. This chart is only designed as a general guide to get the charge of the system close to factory requirements. Prokup Media Inc. assumes no responsibility for the operation and reliability of systems charged using the data provided in this table. The data was gathered by comparing a wide range of superheat charging table data from various manufacturers and then compiling average values for various combinations of outdoor air and indoor air conditions.

Use of this chart requires that the indoor air volume be set to the manufacturer required level and that the metering piston be properly sized.

Subcooling

Liquid Subcooling

Introduction

The condensing of vapor refrigerant into a liquid refrigerant is a change of state that requires the transfer of a large amount of heat energy from the refrigerant to the outdoor air. The refrigerant in the initial circuits of the condenser coil is typically a mixture of refrigerant liquid and vapor. The refrigerant in this state is called saturated. Saturated refrigerant will change state with any addition or removal of heat energy.

As the saturated refrigerant travels further into the condenser circuiting, it will give up additional heat to the cooler outdoor air. As the refrigerant transfers more heat to the outdoor air, its temperature begins to drop below the saturation temperature and enters a pure liquid state, meaning no vapor bubbles are present. Liquid refrigerant that is below saturation temperature and in a pure liquid state is called "subcooled".

Subcooling usually occurs in the independent parallel circuits of the condenser coil. However, some coils have an additional circuit where the liquid refrigerant further circulates and removes more heat from the liquid. This circuit is called a subcooling loop.

In the following example, a temperature probe has been attached to the liquid line at the outlet of the condenser coil. A temperature probe measures the temperature of the liquid line at the outlet of the condenser, which, in this example, is 107°F. The high pressure gauge indicates a liquid line pressure of 250 PSIG, which corresponds to a saturation temperature of 117°F for R-22. The liquid line in this example is 10°F cooler than the saturation temperature. This 10°F difference in temperature is the subcooling.

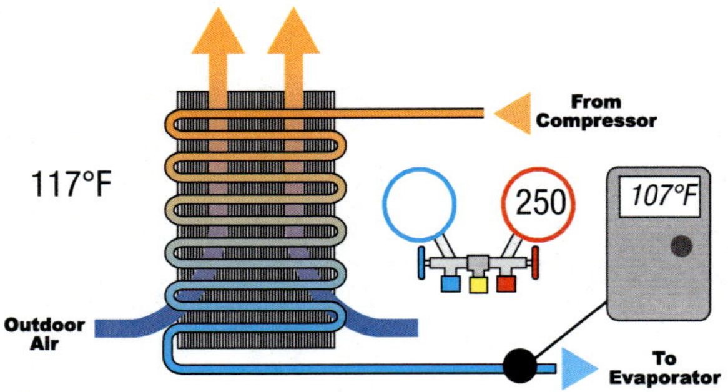

The liquid refrigerant can only be cooled as low as the temperature of the air surrounding the condenser coil. If liquid refrigerant leaves the condenser coil at a temperature that is colder than the air passing through the coil, there is a pressure drop in the coil that is causing the temperature of the refrigerant to fall. For example, if a drier installed in the condensing unit became plugged with debris, the temperature of the liquid at the outlet of the drier would be lower than that of the outdoor air.

This is a problem because, as refrigerant pressure drops, so does the saturation temperature. If the pressure of the refrigerant drops to a level where its temperature equals its saturation temperature, vapor will begin to form in the liquid line.

The Importance of Subcooling

The liquid that exits the condenser coil must travel through the liquid line to the metering device. As the liquid travels through the line, its pressure drops due to friction between the moving liquid and interior surface of the piping. There is also pressure loss that occurs when the liquid travels uphill when the evaporator is located above the condenser coil. As the liquid pressure falls, the corresponding saturation temperature also drops, reducing the temperature difference between the saturation temperature and the temperature of the liquid refrigerant. In other words, the refrigerant subcooling at the condenser coil outlet is greater than it is further down the liquid line.

In worst case conditions, the liquid pressure may drop to a pressure where the refrigerant temperature equals the saturation temperature. At this point, The refrigerant will become a mixture of vapor and liquid, increasing the likelihood of liquid flashing into a vapor.

The net effect of flash gas is poor refrigerant flow into the evaporator circuiting, and wearing of the metering device orifice area.

To prevent this condition from occurring, it is important to charge the system so there is adequate subcooling to overcome the pressure loss that will occur in the liquid line. The amount of subcooling required will be determined by liquid line length, liquid lift, fittings, liquid line size, and overall condenser size.

An R-22 system with an average allowable maximum liquid line pressure drop of 30 PSIG, or an R-410A system with an average allowable maximum liquid line pressure drop of 50 PSIG, need 10°F of subcooling to overcome a system's liquid line pressure drop.

How Subcooling Changes

The amount of condenser subcooling depends on the refrigerant level in the condenser coil. The refrigerant level in the condenser will vary by how much charge is in the system.

If the refrigerant level is high, the pressure in the condenser will be high, and the amount of condensing liquid will be high.

The vapor refrigerant will condense to liquid early in the condenser circuiting, thus the refrigerant travels farther as a liquid in the condenser. The farther it travels as a liquid, the more its temperature will drop.

If the refrigerant level is low, the pressure in the condenser will be low, and less liquid will be condensed from the hot vapor. The vapor refrigerant will condense to liquid late in the condenser circuiting, thus the refrigerant travels farther as a vapor into the condenser circuiting. The farther it travels as a vapor, the less its temperature will drop.

At high charge levels, the subcooling will be high. At low charge levels the subcooling will be low.

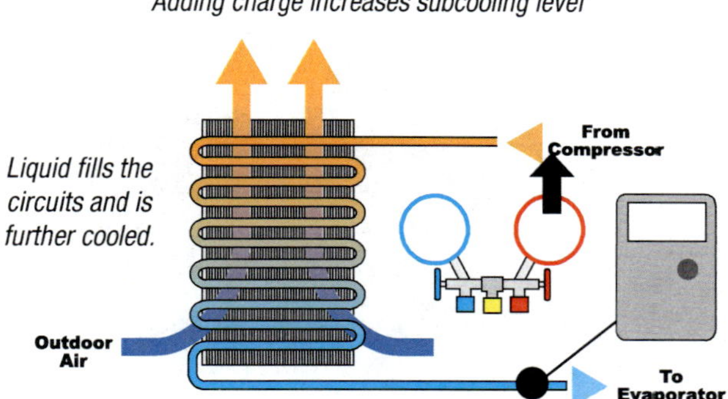

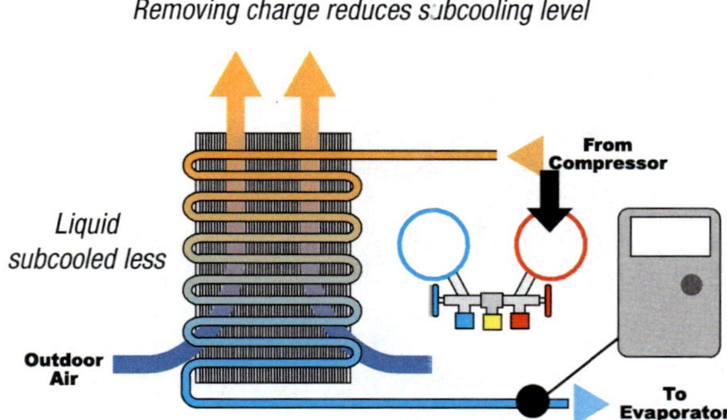

Subcooling and Heat Load

If the metering device is a fixed piston type, changes in the indoor air heat load will also change the subcooling level. As heat load changes and suction pressure changes, the amount of subcooling will fluctuate. For example, at low heat loads on the evaporator coil, the indoor coil will flood and the condenser will starve. The result will be low subcooling.

Condenser Coil Size and Subcooling Design Levels

To reach higher energy efficiency ratings, manufacturers of air conditioning systems keep the condensing pressure as low as possible without reducing liquid line pressure to a level that prevents the metering device from flowing adequate refrigerant into the evaporator circuiting. To lower condensing pressure, the size of the condenser coil is increased. The increase in size of the coil has resulted in higher levels of subcooling due to the large amount of refrigerant in these systems. It is common to see subcooling levels under normal charge levels in the range of 10°F to 20°F. Subcooling requirements are typically printed on the condensing unit dataplate.

Overcharging For More Subcooling May Result In Abnormal Condensing Pressure

The condenser coil has a limited amount of piping to hold refrigerant. If the system liquid line pressure drop is found to be excessive and flash gas is forming in the liquid line, compensating by adding charge to the system to increase the liquid subcooling level may cause the condensing pressure to rise too high. The system will then operate with poor compressor performance and capacity. In that case, the pressure drop in the liquid line will need to be reduced by increasing the line set size, or reducing the number of elbows and fittings if possible.

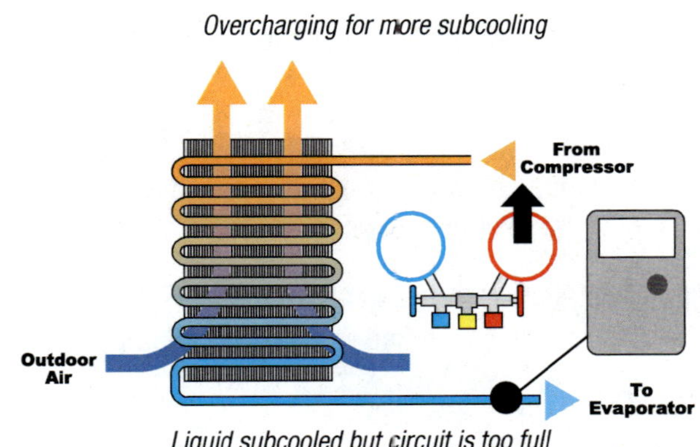

Liquid subcooled but circuit is too full

Excessively large liquid line pressure loss is due to high liquid lifts, under-sized refrigerant lines, and too many liquid line accessories.

Getting the Right Amount of Subcooling

See the example below. This system has a liquid subcooling level requirement of 10°F and uses R-22 refrigerant. The liquid line in this example has three pressure ports installed along its length. Each port will read a lower pressure as the liquid travels down the liquid line. In this example, the liquid line pressure is 220 PSIG at the condensing unit's liquid service valve, corresponding to a saturation temperature of 108°F. The liquid temperature leaving the condenser coil is 98°F. A perfect 10°F of subcooling. Let's see if this subcooling level works in our example:

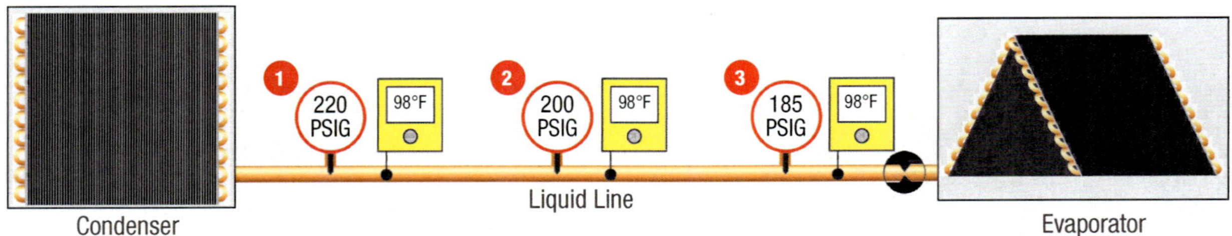

1 At the first gauge port on the liquid line, the pressure is 220 PSIG with a liquid line temperature of 98°F.

- Saturation Temperature @ 220 PSIG = 108°F.
- 108°F - 98°F = 10°F of subcooling.

2 At the second gauge port the pressure has dropped by 20 PSIG to 200 PSIG. The liquid line is still 98°F.

- Saturation temperature @ 200 PSIG = 101°F.
- 101°F - 98°F = 3°F of subcooling.

3 Now, the liquid goes to the last gauge port which has a pressure down to 185 PSIG. The liquid line is still 98°F.

- Saturation temperature @ 185 PSIG = 96°F.
- The liquid is now 2°F above saturation temperature.

Our example may result in vapor (flash gas) forming in the liquid line. The problem with this system is that the liquid line pressure drop is in excess of the maximum 30 PSIG (R-22 limit for 10°F of subcooling.) The solution is to add refrigerant, but this will cause the condensing pressure to rise and the system capacity to drop. The better solution is to reduce the pressure drop by increasing the liquid line size or reducing number of fittings installed on the line.

Condensing Pressure and Liquid Subcooling use in Diagnostics

The condenser coil should operate at the pressure indicated on the system's charging charts. When an abnormal pressure is present, the temperature of the liquid leaving the condenser coil should be measured. At no time should the liquid refrigerant be cooler than the temperature of the outdoor air passing through the condenser coil. If refrigerant is leaving the condenser coil at a lower temperature, check for a pressure drop in the coil.

If the liquid refrigerant is warmer than the air entering the condenser coil, the liquid subcooling level should be calculated. The liquid subcooling level, when matched with other system observations, holds clues to possible system problems.

If the liquid pressure is high, and the liquid temperature is hot with low subcooling, there is a lack of heat transfer at the condenser coil. Check for a dirty condenser coil, a failing condenser fan, or for obstructions that may prevent air from circulating through the condenser coil. A non-condensable gas test may also be required (See Chapter B: Refrigerant & Refrigerant Piping).

If the coil is clean, and the liquid temperature is warm with normal subcooling, It's likely there is a high load on the evaporator coil.

If the condensing pressure is high, and the subcooling is high, the condenser is flooded with refrigerant.

Subcooling

Service Procedure: Subcooling Calculation

Tools:

- Digital Temperature Probe
- Refrigerant Gauges

STEP 1

Attach a high pressure gauge to the pressure port on the liquid line at the outdoor unit service valve.

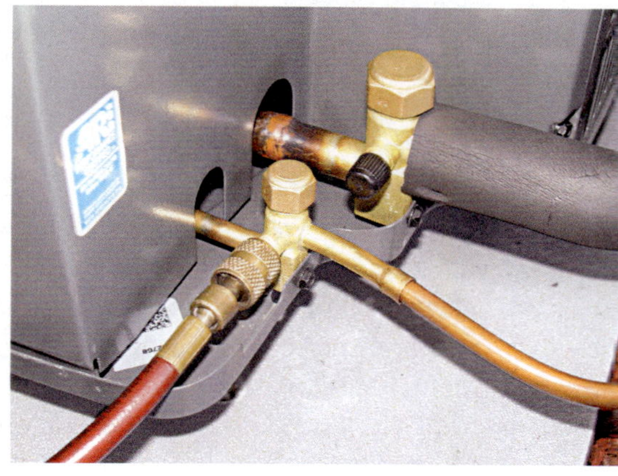

STEP 2

Read the liquid line pressure and convert to saturation temperature. (Use the scale on the gauge or a temperature pressure chart.)

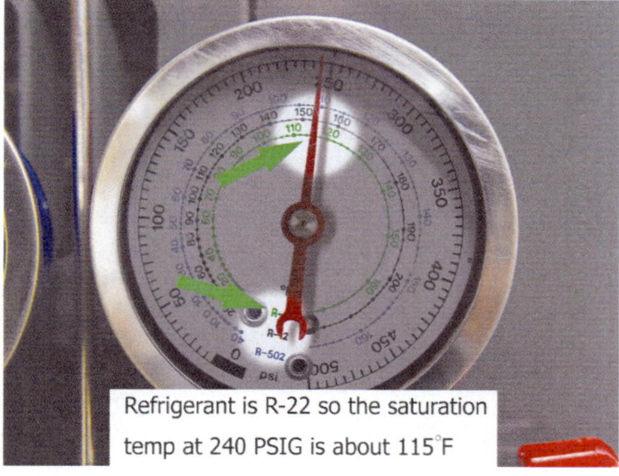

STEP 3

Attach a digital temperature probe to the liquid line near the gauge port. Measure the temperature of the liquid line. It may take up to 20 minutes for the liquid line temperature to become stable following a change in charge level.

STEP 4

Subtract the liquid line temperature from the saturation temperature corresponding to the liquid line pressure. The difference in temperature is the subcooling level.

115.0°F (saturation temp found in step 2)

-102.7°F (temp of liquid line from step 3)

____

12.3°F subcooling temp

Allow up to 20 minutes from charge changes to measure liquid subcooling.

TXV System Charging Using Subcooling

Introduction

A Thermostatic expansion valve (TXV) maintains a constant level of evaporator superheat over a wide range of liquid line pressure changes and indoor air heat loads. It stabilizes superheat by varying the size of its internal orifice in response to changes in the liquid line pressure and indoor air heat load. Superheat may range from a low of about 8°F to a high of around 12°F at normal charge levels and correct system operation. To operate properly, the TXV must receive a solid column of liquid refrigerant from the liquid line. In liquid form, refrigerant is in a subcooled state.

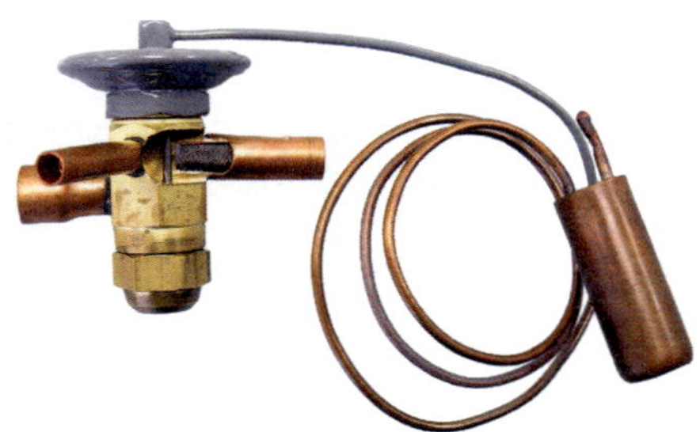

Typical Thermostatic Expansion Valve (TXV)

The subcooling method is used to charge a TXV-equipped air conditioning system. With this charging method, the liquid subcooling level is monitored as refrigerant is added to or recovered from the system. When the subcooling level has been adjusted to the required level, the system will operate at the correct efficiency and capacity.

Technician measures the liquid line temperature with a digital temperature probe.

If charging a system when the air temperature in the home is high, such as a new system start up on a hot day, run the air conditioning system and allow the air temperature in the home to reach a comfortable temperature before making a final charge adjustment. Comfort temperature range is considered 75°F to 80°F dry bulb.

The indoor air volume must also be set to the correct level before attempting to adjust the refrigerant charge. Typically, indoor air volume is set to 400 CFM per ton of cooling capacity. This range may deviate in some instances to a low of 350 CFM per ton of cooling to a high of 450 CFM per ton of cooling. Verify air volume requirements by referring to the system's charging charts. If no reference to required air volume is found, adjust the air volume for 400 CFM per 1 ton of cooling capacity.

The system charging chart will specify the minimum outdoor air temperature at which the subcooling charging method can be performed. If the outdoor air temperature is below the minimum outdoor air temperature specified, the system must be charged by weighing in the correct amount of refrigerant. The minimum outdoor air temperature for charging using the subcooling method is typically around 60°F.

Charge Level and Subcooling

Adding or recovering refrigerant from an air conditioning system will change the amount of liquid subcooling in the condenser coil. When refrigerant is added to the system, the amount of refrigerant condensing in the condenser coil increases. This condensed liquid refrigerant backs up further into the condenser circuiting and results in a high side line pressure increase. This increase in pressure raises the corresponding saturation temperature in the condenser coil, and causes a wider difference between liquid line temperature and saturation temperature. The result is higher subcooling with slightly higher condensing pressure.

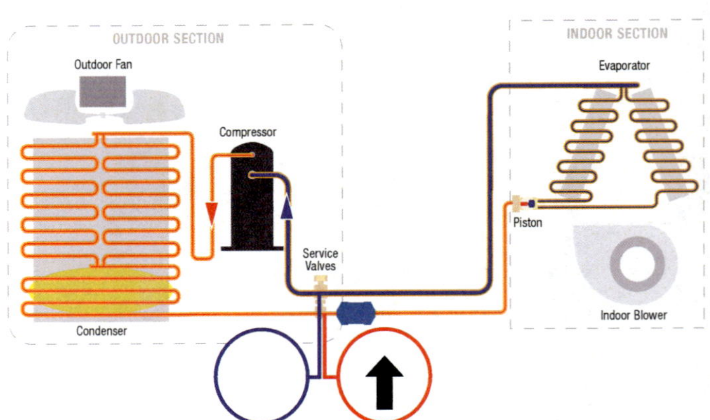

Backing up liquid into the condenser circuits raises liquid line pressure and increases liquid subcooling. The addition of charge increases pressure and the amount of coil circuiting used to subcool liquid.

If refrigerant is removed from the system, the condensing pressure falls and the amount of liquid refrigerant condensing in the condenser circuiting drops. The amount of coil circuiting used for subcooling decreases. The result is less of a difference between saturation temperature and liquid line temperature, meaning lower liquid subcooling.

System Subcooling Level Requirements

Every system requires a different amount of liquid subcooling based on the pressure drop in the liquid line and the design and size of the condenser coil. In some cases, the subcooling level required for correct operation of the system is printed on the nameplate of the condensing unit. As previously stated, 10°F of subcooling will overcome design pressure drop for most properly sized liquid lines. However, some systems may require more subcooling. These systems are typically high efficiency models where the outdoor condenser coil is very large. The large coil may require significant refrigerant charge to reach the required condensing pressure. The high subcooling is a result of the large amount of refrigerant in the condenser coil.

Liquid Subcooling Levels as Charge is Added

In the following series of figures, liquid line pressure and liquid line temperature are shown as refrigerant is added to an R-22 system. An interval of 15 minutes was allowed between each change in charge to allow for stabilization of liquid line temperature and pressure.

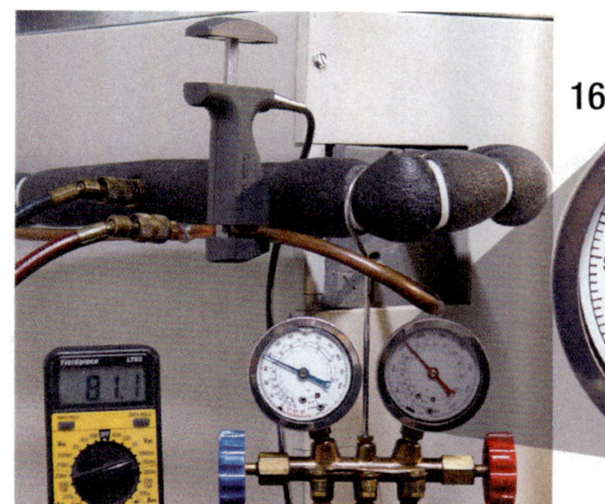
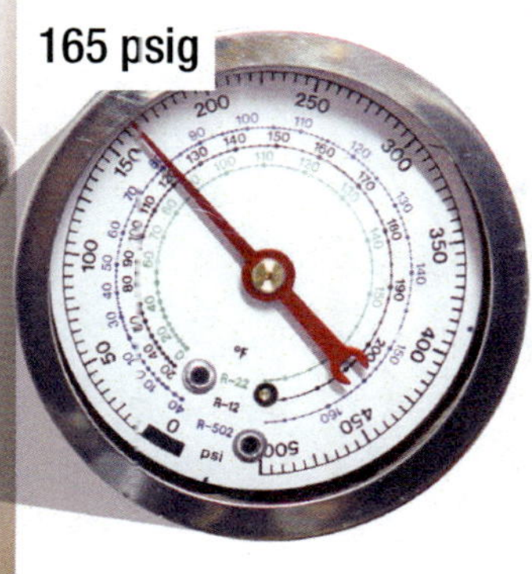

Liquid Line pressure and Liquid Line temperature before charge is added.

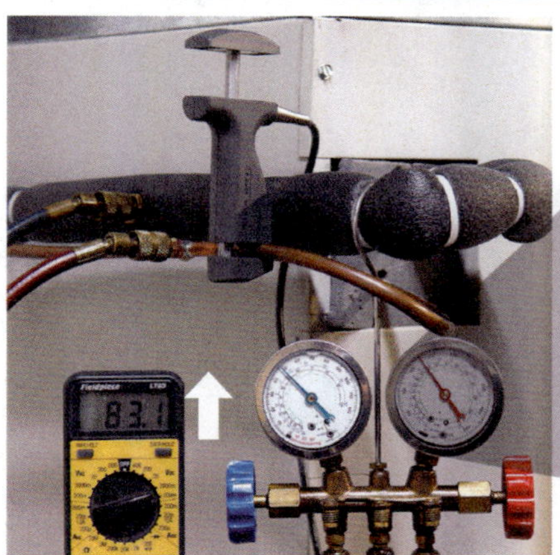
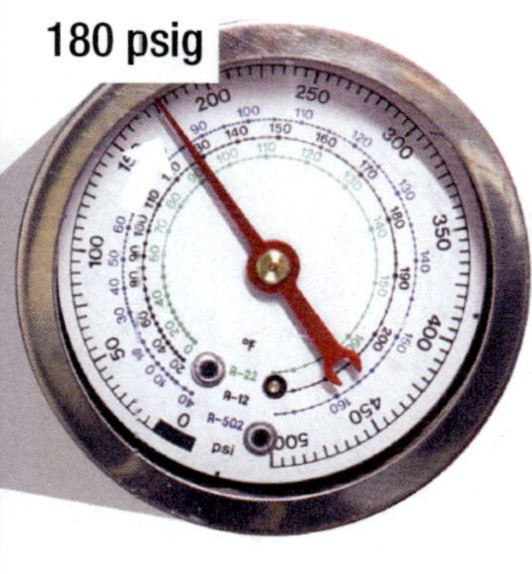

Liquid line pressure and liquid line temperature after addition of 1/2 pound of R-22.

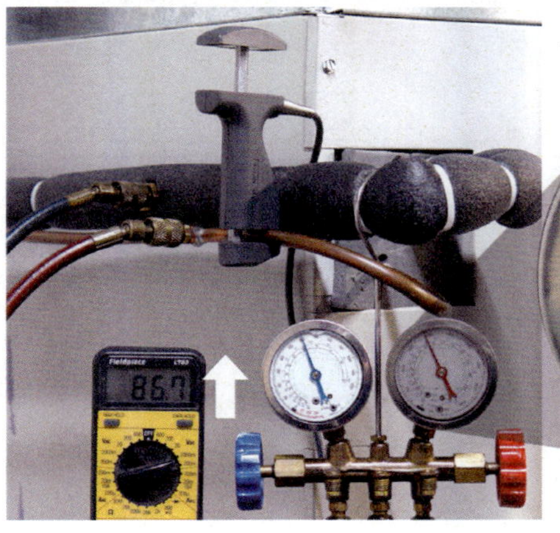
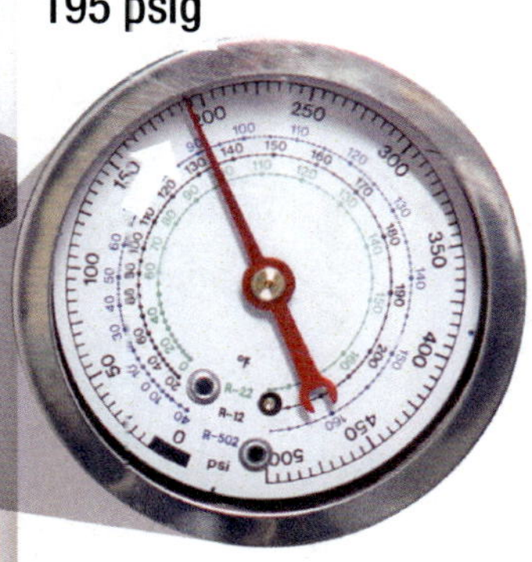

Liquid line pressure and liquid line temperature after addition of 1 pound of R-22.

Subcooling Charging Charts

Some air conditioners will come with a subcooling charging chart that plots the required subcooling based on the length and vertical liquid lift of the liquid line.

Two charts are usually provided, one chart to determine the pressure drop range of the liquid line, and the other to determine the liquid pressure and liquid line temperature requirements.

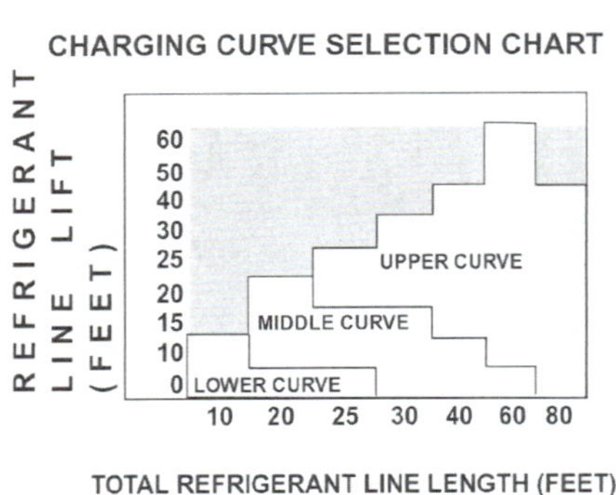

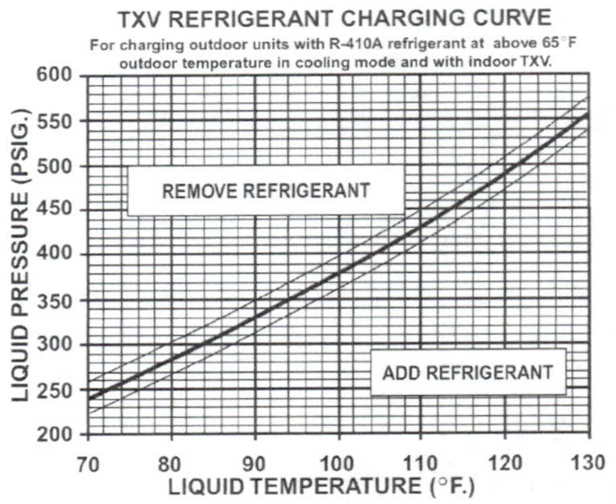

Provided courtesy of The Trane Company

When using these charts, the first step is to plot the length of the liquid line and any vertical lift of liquid to determine which curve range the liquid line application falls into. In this chart (see right), the ranges are labeled lower curve, middle curve, and upper curve. Each range represents a different level of liquid subcooling requirement.

For example, let's plot the subcooling requirement for a liquid line with 10 feet of vertical lift and 20 feet of length onto the chart below. These two values intersect within the boundary for the middle curve, so the middle curve in the next chart will be used.

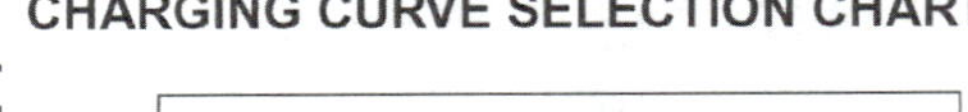

Next, go to the liquid pressure/liquid line temperature chart. There are three diagonal curved lines on this chart: a lower, a middle, and an upper curve. These lines are used to plot the intersection of the liquid line pressure and the required liquid line temperature. Notice the three lines represent the curves referred to on the charging curve selection chart (above).

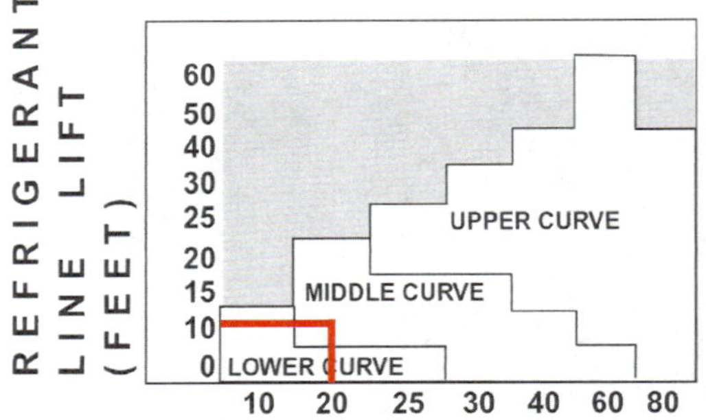

Let's plot an example on the chart shown below: If the liquid line pressure was 350 PSIG, and the lower curve was determined to be the plot line, the required liquid line temperature would be 98°F. If the upper line was determined to be the plot line, the liquid line would need to be 90°F. The difference between the two examples equals a difference in required liquid subcooling levels based on two different liquid line applications. The lower line example equals 9°F of subcooling while the upper line equals 17°F of subcooling.

TXV REFRIGERANT CHARGING CURVE

Notice on the chart that when the liquid line temperature is too warm compared to the required liquid line temperature, the chart directs to add refrigerant into the system. When the temperature of the liquid line is below the required liquid line temperature, the chart directs to recover refrigerant from the system. The addition or removal of refrigerant from the system will change the pressure and liquid line temperature. These changes may take up to 20 minutes to stabilize.

Re-measure both liquid pressure and liquid line temperature and re-compare to the chart. Make charge adjustments as needed until the pressure and temperature are equal to the values indicated on the liquid subcooling charging chart.

Allow time for the liquid line temperature to reach desired temperature. Up to 20 minutes may be required.

Properly Charged TXV Systems

The Evaporator Coil

The evaporator coil will operate with stable suction pressure. The saturation temperature range of the evaporator coil will be slightly higher than a similar system equipped with a fixed piston type metering device. The suction vapor superheat level will be stable at close to 10°F.

During times where the system operates at cooler outdoor air temperatures, the TXV will increase the diameter of its internal metering orifice to maintain suction pressure at a higher level than a piston system. The higher suction pressure will result in better compressor performance.

The Compressor

With the correct air volume heat load, the compressor will not experience any liquid floodback due to the metering action of the expansion valve. The constant suction vapor superheat level will also keep the compressor running cooler than a piston equipped system.

The Condenser Coil and Liquid Subcooling Levels

The condenser coil will have enough refrigerant to maintain the correct liquid subcooling level for the liquid line application. If the liquid line is properly sized, no flash gas will form.

Summary: TXV Metering System & Correct Charge Level

- Normal and stable suction pressure.
- Superheat will be steady at around 10°F.
- Compressor will operate at optimal temperature and liquid floodback will not occur during run operation.
- Liquid subcooling will be adequate to prevent flash gas formation in the liquid line.

Undercharged TXV Systems

The Evaporator Coil

The evaporator coil may operate normally with the correct suction pressure and suction vapor superheat. During periods of high liquid line pressure drop, the system may lack adequate subcooling, resulting in flash gas forming in the liquid line. The flash gas will cause fluctuations in suction pressure and suction vapor superheat.

The Compressor

If the charge level is low enough to cause high suction vapor superheat, the compressor will run hot and may overheat. The compressor motor internal temperature protection may trip and shut down the compressor. High operating temperatures may cause oil breakdown and bearing wear or damage.

The Condenser Coil and Liquid Subcooling Levels

The condenser coil will operate at lower than required pressure. The lack of charge will result in lower than required liquid subcooling level. The low subcooling level may cause flash gas to form in the liquid line.

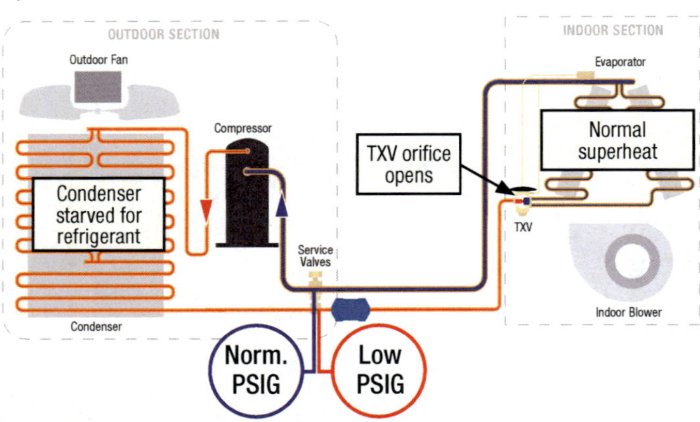

Summary: TXV Metering System & Undercharge

- Normal and stable suction pressure if charge is slightly low. Suction pressure will fall to low levels if charge is excessively low.
- Superheat may be normal or high.
- Compressor will be hot.
- Condensing pressure will be low. Liquid subcooling will be low.

Overcharged TXV Systems

The Evaporator Coil

The evaporator coil may operate normally with the correct suction pressure and suction vapor superheat. The expansion valve will throttle back flow of refrigerant in an attempt to maintain the correct suction vapor superheat levels.

The Compressor

The compressor will lose capacity due to elevated condensing pressure caused by an overcharge condition in the condenser. The expansion valve will prevent excessive flooding of the evaporator coil.

The Condenser Coil and Liquid Subcooling Levels

The condenser coil will operate at higher than required pressure. The excess charge will cause high condensing pressure and low compressor capacity. The liquid subcooling level will be abnormally high.

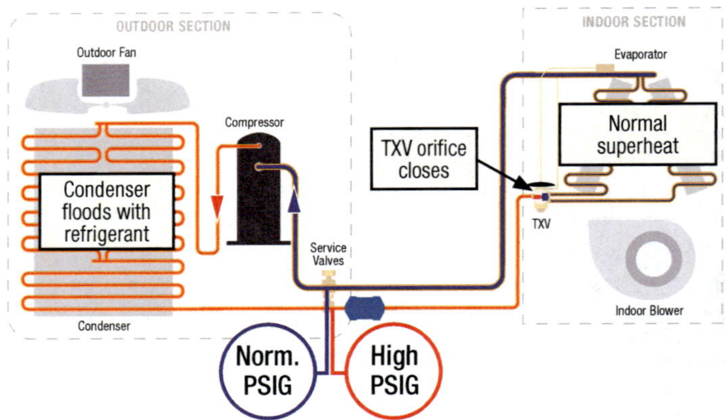

Summary: TXV Metering System & Overcharge

- Suction pressure will be normal or slightly elevated.
- Superheat will be normal.
- Compressor capacity will be low.
- Condensing pressure and liquid subcooling will be high.

Service Procedure: TXV System Charging

Tools:

- Digital Temperature Probe
- Refrigerant Gauges

STEP 1

Make sure that indoor air volume has been adjusted to the correct level. See the charging chart for CFM specification. If it is not present, adjust the indoor air volume for 400 CFM per 1 ton of cooling capacity.

The outdoor air temperature should be above 60°F. If it is colder, weigh the charge in using a charging scale. Return when outdoor air temperature is above 60°F to make final charge adjustment.

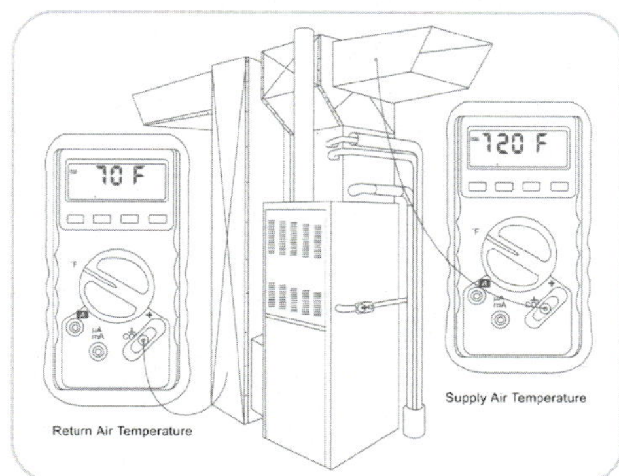

STEP 2

Make certain the evaporator coil and condenser coil are clean. Confirm correct suction and liquid line sizing.

STEP 3

Attach refrigerant gauges to the system.

Attach a digital temperature probe to the liquid line at the outlet of the condensing unit.

STEP 4

Run the system and add refrigerant. Allow system pressures to stabilize. Measure the liquid line pressure and convert to saturation temperature. Measure the liquid line temperature. Subtract liquid line temperature from the saturation temperature to determine liquid subcooling level.

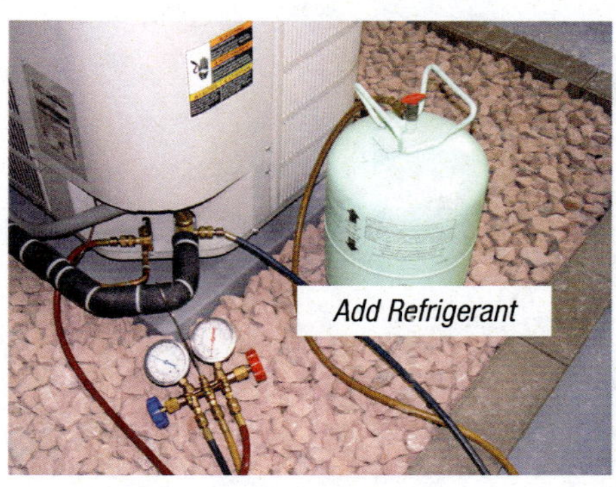

STEP 5

If subcooling charging chart is present, calculate the required liquid subcooling level for the liquid line application. If the chart is not available, check the condensing unit nameplate for possible liquid subcooling charge data. If this is not present, charge the system for a subcooling level of 10°~12°F.

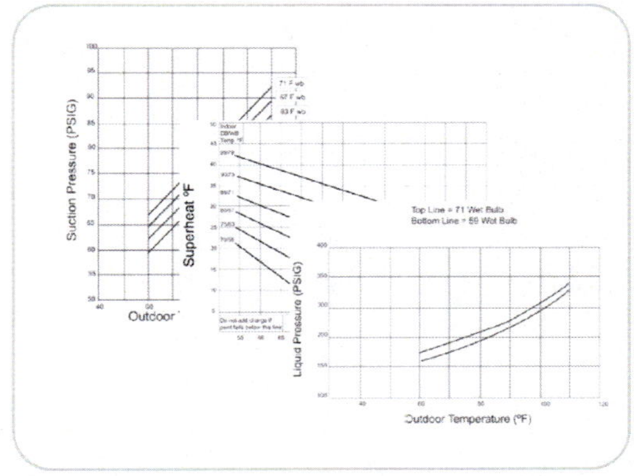

STEP 6

If the liquid subcooling level is below the required value, add some refrigerant to the system. If the liquid subcooling level is too high, recover refrigerant from the system. Re-check subcooling level to confirm the correct charge. (Allow up to 20 minutes for the system conditions to stabilize.)

STEP 7

Confirm correct operation of the expansion valve by measuring suction pressure, suction line temperature, and calculating suction vapor superheat. The measured superheat should be in a range of 8°F~15°F. (Some systems where large evaporator coils are matched to smaller condensing units may show higher levels of superheat. 2 stage capacity air conditioning systems operating at first stage may also show high suction vapor superheat levels.)

NOTE: R-410A SYSTEM CHARGING INFORMATION

When adding refrigerant charge to an R-410A system, add the refrigerant in a liquid state. See the refrigerant cylinder information for details.

Refrigeration Circuit Diagnostics

Troubleshooting the Refrigeration Cycle

Introduction

To properly troubleshoot refrigeration circuit problems, a technician must understand, among other things, how the system works, why refrigerant pressures rise and fall, and how to calculate - and the importance of - suction superheat and liquid subcooling. It is recommended that readers have a firm understanding of the topics, lessons, and procedures covered in the previous chapters of this book before exploring the topics and procedures discussed in this chapter. This chapter references terms, concepts, and procedures covered earlier in the book. Referring back to the appropriate sections, as needed, might prove to be helpful.

Refrigeration cycle troubleshooting begins with determining what the system should be doing when it is operating properly. This includes using charging charts to determine what the optimal pressures should be for a given environment. Once the optimal operating conditions are known, the system can be analyzed by comparing what should be happening to what is actually happening. Be aware that for every one-pound deviation from the required suction pressure, the system capacity will be reduced by approximately 1% if R-22, and by 0.6% if R-410A.

Diagnosing Fixed Metering System Problems

Fixed metering refrigeration cycle problems can exhibit the following symptoms:

- Low suction pressure.
- High suction pressure.

The following diagnostic procedures will help identify the cause of low suction or high suction pressure in a fixed metering system.

For fixed metering systems, the suction vapor superheat requirement should be calculated, and liquid subcooling should be 10°F to 15°F degrees, depending on the specific piece of equipment.

Low Suction Pressure

If the system suction pressure is too low, there are three potential causes:

- Excessive pressure loss in an undersized or kinked suction line.
- Low heat load (this could be a dirty evaporator coil, a lack of indoor air volume, or simply a cold return air temperature).
- A starved evaporator coil due to a restriction or charge-related problem.

Note that low suction pressure may cause ice or frost to form on the surface of the evaporator coil.

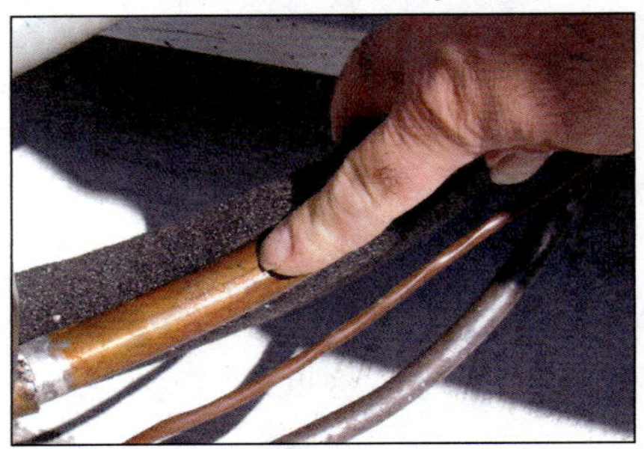

Tools:

- Refrigeration Gauges
- Thermometer

STEP 1

Inspect the suction line for any obvious kinks. If no kinks are seen, the line size may be the problem, install a pressure tap on the suction line at the outlet of the evaporator circuit.

STEP 2

Run the system and compare the pressure at the outlet of the evaporator circuit to the suction pressure at the outdoor condensing unit.

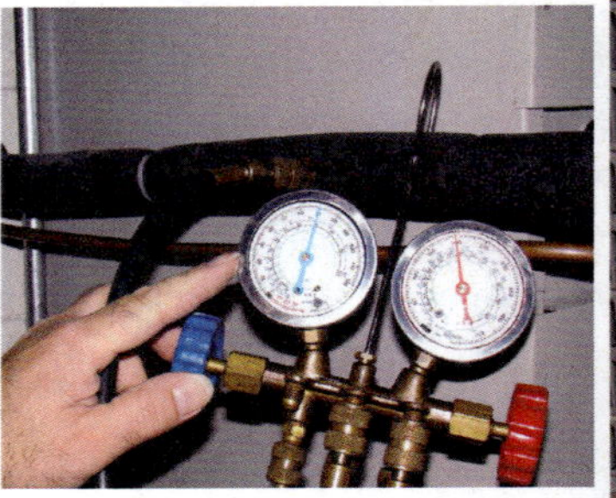

If there is a pressure drop in excess of 3 PSIG for R-22 (5 PSIG for R-410A), the line is either undersized, or may be kinked. Evaluate the suction line to determine if it is undersized.

STEP 3

To check for an unseen kink or obstruction in the suction line, measure the temperature of the suction line at various points on the line. If a temperature drop in excess of 3°F is detected across two points, there is a pressure drop between those points. Locate the problem and repair the line as needed.

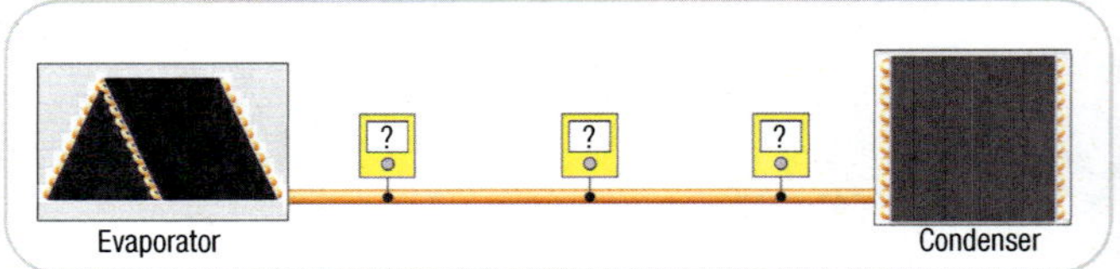

Check the temperature at different points along the suction line in order to find an excessive drop in temperature.

Service Procedure:
Diagnosing Low Heat Load

Tools:

- Refrigeration Gauges
- Psychrometer

STEP 1

Measure the return air wet bulb and dry bulb temperature at the inlet of the air handler/furnace. If the air handler is not accessible, readings can be taken at the filter grill, but there is the possibility of inaccurate readings if the ductwork is not adequately and properly sealed. Plot the required suction vapor superheat on the factory-charging chart. If a chart is not available, use the generic chart located in the Charging chapter.

STEP 2

With the system operating in cooling mode, measure the suction pressure and suction vapor superheat. If the suction pressure is low, and the superheat is lower than required, the system lacks air volume or heat.

STEP 3

Check to ensure the evaporator coil and air filter are clean, and that the indoor blower wheel is clean and spins freely. Replace air filter and clean the coil and/or blower wheel as necessary, and remove or repair any obstructions to the blower wheel.

STEP 4

Perform an air volume calculation and correct any air volume issues.

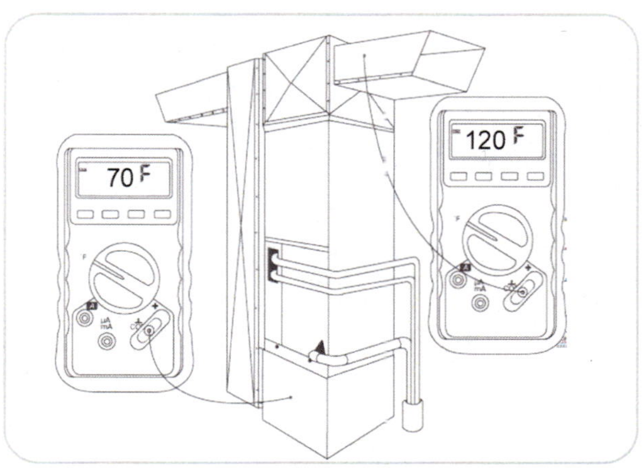

Service Procedure:
Diagnosing a Starved Evaporator Coil

Tools:

- Refrigeration Gauges
- Psychrometer

STEP 1

Measure the return air wet bulb and dry bulb temperature at the inlet of the air handler/furnace. If the air handler is not accessible, readings can be taken at the filter grill, but there is the possibility of inaccurate readings if the ductwork is not adequately and properly sealed. Plot the required suction vapor superheat on the factory-charging chart. If a chart is not available, use the generic chart located in the Charging chapter.

STEP 2

Measure and adjust air volume to 350-450 CFM for each ton of cooling.

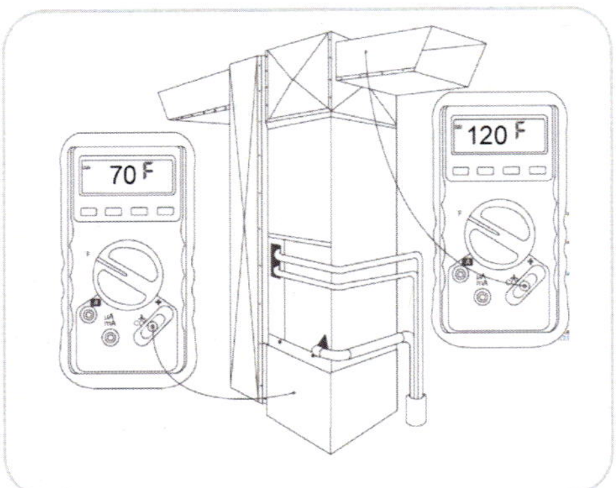

STEP 3

With the system operating in cooling mode, measure the suction pressure, suction line temperature, and then calculate the suction vapor superheat. If the suction pressure is low and the superheat is higher than required, the evaporator coil is not getting enough refrigerant (starved).

STEP 4

Check the discharge pressure to ensure it is at the required level. If it is lower than required, add refrigerant to the system.

If refrigerant is added to the system, and both the suction and discharge pressures increase, and superheat decreases, the system is low on charge. If both pressures stay low despite the addition of refrigerant, and the compressor amp draw increases, there is a restriction in the refrigeration system. If the addition of refrigerant causes the discharge pressure to rise rapidly, yet the suction pressure stays low, there is now an overcharge in the condenser and a liquid circuit restriction.

STEP 5

If a restriction is present, it can be in the condenser coil, the liquid line drier, liquid line, or metering device. Check the temperature of the liquid line at the outdoor unit service valve. The liquid temperature should be above or at the temperature of the outdoor air. If the liquid line is cooler than the outdoor air, the restriction is in the condenser circuits or a drier inside the outdoor unit.

STEP 6

Measure the temperature of the refrigerant leaving the condenser circuits. They should all be relatively close in temperature. If one of these points is disproportionately cooler than the others, there is a restriction in that circuit of the condenser coil. Locate and repair the restriction.

Make sure to check the temperature at all exits of the parallel circuits.

STEP 7

Check the temperature of the liquid entering and leaving the liquid line service valve and drier. There should be very little temperature change through the service valve. The drier may have up to a 2°F difference in temperature between the inlet and outlet of the drier. If there is an excessive temperature drop, the drier is plugged and must be replaced.

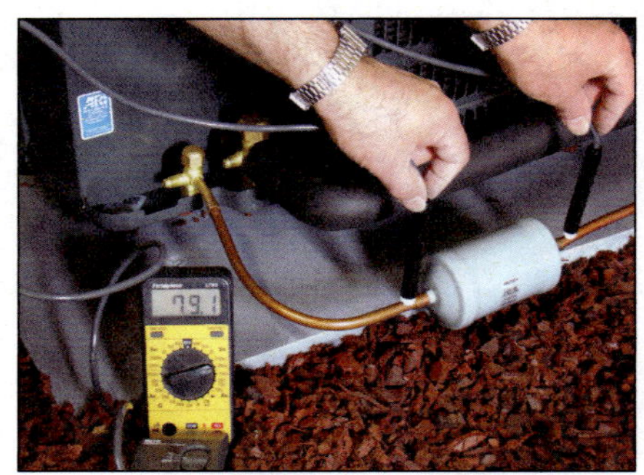

STEP 8

Check the temperature of the liquid line at the outlet of the liquid line drier, and at about 1 foot before the metering device.

There should be no noticeable difference in liquid line temperature between these two points. If there is a notable temperature drop, the liquid line may be undersized, or may be kinked. Evaluate the liquid line to determine if the line is undersized.

To check for a kink or obstruction in the liquid line, measure the temperature of the liquid line at various points on the line. If a temperature drop is detected across two points, there is a pressure drop between those points. Locate the problem and correct as needed.

STEP 9

If all components in the liquid line have checked out OK, the restriction must be in the metering device, or metering device distributor tubes. Recover the charge and inspect the metering device for debris and check that the piston is properly sized for the system. Also, check the screen assembly protecting the distributor tubes for debris. Re-install the orifice and re-charge the system as specified by manufacturer charging charts.

High Suction Pressure

If the system suction pressure is too high, there are four potential causes:

- A flooded evaporator coil. This can be caused by an oversized metering device, overcharge, or the metering device is not properly seated and bypassing liquid into the evaporator circuits.
- High evaporator heat load. This can be caused by the infiltration of hot air into the return duct, or if there is a high air temperature in the home.
- Dirty condenser coil or non-concensable gases in system.
- Compressor valve failure. (reciprocating compressor).

The following diagnostic procedures will help identify the cause of high suction pressure in a fixed metering system.

Service Procedure: Diagnosing a Flooded Evaporator Coil

Tools:
- Refrigeration Gauges
- Psychrometer

STEP 1

Measure the return air wet bulb and dry bulb temperature at the inlet of the air handler/furnace. If the air handler is not accessible, readings can be taken at the filter grill, but there is the possibility of inaccurate readings if the ductwork is not adequately and properly sealed. Plot the required suction vapor superheat on the factory charging chart. If a chart is not available, See Chapter C: Superheat.

STEP 2

Measure and adjust air volume to 350-450 CFM for each ton of cooling.

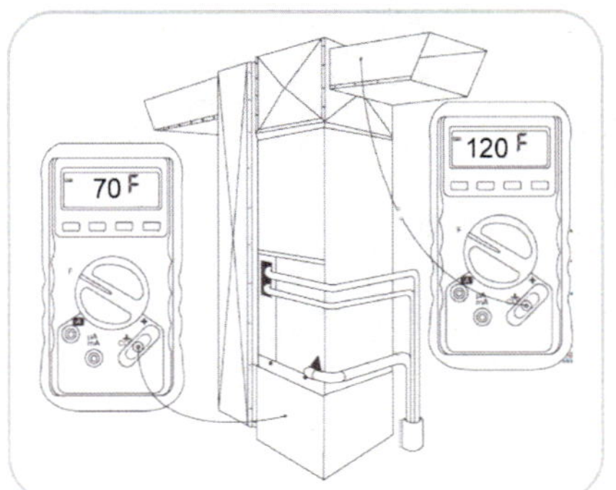

STEP 3

With the system operating in cooling mode, measure the suction pressure, suction line temperature, and then calculate the suction vapor superheat. If the suction pressure is high and the superheat is lower than required, the evaporator coil is being fed too much refrigerant.

STEP 4

If the discharge pressure is lower than it should be, go to step 5.

If the discharge pressure is high, check the temperature of the liquid line at the outdoor unit service valve. If the line is hot, clean the condenser coil and repeat step 3. If the liquid line is warm, recover charge until the suction and discharge pressures are within normal ranges.

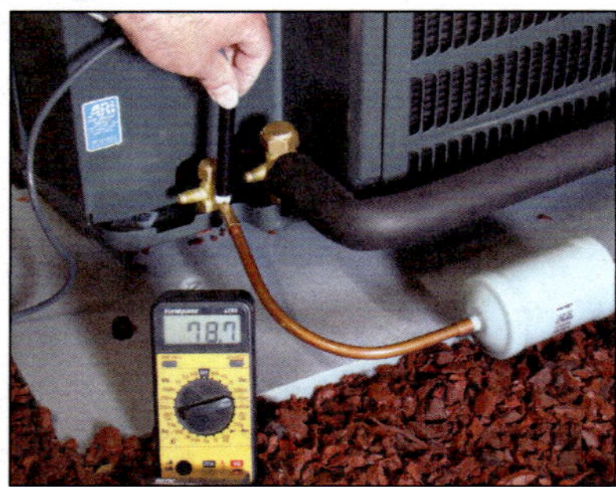

STEP 5

Recover the charge and check that the piston is properly sized for the system and seating firmly.   Re-install the orifice and re-charge the system as specified by manufacturer charging charts.

If the pressures are still not correct, test the compressor for proper operation.

Service Procedure: Diagnosing High Evaporator Heat Load

Tools:

- Refrigeration Gauges
- Psychrometer

Step 1

Measure the return air wet bulb and dry bulb temperature at the inlet of the air handler/furnace. If the air handler is not accessible, readings can be taken at the filter grill, but there is the possibility of inaccurate readings if the ductwork is not adequately and properly sealed. Plot the required suction vapor superheat on the factory-charging chart. If a chart is not available, use the generic chart located in the Charging chapter.

STEP 2

Measure and adjust air volume to 350-450 CFM for each ton of cooling.

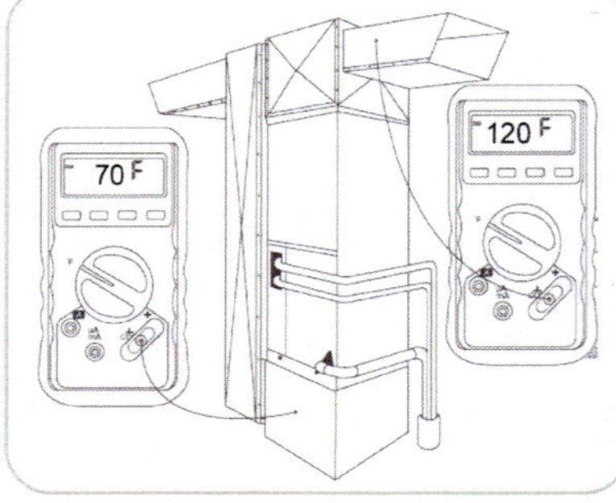

STEP 3

With the system operating in cooling mode, measure the suction pressure, suction line temperature, and then calculate the suction vapor superheat. If the suction pressure is high, and superheat is higher than required, there may be a high heat load on the evaporator coil. Confirm this by checking the discharge pressure and subcooling level. If the discharge pressure is high, and subcooling normal, the system has a high heat load. If the discharge pressure is high, and the subcooling is low, check for a dirty condenser and/or potential non-condensable gas in the system.

If the system indicates a high evaporator load problem, go to the next step.

Measure the room air temperature near the thermostat, and the return air temperature at the air handler/furnace filter rack. The air temperature at the thermostat should be the same, or very near, the air temperature at the filter rack. If the room air is significantly cooler than the air at the filter rack, warm air is infiltrating the return air duct. Find and repair the source of the air leak and repeat steps 1-4. If the return air ducting checks out OK, proceed to the next step.

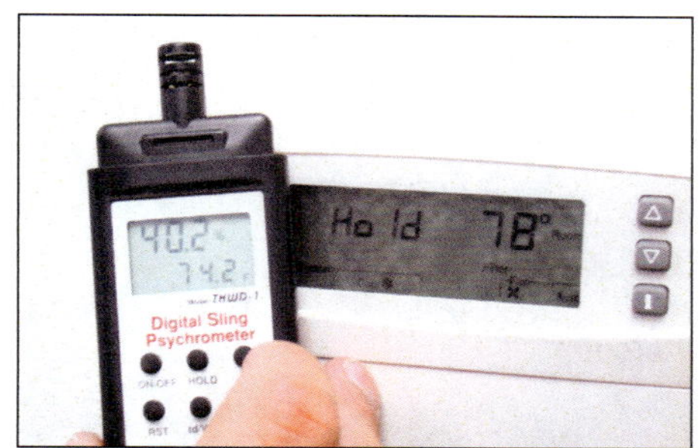

STEP 5

Allow the system to run. As the return air temperature falls, the suction pressure and suction vapor superheat will drop. If the room air temperature does not decrease, the system may not have enough capacity to cool the structure.

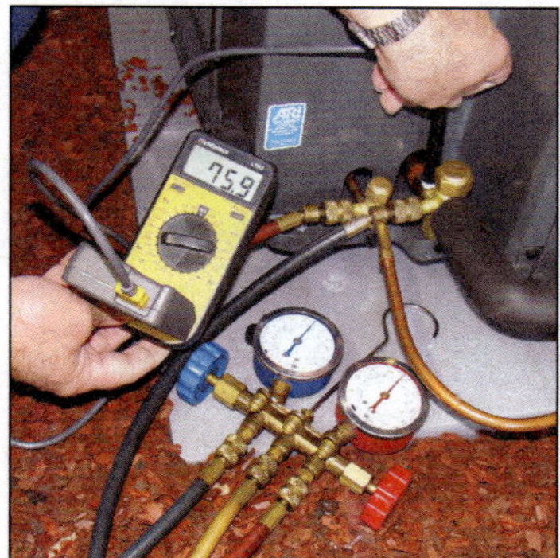

As the system runs the suction pressure and suction vapor superheat should fall.

Service Procedure: Diagnosing Non-Condensable Gasses in the System

See Chapter B: Refrigerant & Refrigerant Piping for non-condensable gas test procedure.

Service Procedure: Diagnosing a Bad Compressor

See Chapter G: Compressors for compressor test procedures.

Fixed Metering System Diagnostics Summary

- Check the suction line and liquid for undersizing and kinks. Because these are field installed components, the likelihood of error is high.

- Determine if low or high heat load is present. Ducting can be incorrectly sized or poorly installed. Undersized or leaking duct can cause heat load problems.

- Check charge levels and verify the piston is the correct size.

- Check system for the presence of a non-condensable gas.

- Perform compressor tests.

Diagnosing TXV System Problems

TXV system problems can exhibit the following symptoms:

- Normal suction pressure & superheat, with low or high discharge pressure & subcooling.
- Suction pressure hunting with fluctuating superheat.
- Low suction pressure with high superheat.
- High suction pressure with low superheat.
- High suction pressure with high superheat.

The following diagnostic procedures will help identify and correct the causes of these symptoms.

Normal Suction Pressure & Superheat, with Low or High Discharge Pressure & Subcooling

Possible Causes:

- Overcharge in condenser if discharge pressure and subcooling are too high.
- Undercharge if discharge pressure is too low and subcooling is too low.

See Chapter D: Subcooling, "TXV System Charging Using Subcooling" to measure and adjust the system charge level.

Suction Pressure Hunting with Fluctuating Superheat

Possible Causes:

- Liquid line flash gas.
- Low air volume/low heat load on evaporator coil.
- Oversized or overfeeding TXV.

Tools:
- Refrigeration Gauges
- Thermometer

STEP 1

Determine the required suction pressure, discharge pressure and subcooling using the factory charging charts. The suction vapor superheat level should be close to 10°F for both R-22 and R-410A systems. Typically, the subcooling requirement is 10°F, but newer systems may be higher. In many cases, the target subcooling level is often printed on the nameplate of the condensing unit.

Check the outdoor condensing unit for the factory required subcooling temperature

STEP 2

Check and clean the condenser coil as needed, then calculate the liquid subcooling at the condenser. If subcooling is low, or below factory specifications, check for a restriction in the condenser coil circuits or factory installed liquid line drier (if present). If no restrictions are found, add refrigerant charge to attain proper subcooling. Proceed to the next step if subcooling level checks out OK.

STEP 3

Check the temperature of the liquid line at the inlet and outlet of the field installed liquid line drier, and 1 foot before the TXV device.

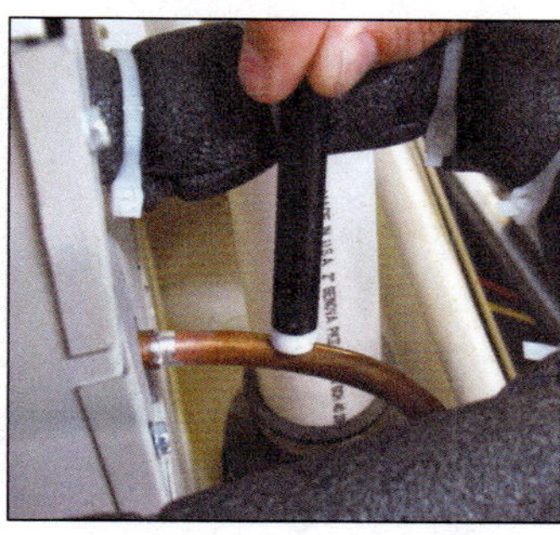

There should be no noticeable difference in liquid line temperature between these points. If there is a noticeable temperature drop, flash gas may be forming in the liquid line. The drier may be restricted, or the liquid line may be undersized or kinked. Evaluate the liquid line to determine if the line is undersized, and check for and repair any kinks or restrictions in the liquid line. Proceed to the next step if the liquid line checks out OK.

STEP 4

Inspect evaporator coil for dirty surface and clean if needed. Adjust air volume to 350-450 CFM for each ton of cooling.

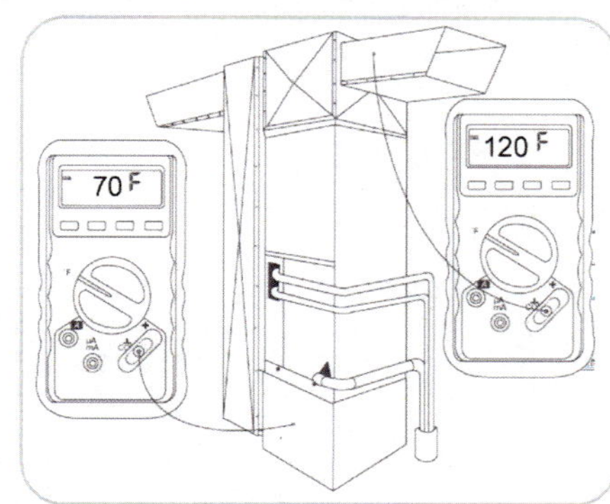

Proceed to the next step if air volume and heat load check out OK.

STEP 5

If the TXV continues to hunt, even after the air volume, a clean evaporator coil, and a properly sized and unobstructed liquid line have been verified, The TXV may be oversized or overfeeding. Ensure the TXV sensing bulb is properly positioned and insulated as per manufacturer requirements. Check the size of the TXV and compare to the manufacturer's specified size requirement. Replace the TXV with one that matches the manufacturer specifications.

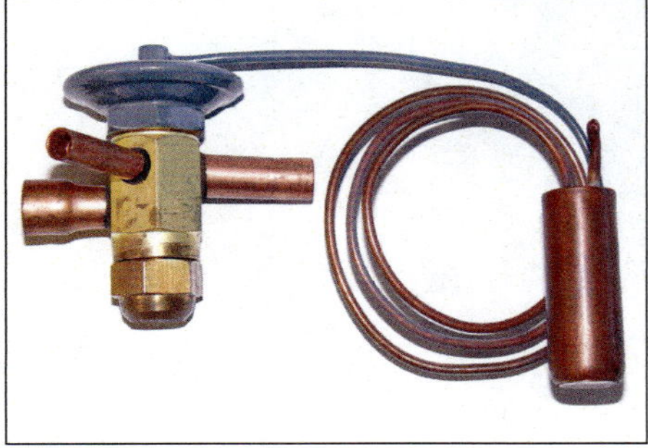

Low Suction Pressure with High Superheat

Possible Causes:

- Low charge.
- Restriction in the refrigeration system.
- Undersized or Underfeeding TXV.

Systems operating in this condition lack refrigerant in the evaporator circuits. First, confirm the system is properly charged. If the addition of charge does not significantly change system pressures, or if the discharge pressure rapidly rises, there is likely a restriction in the system or a problem with the TXV.

Service Procedure: Correcting Low Suction Pressure with High Superheat

Tools:

- Refrigeration Gauges
- Thermometer

STEP 1

See Service Procedure: Correcting Pressure Hunting with Fluctuating Superheat, and perform Steps 1 through 3 to determine if any line restrictions are present.

STEP 2

If no restrictions are present, remove the TXV sensing bulb from the suction line. With the system running, warm the TXV sensing bulb in your hand. The heat from your hand should momentarily cause an increase in the flow of refrigerant into the evaporator coil as noted by an increase in suction pressure. You might also be able to hear the intensity of refrigerant flow into the evaporator increase. Note: A line pressure tap may need to be installed at the evaporator coil to observe this pressure fluctuation.

If the suction pressure increases, ensure the TXV sensing bulb is properly positioned and insulated as per manufacturer requirements. If no pressure change is detected, replace the TXV with one that matches the manufacturer specifications.

High Suction Pressure with Low Superheat

Possible Cause:

- Liquid bypassing the TXV.

If the suction pressure is too high and the suction vapor superheat too low, the evaporator circuits are operating in a flooded condition. The TXV sensing bulb may be improperly positioned or poorly insulated.

Service Procedure: Correcting High Suction Pressure with Low Superheat

STEP 1

Check the position of the TXV sensing bulb. It should be installed at the 2 o'clock or 10 o'clock position on the suction line. Ensure The bulb is in full contact with the suction line. Also ensure the bulb is completely insulated. Correct the bulb position and insulation as needed.

The TXV sensing bulb should be covered with insulation. The insulation has been removed in this image to show proper placement of the bulb.

STEP 2

The TXV equalizer line exerts a closing pressure on the TXV diaphragm. Confirm the equalizer line is not kinked, which would prevent the closing pressure from reaching the TXV. If the equalizer line checks out OK, replace the TXV.

High Suction Pressure with High Superheat

Possible Cause:

- High evaporator heat load.

If the suction pressure is too high and the suction vapor superheat is too high, the refrigerant in the evaporator circuit is absorbing a large amount of heat. The expansion valve is wide open but cannot feed enough refrigerant into the evaporator circuit to compensate.

Service Procedure: Correcting High Suction Pressure with High Superheat

Tools:

- Thermometer

STEP 1

With the system running, measure the room air temperature near the thermostat, and the return air temperature at the air handler/ furnace . If the air handler is not accessible, readings can be taken at the filter grill, but there is the possibility of inaccurate readings if the ductwork is not adequately and properly sealed.

The air temperature at the thermostat should be the same, or very near, the air temperature at the filter rack. If the room air is significantly cooler than the air at the filter rack, warm air is infiltrating the return air ducting. Find and repair the source of the air leak.

If the air temperatures are equal and the system cannot cool the space, the system may be too small for the heat load. Take any possible measures to reduce the heat load. If the heat load cannot be lowered to a level within the capacity of the system, the system will need to be replaced.

Quick Reference Diagram

Low Evaporator Heat Load: Fixed Metering Device and TXV

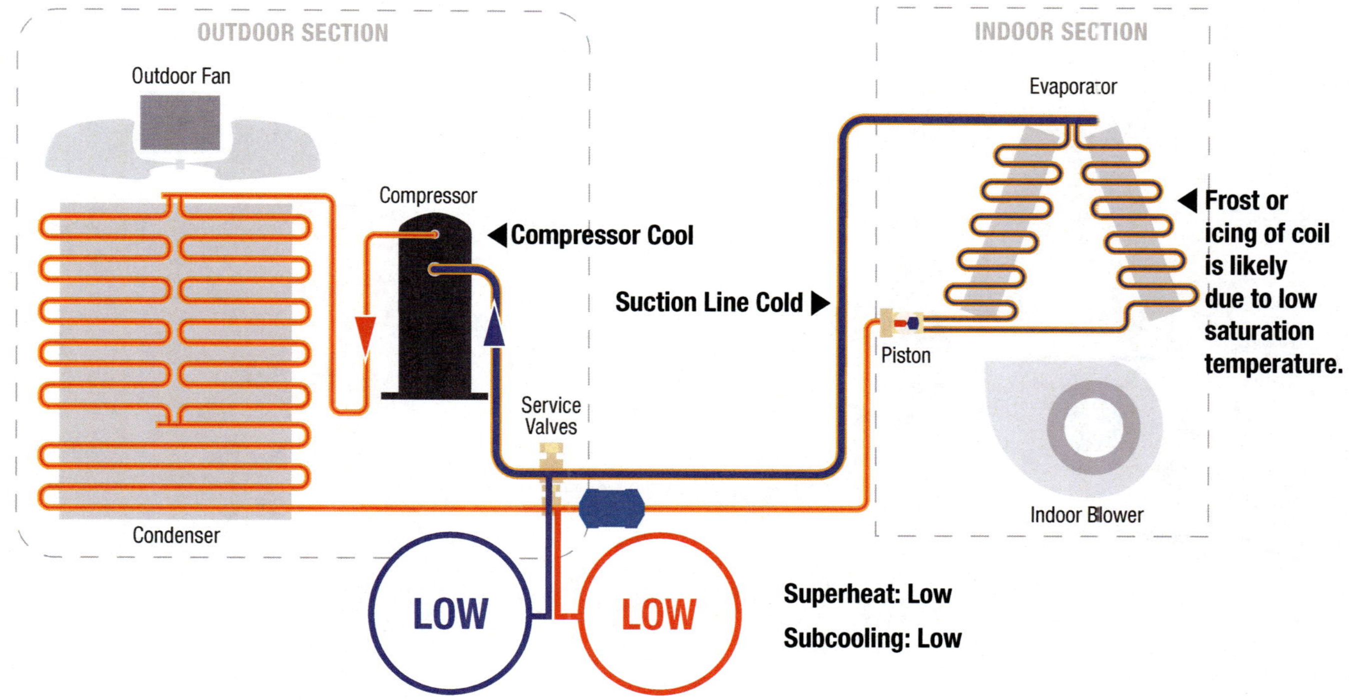

Problem Overview

Low evaporator heat loading occurs when there is not enough heat being absorbed by the refrigerant in the evaporator coil.

Symptoms (Fixed Metering)

Systems running with low evaporator heat load will run at pressures below factory required levels. Abnormally low superheat and subcooling levels will be present when the evaporator coil lacks heat load.

The low heat load condition may cause the evaporator coil temperature to be very cold. Ice or frost may form on the evaporator surface and suction line surface. The compressor shell may sweat and the hot gas temperature will be cool. In extreme cases, superheat may fall to zero degrees and liquid can flood back to the compressor.

Symptoms (TXV Metering)

Same as above but TXV may hunt in an attempt to maintain proper superheat.

Causes

- Dirty return air filter.
- Dirty indoor fan blower assembly.
- Indoor fan motor not on high speed (incorrect air volume).
- Undersized ductwork.
- Dirty evaporator coil.
- Thermostat setpoint too low.

Recommended Procedures:

Diagnosing Low Head Load (See page E3)

Air Volume Measurement (See Chapter J: Air Volume)

High Evaporator Heat Load: Fixed Metering Device and TXV

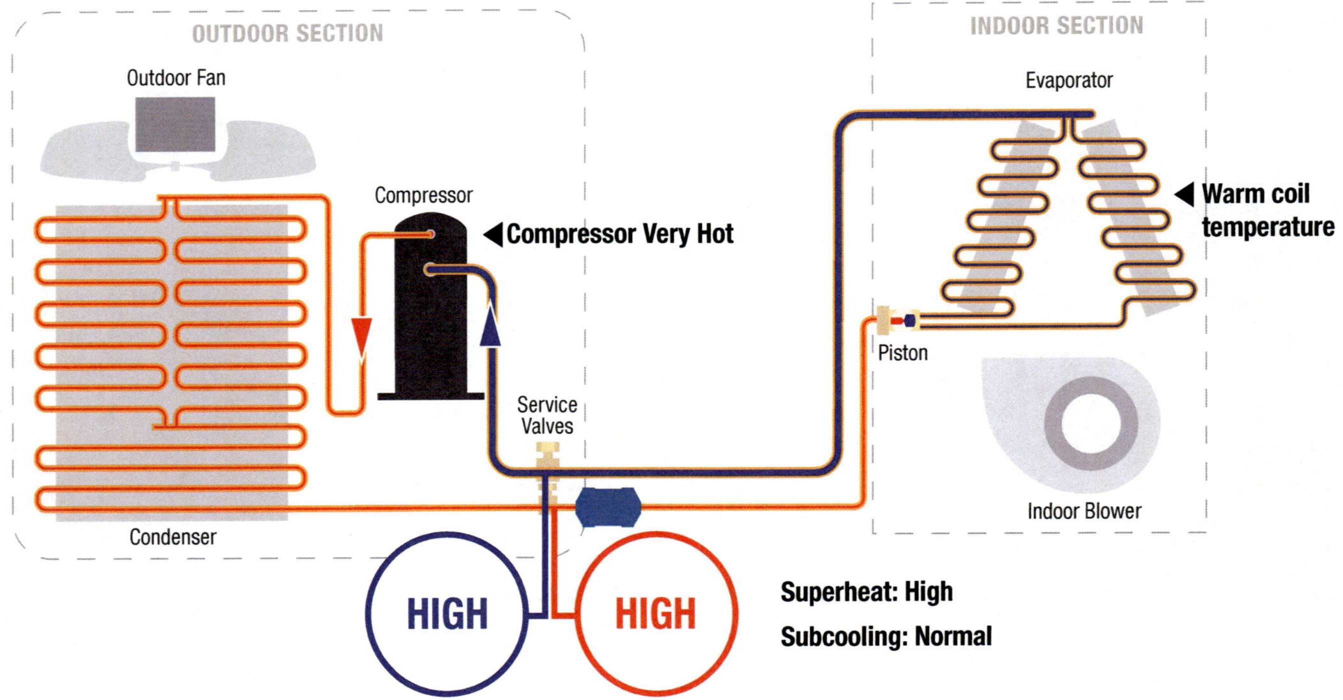

Problem Overview

High evaporator heat load occurs when there is excessive heat absorbed by the refrigerant in the evaporator coil. This condition can cause the compressor to shut off on its internal overload, or trip the electrical circuit breaker.

Symptoms (Fixed Metering)

Systems running with high evaporator heat load will run at pressures above factory required levels. Abnormally high superheat and normal subcooling levels will be present when the evaporator coil is exposed to higher than allowed heat load.

The high heat load condition may cause the evaporator coil to be very warm. The warm evaporator temperature will reduce the amount of moisture being removed from the air.

The compressor shell and refrigerant vapor temperatures will be extremely high.

Symptoms (TXV Metering)

Same as above but the TXV may keep superheat level close to the required range of 10-15°F. In extreme cases of high heat load, the superheat may go much higher.

Causes

- Infiltration of unconditioned air into the return duct.
- New systems started on a very hot day.
- Periods of high occupancy in the conditioned space.
- System undersized for structure heat gain.

Recommended Procedures:

Diagnosing High Evaporator Heat Load (See page E6)

Low System Charge: Fixed Metering Device and TXV

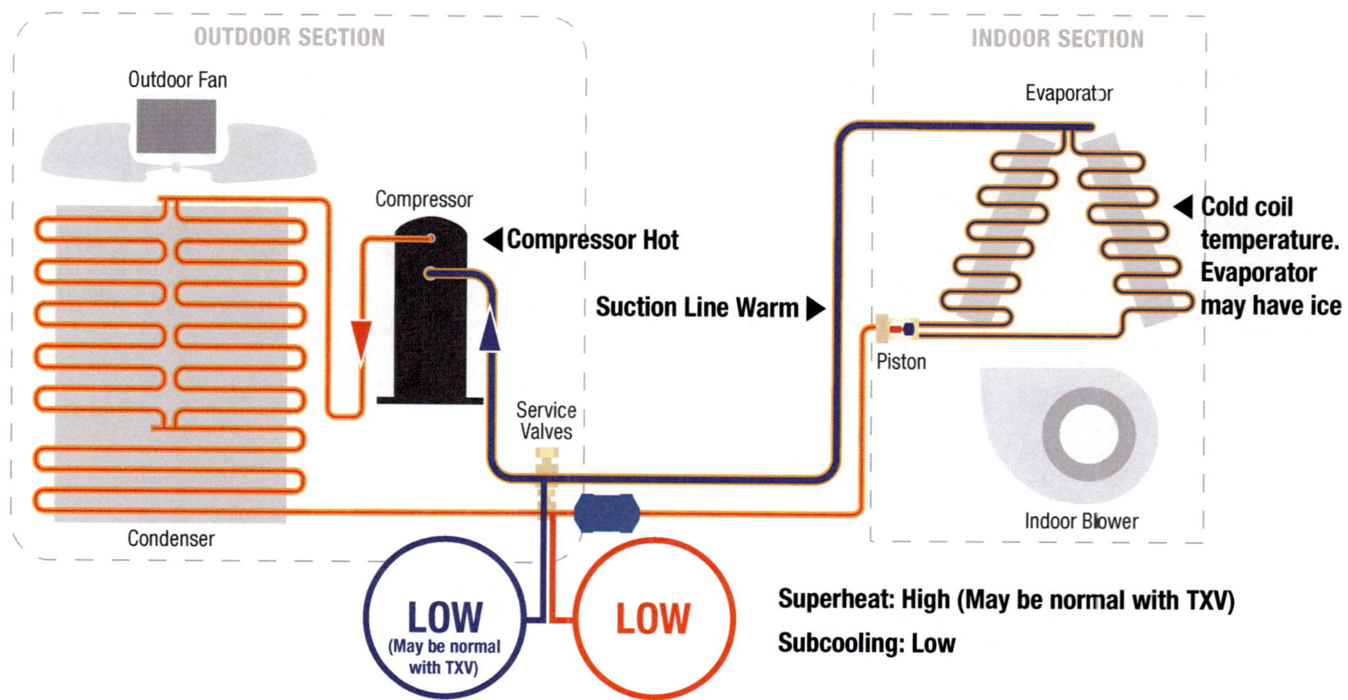

Problem Overview

Low system charge can be caused by a leak in the refrigerant piping or equipment, or poor installation or service skills. A system running with a low charge will have low capacity, low efficiency, and a hot compressor motor.

Symptoms (Fixed Metering)

Systems with low charge will run at pressures below factory required levels. Abnormally high superheat and low subcooling levels will be present. The compressor motor will run hot due to a lack of refrigerant vapor to cool the motor windings. The low suction pressure will create a cold evaporator coil and may form a coating of ice.

Symptoms (TXV Metering)

When a system is undercharged with a TXV, the TXV may open enough to maintain adequate suction pressure and suction vapor superheat.

The condensing pressure and subcooling will be low. If the system charge is very low, the system will exhibit the same symptoms as a fixed metering type system.

Causes

- Poor installation or service skills.
- Refrigerant leaks.

Recommended Procedures:

Fixed Metering Charging Using Superheat (See Chapter C: Superheat)

TXV System Charging Using Subcooling (See Chapter D: Subcooling)

Note: The system should be inspected for refrigerant leaks using industry approved methods and devices prior to charging.

High System Charge: Fixed Metering Device and TXV

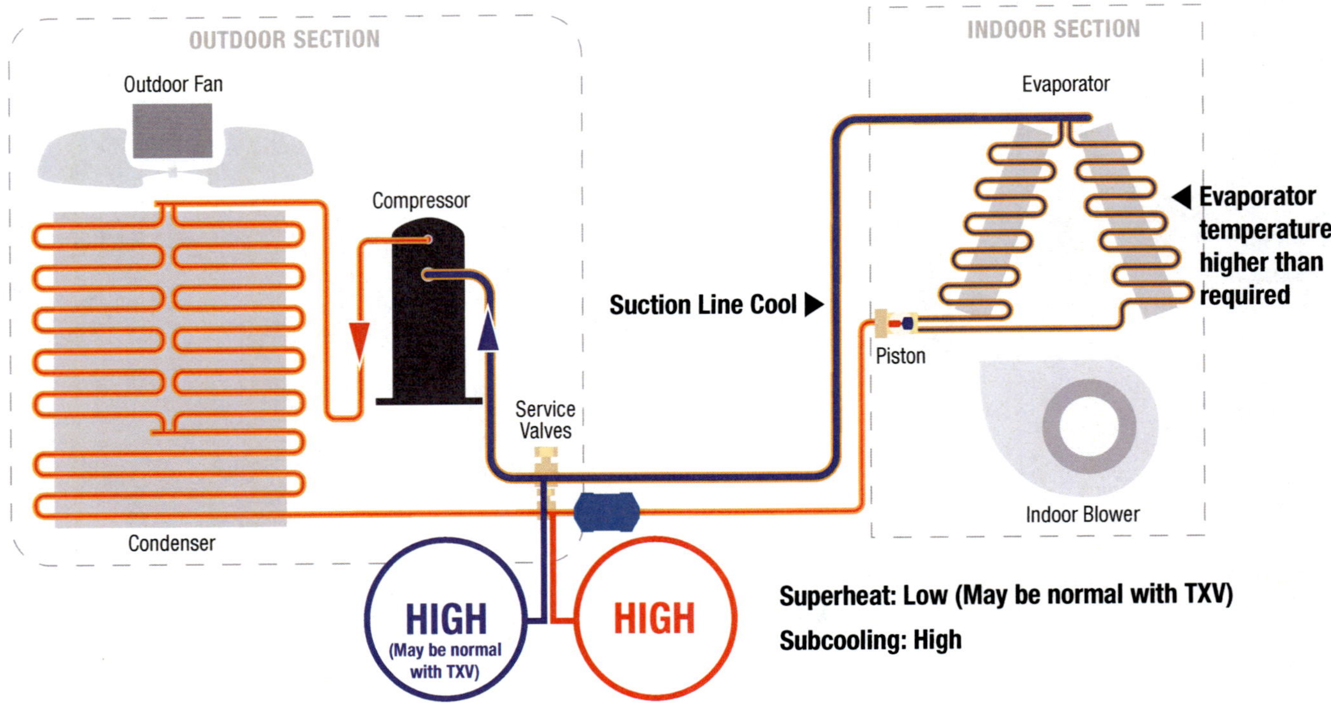

Problem Overview

High system charge (overcharge) is caused by poor installation or service skills. A system running with an overcharge may experience excessive energy consumption, starting problems, internal overload tripping, indoor air humidity complaints, compressor failure due to liquid flood back.

Symptoms (Fixed Metering)

Systems running with an overcharge will run at pressures above factory required levels. Abnormally low superheat and high subcooling levels will be present. The compressor motor may experience liquid flood back problems.

Symptoms (TXV Metering)

When a TXV system is overcharged, the suction pressure and superheat may appear normal.

The appearance of normal conditions are a result of the TXV throttling the excess refrigerant into the evaporator coil. The excess charge is then stored in the condenser coil, which elevates the head pressure and liquid subcooling levels.

If the charge is high enough, the suction pressure may be too high with slightly low superheat present.

Causes

- Poor installation or service skills.

Recommended Procedures:

Fixed Metering Charging Using Superheat (See Chapter C: Superheat)

TXV System Charging Using Subcooling (See Chapter D: Subcooling)

Non-Condensable Gases: Fixed Metering Device and TXV

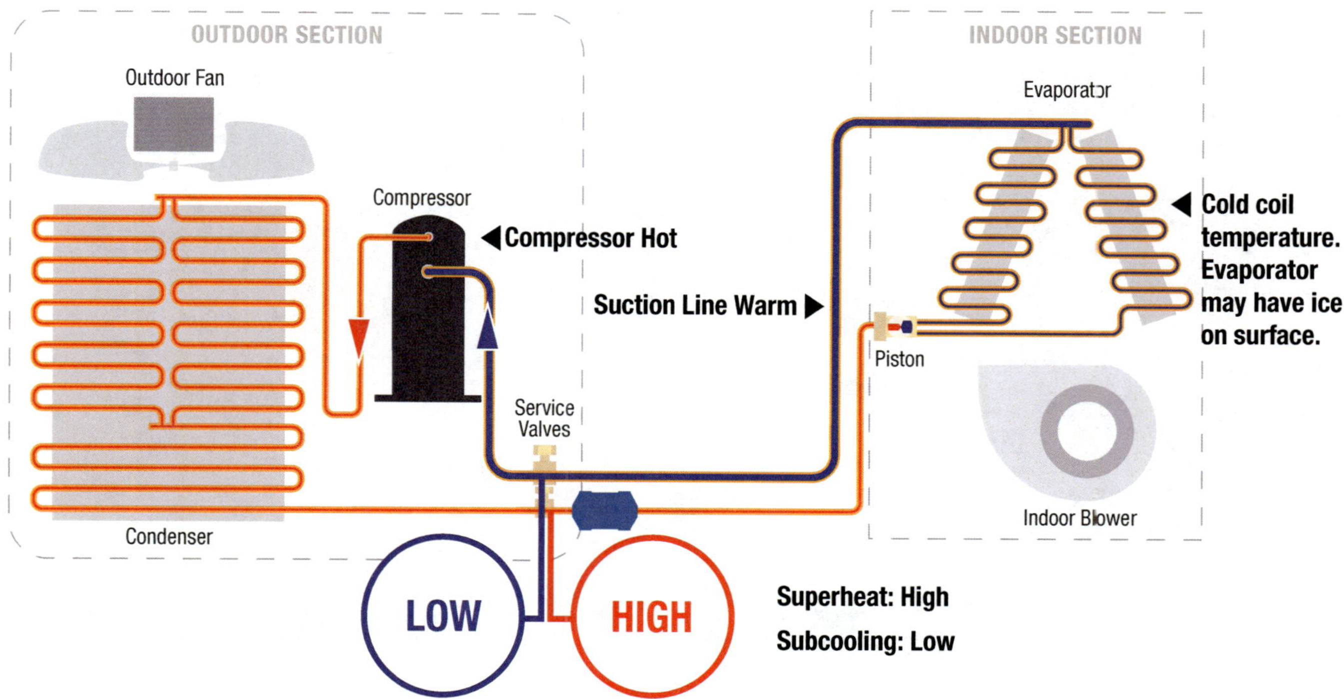

Problem Overview

Air, Nitrogen, Hydrogen and other foreign gases present in a refrigerant system are referred to as non-condensables since they will not condense into a liquid at the pressures encountered in the refrigeration system. These gases will accumulate in the condenser coil, reducing its effectiveness.

Symptoms

Systems running with non-condensables present will experience higher than normal head pressure. The suction pressure will be low if there is no subcooling present. If subcooling is present, the suction pressure may be slightly high with fixed metering systems but may appear normal in TXV systems.

The subcooling level with non-condensables will be lower than required. If no subcooling is present, there will be flash gas in the liquid line ahead of the metering device.

The superheat level will be high when flash gas is present. If there is subcooled liquid entering the metering device, and the suction pressure is high with a fixed metering system, the superheat may be low. If the system is using a TXV, the superheat may appear normal

Causes

- Poor installation or service skills.
- Failure to properly evacuate the system after the system has been opened.

Recommended Procedures:

 Detecting Mixed Refrigerants & Non-Condensable Gases
(See Chapter B: Refrigerant and Refrigerant Piping)

Quick Reference Diagram

Suction Line Restriction: Fixed Metering Device and TXV

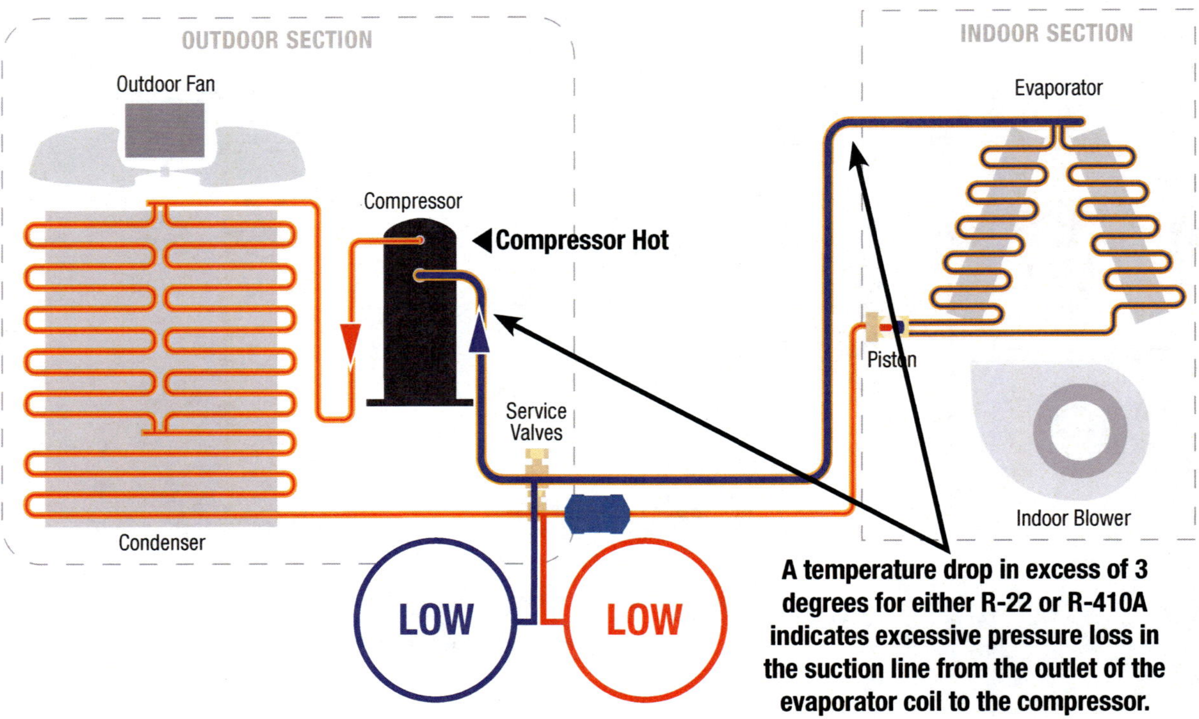

Problem Overview

Excessive pressure loss in the suction line will cause low cooling capacity and potential evaporator coil freezing.

Symptoms

Systems running with a large pressure loss in the suction line will run at pressures below factory required levels. Because the pressure is being read downstream from the restriction, the actual calculated superheat will be inaccurate. Find the problem by checking the temperature of the suction line at the outlet of the evaporator coil and then comparing it to the temperature of the suction line at the inlet to the compressor. There should be a maximum temperature drop of 3°F degrees. If the actual temperature drop is higher, find the cause of the pressure drop.

Causes

- Undersized suction lines.
- Kinks in the suction line tubing.
- Restricted suction line drier.
- Excessive use of suction line pipe fittings.

Recommended Procedures:

Diagnosing Excessive Suction Line Pressure Loss (See page E2)

Evaluating Potential Suction Line Pressure Drop (See Chapter B: Refrigerant and Refrigerant Piping)

Liquid Line Restriction: Fixed Metering Device and TXV

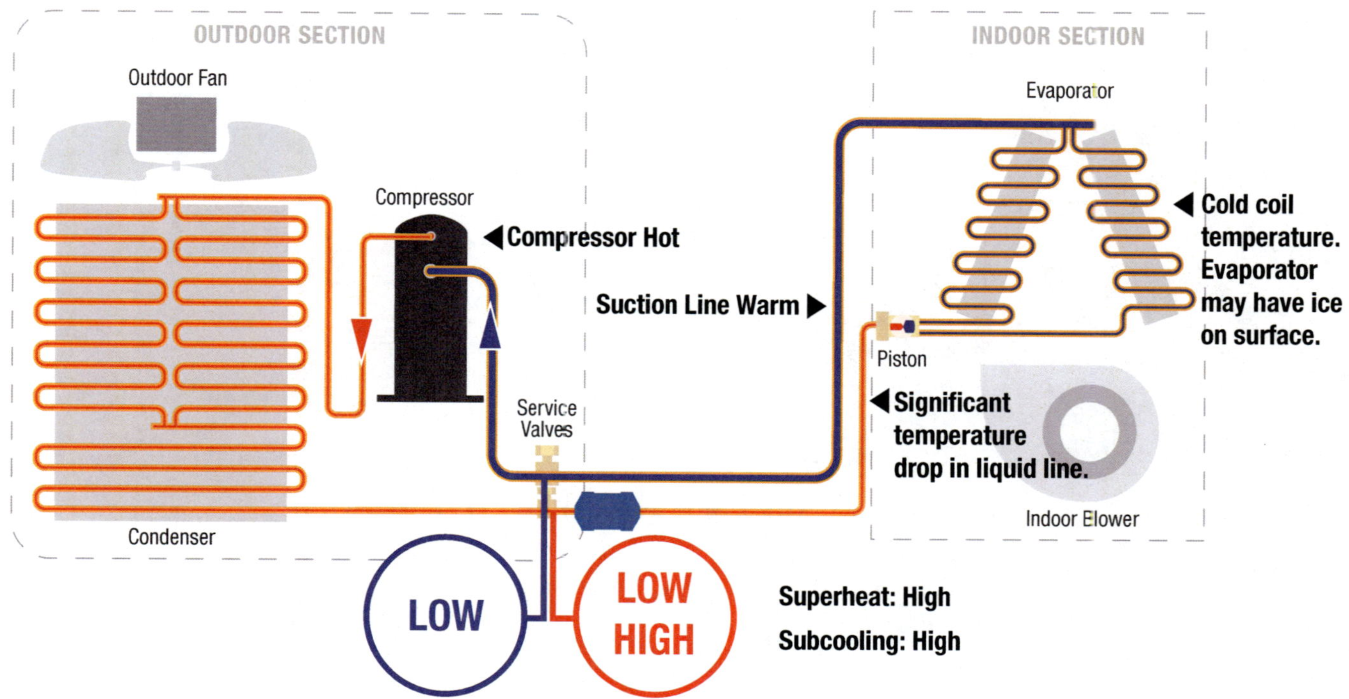

Problem Overview

Restrictions in the liquid line will prevent the evaporator coil from receiving the proper amount of refrigerant. The restriction will cause a large pressure drop between the condenser coil and the metering device. This large pressure drop can be detected by checking the temperature of the liquid line at the outlet of the condenser coil and at the inlet to the metering device.

Symptoms

Systems running with a liquid line restriction will have low suction pressure along with high superheat. The liquid pressure will be initially low, which will make the system appear undercharged. When refrigerant is added to the system, the liquid pressure will rise but the suction pressure will remain low. At this point, a problem in the liquid circuit can be identified. If subcooling is measured at this time, it will be high. The high subcooling level is an indication that excess refrigerant is being stored in the condenser coil. Look for frosting or sweating of the liquid line or liquid line drier.

Causes

- Restricted liquid line drier.
- Undersized liquid line.
- Excessive liquid line pipe fittings.
- Kinked liquid line.

Recommended Procedures:

Fixed Metering Charging Using Superheat (See Chapter C: Superheat)*

TXV System Charging Using Subcooling (See Chapter D: Subcooling)*

Diagnosing a Starved Evaporator Coil (See page E3, steps 5 through 9.)

Evaluationg Liquid Line Size (See Chapter B: Refrigerant and Refrigerant Piping)

*If the system is undercharged and charge is added, both pressures will begin to rise. If the liquid pressure rises significantly, but the suction pressure remains low, the liquid circuit has a restriction.

Restricted Metering Device: Metering Device and TXV

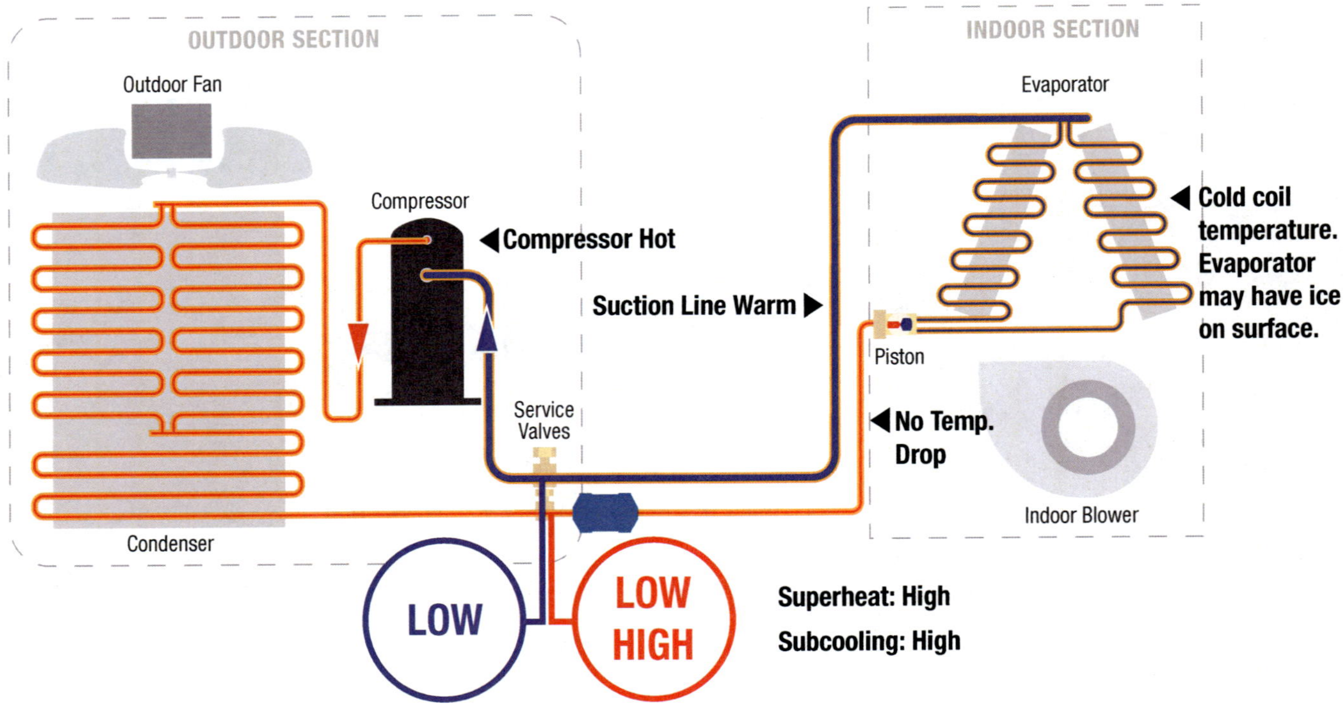

Problem Overview

Restrictions in the metering device will prevent the evaporator coil from receiving the proper amount of refrigerant. The restriction will cause the suction pressure to be too low. With low suction pressure present, the saturation temperature of the refrigerant in the evaporator is very low. Because of the low saturation temperature, freezing of the evaporator surface is likely.

Symptoms

Systems running with a metering restriction will have low suction pressure and high superheat. The liquid pressure will be initially low, which will make the system appear undercharged. When refrigerant is added to the system, the liquid pressure will rise but the suction pressure will remain low. At this point, a problem in the liquid circuit can be identified. If subcooling is measured at this time, it will be high.

The high subcooling level indicates that excess refrigerant is being stored in the condenser coil. There will be no obvious sign of frosting between the condenser coil and the metering device.

Causes

- Undersized fixed metering device (piston).
- Failed expansion valve sensing bulb.
- Debris in metering device (poor installation skills).
- Moisture freezing at metering device (not evacuated properly).

Recommended Procedures:

Fixed Metering Charging Using Superheat (See Chapter C: Superheat)*

TXV System Charging Using Subcooling (See Chapter D: Subcooling)*

Diagnosing a Starved Evaporator Coil (See page E3, steps 5 through 9)

Correcting Low Suction Pressure with High Superheat (see page E9)

Evaluating Liquid Line Size (See Chapter B: Refrigerant and Refrigerant Piping)

*If the system is undercharged and charge is added, both pressures will begin to rise. If the liquid pressure rises significantly, but the suction pressure remains low, the liquid circuit has a restriction.

Condenser Circuit Restriction: Fixed Metering

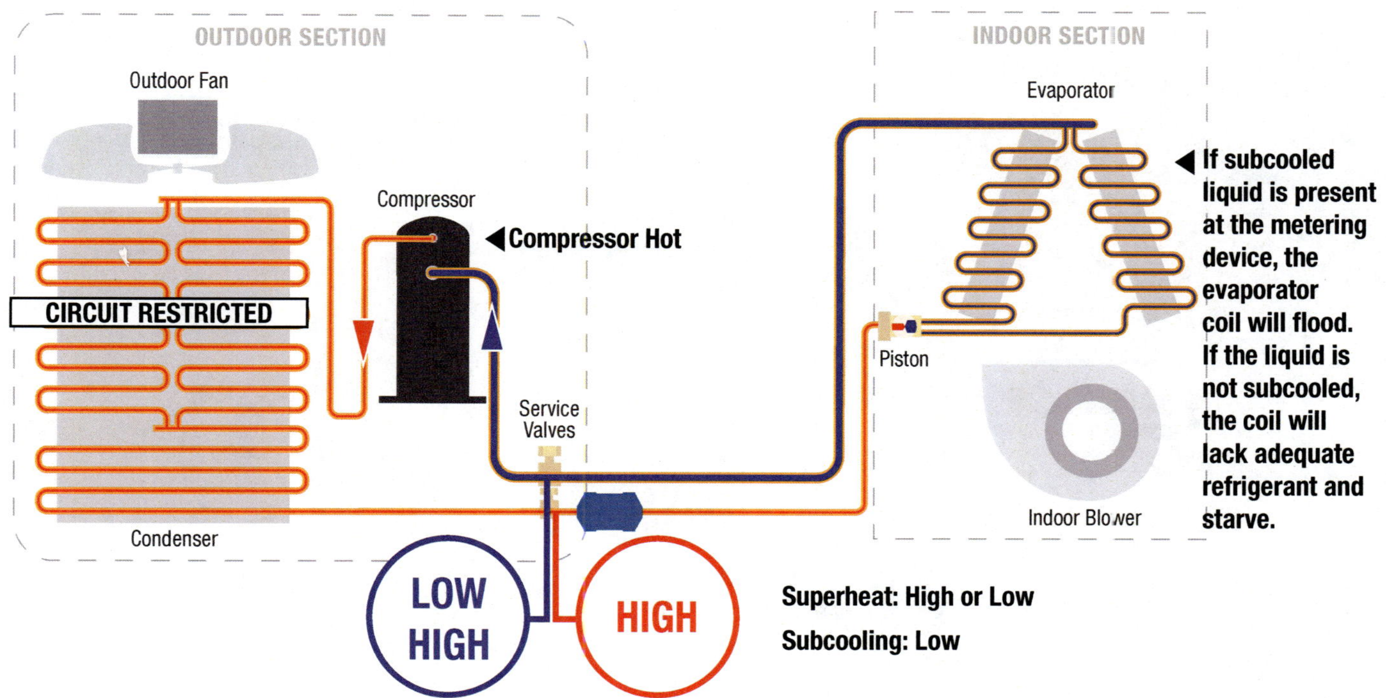

Problem Overview

When a condenser circuit is restricted, its surface area is decreased, and its ability to reject heat is reduced.

Symptoms

Systems running with a partially or fully restricted parallel circuit will exhibit liquid pressures above factory required levels. The suction pressure could be high or low, depending on whether or not there is enough condenser coil surface area to provide adequate subcooling for the amount of liquid refrigerant present. If there is adequate subcooling, the high liquid pressure will flood the evaporator coil. In a flooded state, the evaporator suction pressure will be high and the superheat level will be low. If there is not enough subcooling to overcome the pressure loss in the liquid line, flash gas will be present in the liquid line ahead of the metering device. This condition will starve the evaporator coil. Low suction pressure and high superheat will be present. The restriction in the circuit will cause a corresponding drop in refrigerant temperature.

The refrigerant exiting the restricted circuit will be much colder than the refrigerant exiting other circuits.

Causes

- Debris.
- Handling damage.

Recommended Procedures:

 Checking for Restrictions in Condenser Circuits (See Chapter B: Refrigerant and Refrigerant Piping)

Quick Reference Diagram

Condenser Circuit Restriction: TXV Metering

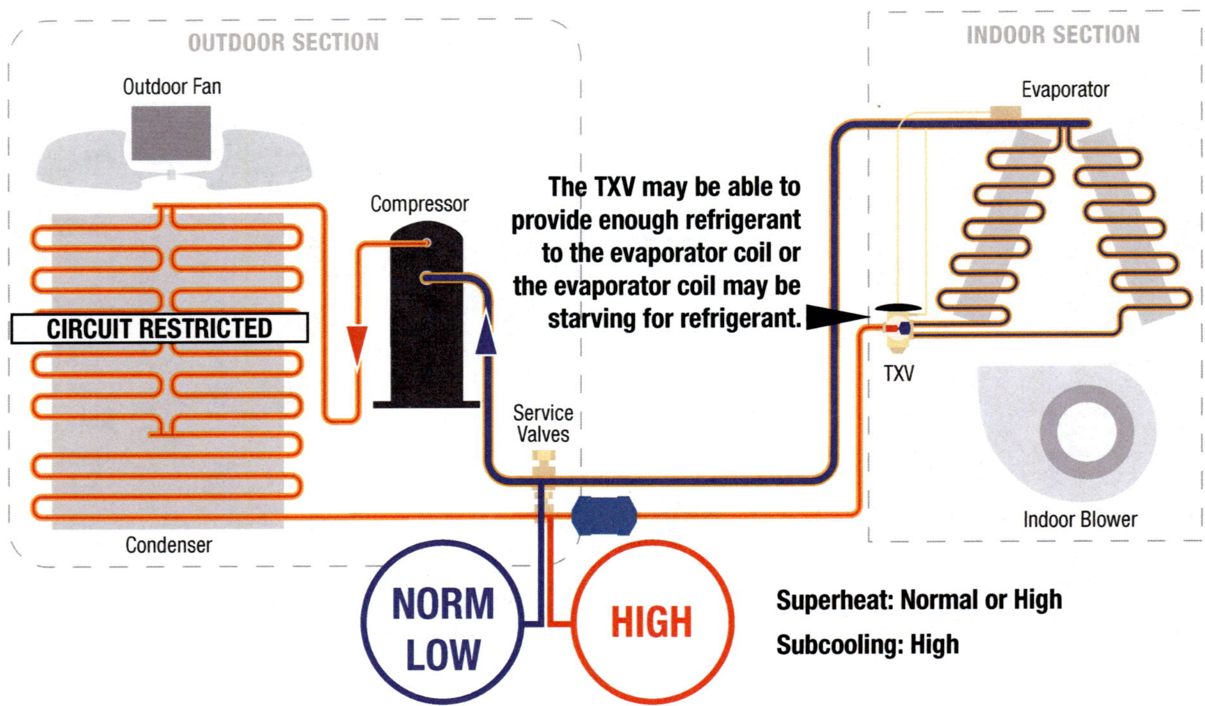

Problem Overview

When a condenser circuit is restricted, its surface area is decreased, and its ability to reject heat is reduced.

Symptoms

Systems running with a partially or fully restricted parallel circuit will run at liquid pressures above factory required levels. The system may have adequate subcooling for the liquid line pressure loss, or it may not. If there is enough subcooling to maintain a subcooled liquid at the metering device, the evaporator will perform normally with the TXV. If flash gas forms in the liquid line due to inadequate liquid subcooling, the TXV will try to open and provide more refrigerant to the evaporator coil. If the TXV cannot feed enough refrigerant into the evaporator coil it will starve.

The refrigerant exiting the restricted circuit will be much colder than the refrigerant exiting other circuits.

Causes

- Debris.
- Handling Damage.

Recommended Procedures:

 Checking for Restrictions in Condenser Circuits (See Chapter B: Refrigerant and Refrigerant Piping)

Condenser Subcooling Circuit Restriction: Fixed & TXV Metering

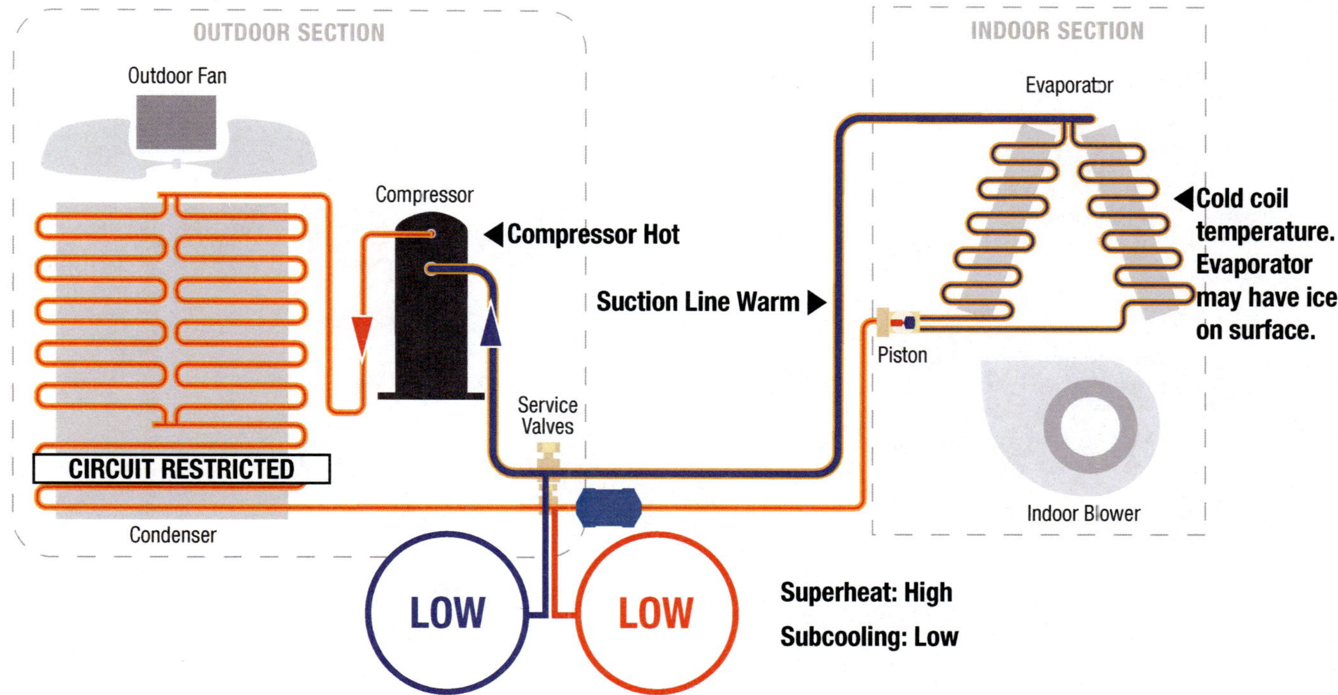

Problem Overview

When a subcooling circuit is restricted, metering of the refrigerant will take place in the condenser coil instead of the metering device.

Symptoms

When the subcooling circuit is restricted, the refrigerant will be metered to a low pressure within the condenser. Since the liquid pressure gauge is downstream from this restriction, the liquid pressure will read low. The temperature of the liquid line will correspond to the saturation temperature equal to the pressure being read on the liquid line gauge.

The refrigerant leaving the condensing coil will be below the temperature of the outdoor air entering the condenser coil. It is likely that frosting or sweating of the liquid line will be present.

Causes

- Debris.
- Handling damage.

Recommended Procedures:

 Checking for Restrictions in Condenser Circuits
(See Chapter B: Refrigerant and Refrigerant Piping)

Over-Feeding Metering Device: Fixed Metering Device and TXV

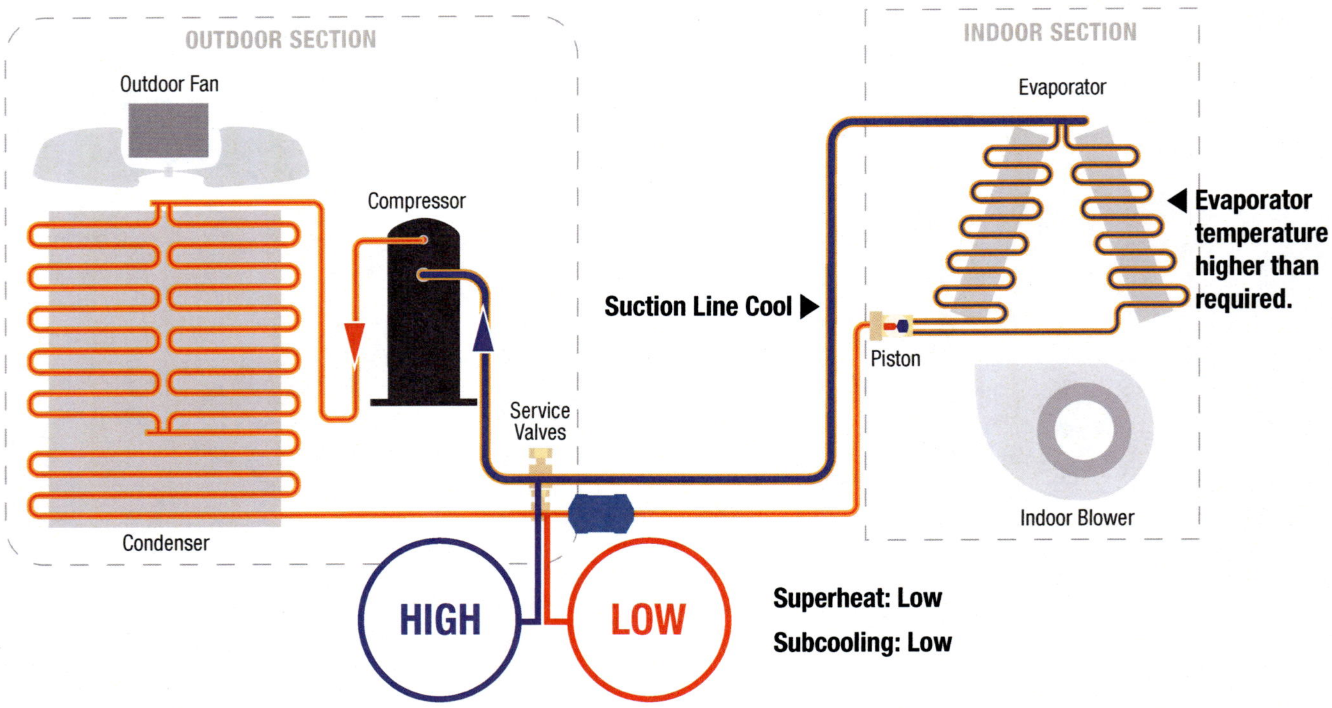

Problem Overview

An evaporator coil that is receiving a higher than normal flow of refrigerant through the metering device will operate in a flooded condition. The system may experience high humidity complaints, compressor starting problems, or compressor failure.

Symptoms

The metering device separates the high-pressure side of the system from the low-pressure side of the system. When an overfeeding condition is present, the low side of the system operates in a flooded state, while the high side of the system operates in a starved state. The suction pressure will be high and the superheat low. The liquid pressure and subcooling will be low.

If the system is using a reciprocating type of compressor, it may appear that the compressor valves are leaking.

Causes

- Oversized piston (Fixed Metering).
- Dirty piston seat not allowing metering device to seat properly.
- TXV sensing bulb not insulated or in good contact with suction line.

Recommended Procedures:

1. Call for cooling.
2. Close the liquid line service valve and pump the system down. If the system does not hold a pump down, and the liquid line service valve is properly closed, replace the compressor. (Reciprocating Only)
3. If the system holds a pump down, inspect the metering device to determine if it is seated properly or too large. (Fixed Type)
4. If the system is using a TXV, make sure the sensing bulb is properly insulated and in good contact with the suction line. If it is not, correct the problem. If it is, replace the TXV.

Evaporator Conditions

Fixed Metering Device

System Condition: Normal

Suction Pressure: **Normal**

This coil is performing to factory specifications. The refrigerant liquid is boiling off to vapor at various points within the circuiting based on the temperature of the outdoor air and indoor heat load.

System Condition: High Heat Load

Suction Pressure: **High**

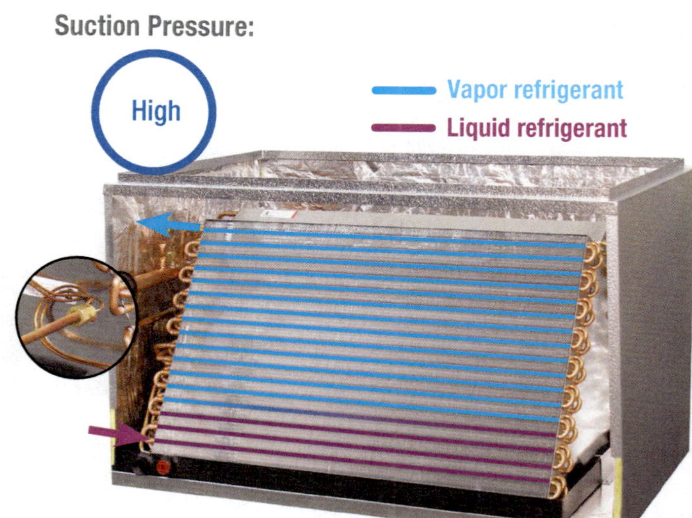

This coil is vaporizing liquid refrigerant earlier in the circuit than it should. The coil is not undercharged because the suction pressure is high. The suction vapor superheat level will be higher than charging chart requirement. This condition is caused by excess air volume or heat load.

System Condition: Flooded Evaporator Coil

Suction Pressure: **High**

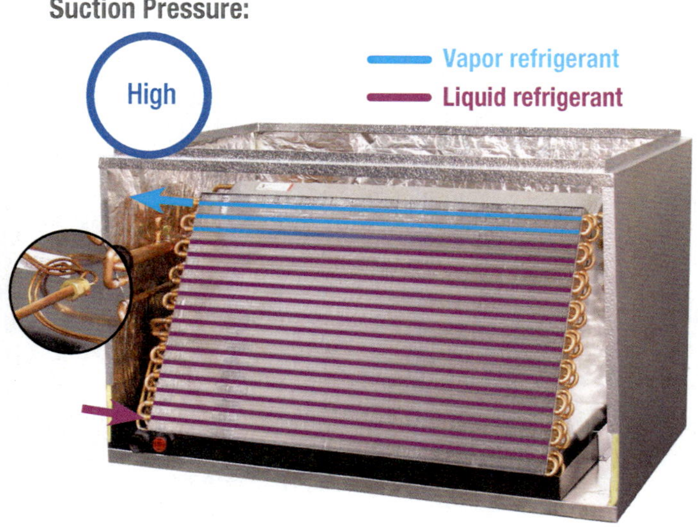

This coil is flooded with liquid refrigerant. The coil pressure is high, and liquid is present very late in the circuiting. If the system were operating at full load conditions, this may appear as normal operation, except for the fact that suction pressure is higher than it should be. Causes include overcharging, and oversizing of the metering piston.

System Condition: Starved Evaporator Coil

Suction Pressure: **Low**

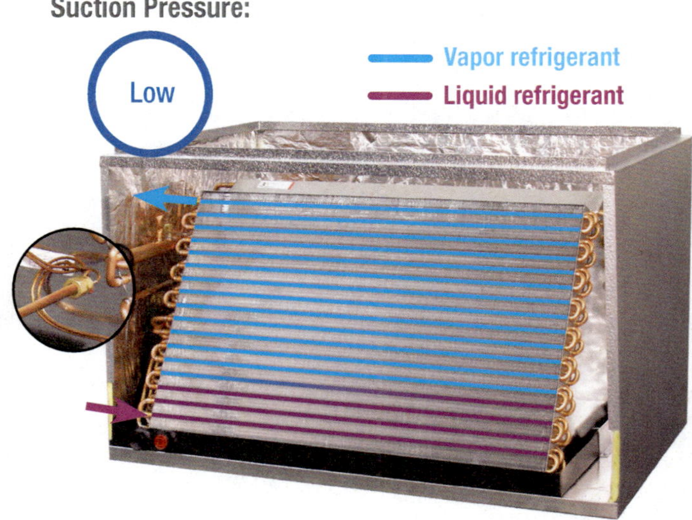

This coil is lacking refrigerant. The saturated liquid is vaporizing early in the circuit and the suction pressure is too low. Superheat is above the required chart value. Causes include undercharging, and liquid restrictions.

TXV Metering Device

System Condition:
Low Heat Load

Suction Pressure:

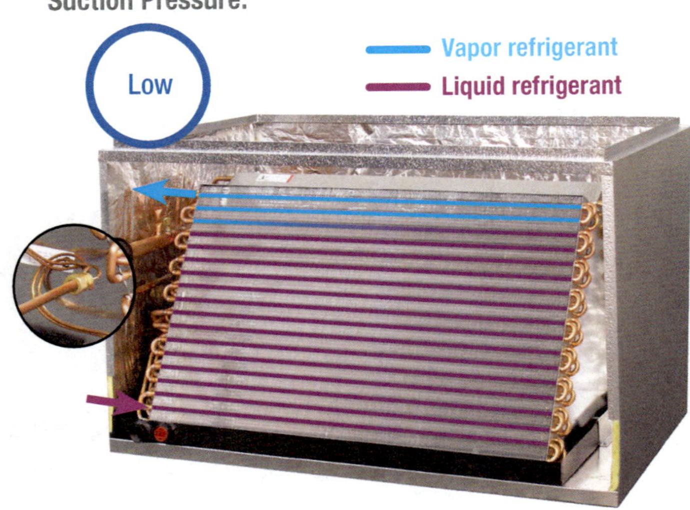

System Condition:
Normal

Suction Pressure:

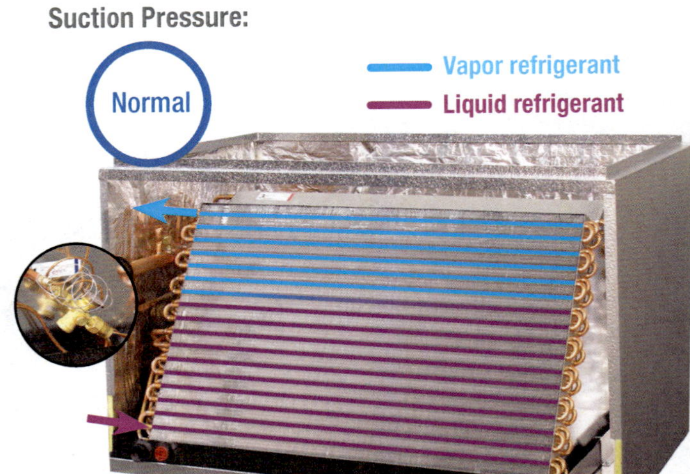

This coil is flooded with liquid refrigerant. The coil pressure is low. This can be due to low air volume, or a lack of heat in the air.

Coils with a TXV operate at 10°F of suction vapor superheat over a wide range of operating heat load and outdoor air temperatures. Be aware that a TXV can mask undercharged conditions by maintaining normal evaporator conditions under certain combinations of indoor heat load and outdoor air temperature. The problem will appear intermittently.

System Condition:
High Heat Load

Suction Pressure:

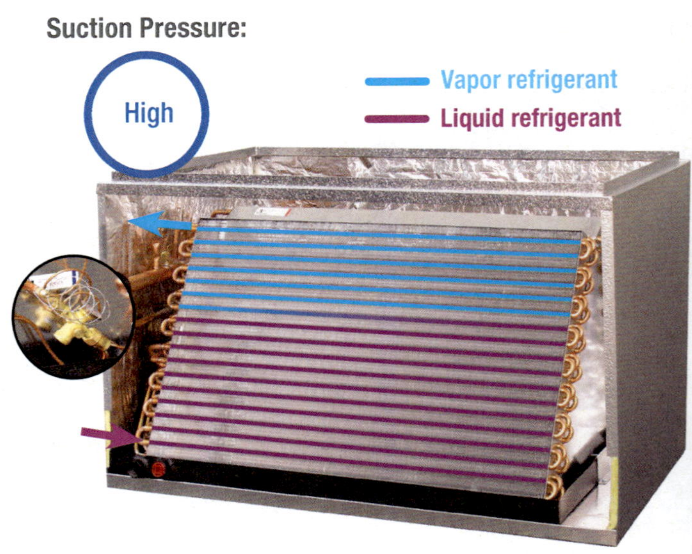

If the heat load on a TXV coil is excessive, the valve will open to try and maintain proper superheat. The suction pressure will be high. If the heat load is high enough, the TXV may not be able to maintain adequate superheat and the superheat will rise to a level above normal.

System Condition:
Flooded Evaporator Coil

Suction Pressure:

(High)

If the TXV over-feeds refrigerant into the circuiting, the suction pressure will be too high and the suction vapor superheat too low. Causes include sensing bulbs that are not insulated, and an improperly functioning TXV.

System Condition:
Starved Evaporator Coil

Suction Pressure:

(High)

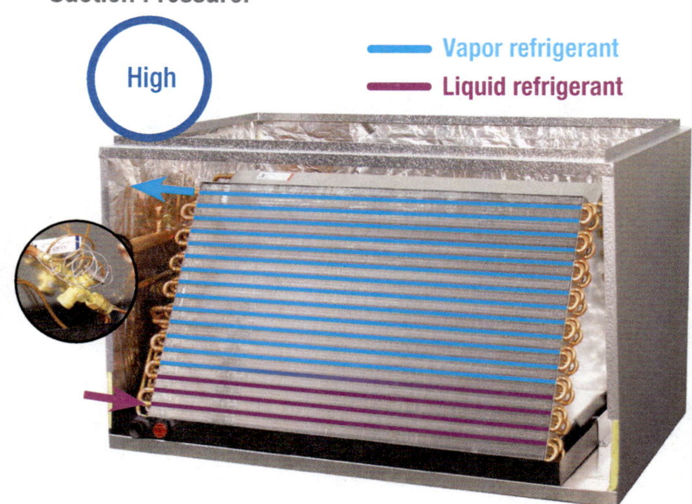

If the coil lacks refrigerant, the suction pressure will be low and the suction vapor superheat high. Causes include low charge, refrigerant circuit restrictions, and failed TXV.

System Condition:
Low Heat Load

Suction Pressure:

(Hunting)

If the heat load on a TXV coil is too low, the TXV will struggle to maintain adequate superheat. The TXV will open and close as it tries to find a balance of refrigerant (hunting). Causes are low air volume, and low heat content in the air.

High Voltage Circuit

Sequence of Operation: Single Phase (Single Speed Condenser Fan)

Single Speed Fan Circuit: Call for Cooling (Refer to the wiring diagram provided below)

When the system calls for cooling, the thermostat makes an electrical connection between the thermostat's subbase terminals R and Y. The field-installed 24VAC low voltage Y wire carries this signal voltage to the indoor unit's low voltage terminal strip Y terminal, which allows the indoor unit to run the indoor blower motor at high speed. At the same time, the thermostat also makes an electrical connection between the subbase terminals R and G. This signal is carried by the field-installed thermostat G wire to the indoor unit's G terminal on the low voltage terminal strip, energizing the fan relay coil, which controls the indoor blower.

From the indoor unit, the Y signal is then carried to the condensing unit, where 24VAC is supplied to the condensing unit's low voltage 24VAC contactor solenoid. When the solenoid coil is energized, it creates an electromagnet, causing the contactor's normally open contacts to close.

When the contacts close, 120VAC from the L1 power supply leg passes to the compressor and its starting circuitry to start the compressor motor. The L2 120VAC potential, which is already present at the compressor's common terminal on systems with a single pole contactor,, places a total potential of 240VAC across the compressor motor circuit. The contactor contacts also supply line voltage to the outdoor fan motor, energizing the motor.

The outdoor fan motor and compressor motor will continue to run as long as there is line voltage to the condensing unit and the contactor solenoid coil remains energized.

The sequence of operation described above can be followed on the schematic diagram shown below. The low voltage circuit is shown in dashed lines on the lower portion of the diagram labeled "A".

Schematic Diagram Reference: Single-Phase Condensing Units Call For Cooling (Single Speed Condenser Fan)

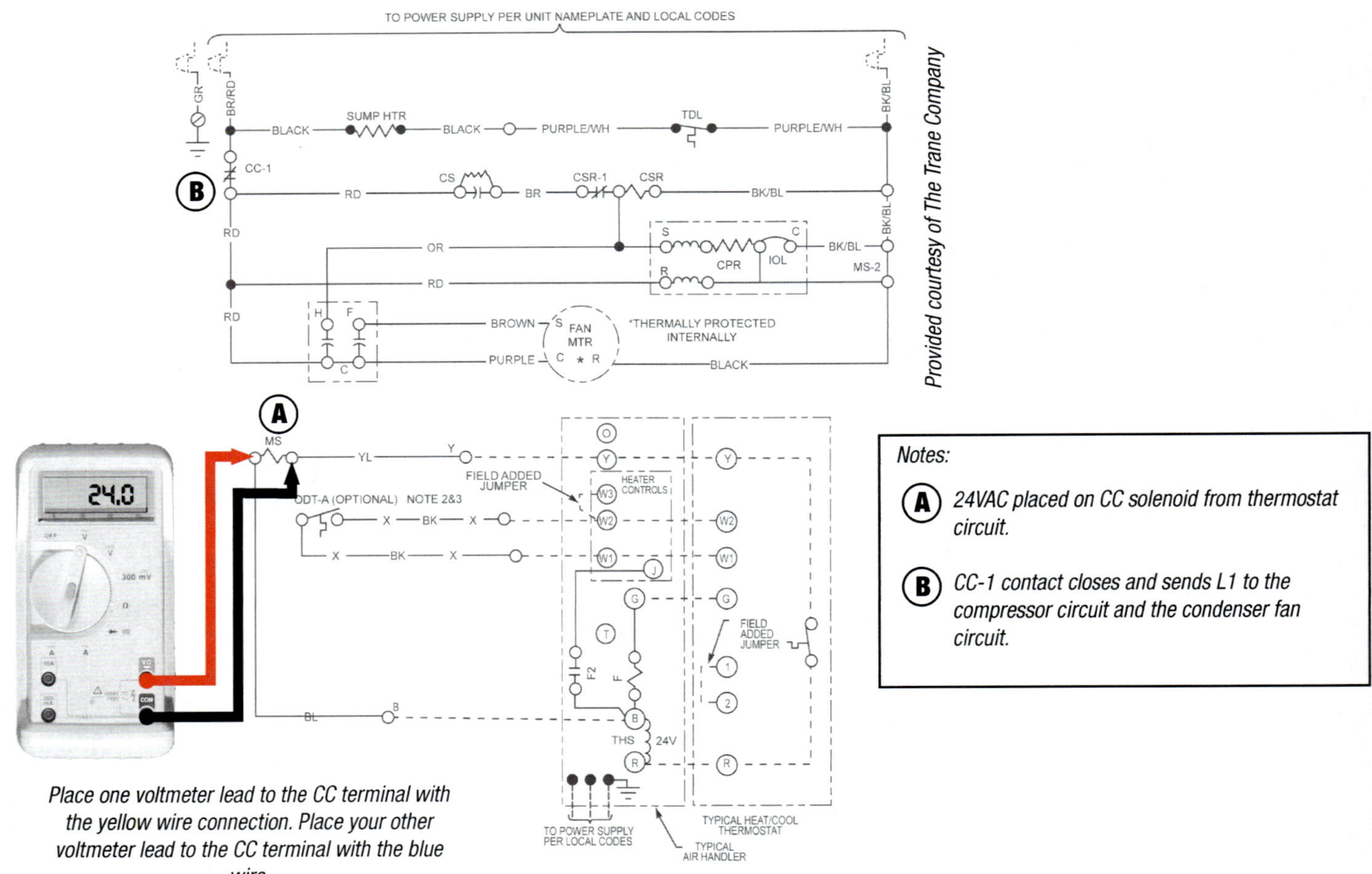

Place one voltmeter lead to the CC terminal with the yellow wire connection. Place your other voltmeter lead to the CC terminal with the blue wire.

Notes:

A 24VAC placed on CC solenoid from thermostat circuit.

B CC-1 contact closes and sends L1 to the compressor circuit and the condenser fan circuit.

High Voltage Circuit

Service Procedure: Checking Line Voltage & Contactor Solenoid Coil

If there is a call for cooling and neither the condenser fan motor nor compressor run, there may be a problem with either the line voltage or the contactor.

Tools:

- Digital Multimeter

STEP 1

Initiate a call for cooling and check to make sure the contactor contacts are pulled in. If not, check for 24VAC across the contactor 24VAC solenoid coil by placing a meter lead to each of the two solenoid coil terminals. If 24VAC is not present, troubleshoot the low voltage circuit to determine the problem. If 24VAC is present, yet the contacts are not pulled in, replace the contactor.

STEP 2

If the contacts are pulled in, place one voltmeter lead to the L1 line terminal of contactor. Place the other lead to the L2 line terminal of contactor. Near 240VAC should be measured. If not, there is a problem with the line voltage supply. Check for an open fuse or tripped circuit breaker. If 240VAC is measured, the problem is not with the line voltage supply.

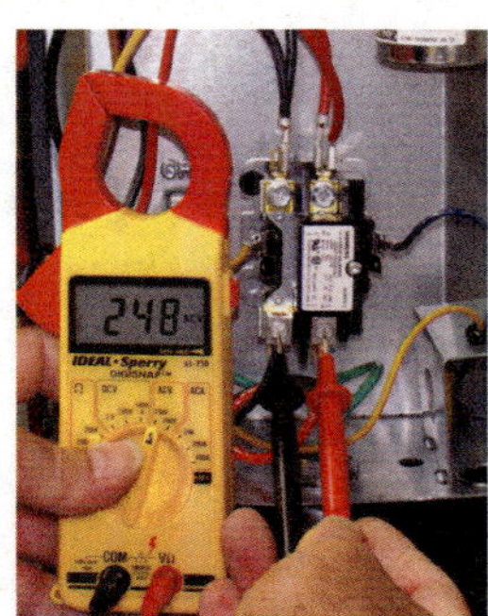

If the circuit breaker has tripped, *DO NOT RESET THE CIRCUIT BREAKER WITHOUT FIRST CHECKING FOR A SHORT TO GROUND IN THE COMPRESSOR MOTOR (See Chapter G: Compressors for compressor troubleshooting).*

Service Procedure: Checking Line Voltage on Contactor Load Terminals

If the contactor solenoid is working properly, and line voltage to the contactor is present, yet the unit fails to energize, the contactor contacts may be damaged, obstructed or pitted.

Tools:

- Multimeter

STEP 1

To check the condition of the contacts, place one voltmeter lead to the contactor line terminal (L1) and the other voltmeter lead to the contactor load terminal (T1).

With power applied to the unit, the contactor solenoid energized and the contacts pulled in, the meter should read 0VAC.

If voltage is read across these two terminals, there is resistance present between the two electrical sets of contacts. When working properly, these contacts should not drop any voltage.

Disconnect power to the unit, properly discharge all capacitors, and inspect the contacts for damage, pitting or obstruction. If the contacts are pitted, replace the contactor. Contactor obstructions are often caused by insect debris. Clean the contactor.

Sump(Crankcase) Heater Circuit

Introduction

Refrigerant will move to the coldest point in a refrigeration system when the system is not calling for compressor operation. If the system is off and the outdoor air temperature is cooler than the temperature of the air at the evaporator coil, the refrigerant migrates to the cold compressor. The liquid refrigerant must be boiled out of the compressor shell prior to the compressor starting. If liquid refrigerant is present during compressor start-up, damage to the compressor could occur.

The sump (crankcase) heater circuit prevents liquid refrigerant from migrating to the compressor shell during the off cycle. The circuit protects the compressor by energizing a high voltage heater that warms the compressor shell during periods when the compressor is not running.

Circuit Operation

This schematic diagram of a typical single-phase condensing unit shows a highlighted area for the sump heater circuit.

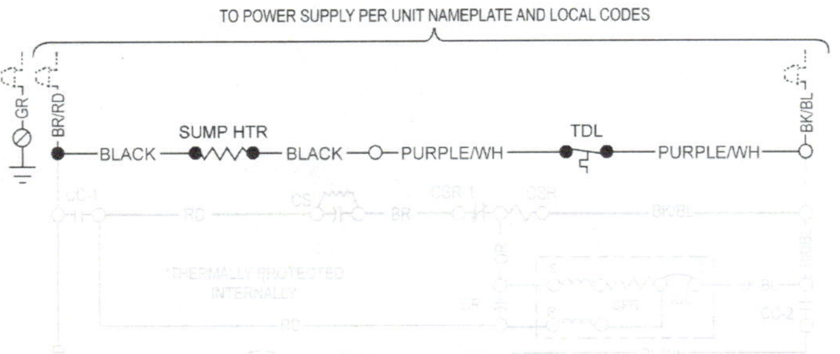

The circuit consists of two components, a heater labeled SUMP HTR, and a normally closed heat activated switch labeled TDL. The TDL switch is a discharge line thermostat that closes its contacts when the compressor is off, and the discharge line is cool, to energize the heater. When the compressor is operating, the discharge line gets hot and the TDL opens its contacts to de-energize the heater. Electrically, the heater and temperature-activated switch are wired in series with each other.

The circuit operates the heater by applying both L1 and L2 power to each side of the heater when switch TDL closes. This means the heater is a 240VAC heater, since the voltage between L1 and L2 is 240VAC. When switch TDL is open, the electrical circuit to the heater is broken and the heater remains off.

The wires that connect the heater and switch are identified by color. The heater has two black wires at either side connected to a wire labeled PURPLE/WH (purple wire with a white stripe). The other side of the switch also has a purple wire with white stripe. This wire is connected to wire BK/BL (black wire with a blue stripe). On the L1 side of the circuit the sump heater's black wire is connected to a wire labeled BR/RD (brown wire with a red stripe).

Notice that L1 is supplied to the circuit via the BR/RD wire. By following the wire color code and schematic diagram shown, the circuit can be easily traced out.

L1 power is applied directly to the left side black heater lead via the BR/RD wire. L2 is supplied to the right hand side of switch TDL via the PURPLE/WH wire. With power present at both L1 and L2, the heater is ready to operate when switch TDL closes.

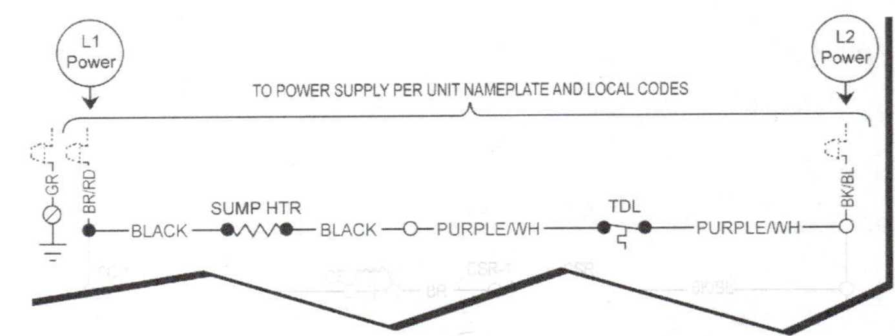

In this diagram the brown wire with red strip can be found supplying L1 power on the left side.

If this switch were to stick and remain in the closed position, the heater would remain on during compressor operation. The compressor would likely overheat and would open its motor winding temperature protector.

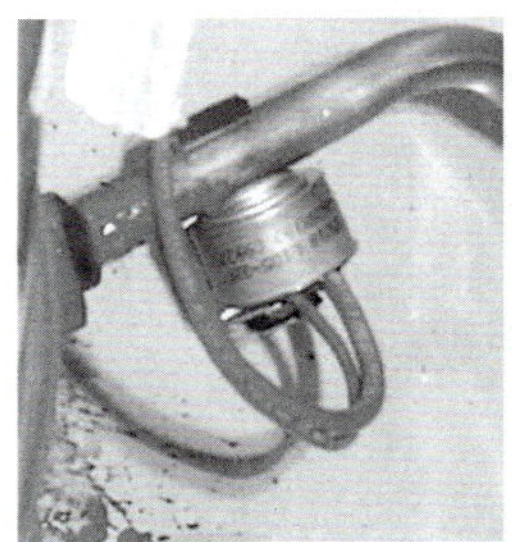

The TDL shown here is used to shut the heater off when the discharge line reaches a specific temperature during compressor operation.

Single-Phase Compressor Circuit

Introduction

A single-phase compressor electrical circuit provides power to the compressor's motor whenever a call for cooling operation is received from the thermostat circuit.

Some motors used in single-phase compressors are Permanent Split Capacitor type motors, or more commonly referred to as PSC motors. The PSC motor is hermetically sealed in the compressor shell. The motor's capacitor is located in the condensing unit's electrical box. The other component used in the circuit is the electrical contactor, which is also located in the unit's electrical control box.

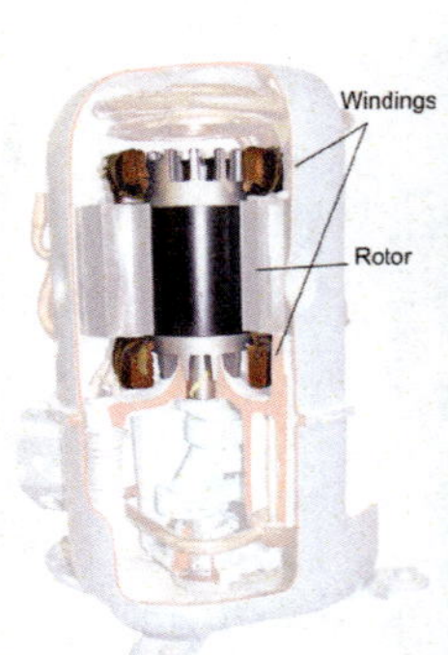

The PSC motor of the compressor is hermetically sealed and is inaccessible to service technicians.

Circuit Operation

Wiring Diagram A (below) illustrates a typical single-phase 240VAC compressor circuit. At the right side of the circuit L2 power is connected to the common terminal of the compressor motor via a connection point at the compressor contactor (single pole). Because this point is tied directly to L2, a voltage of 120VAC to ground will be present anytime there is power supplied to the unit. The color of the wire carrying the L2 voltage is black with a blue stripe. In this circuit, there is no switching of L2 line.

In order for the compressor motor to run, L1 voltage must be supplied to the compressor's run and start terminals. The L1 power leg is connected to the contactor's normally open contacts labeled CC-1 via the BR/RD wire (brown with red stripe) (Diagram B, below). The CC-1 contacts are normally open so they will only close when power is delivered to the contactor's 24VAC electrical solenoid.

The load side of the contacts CC-1 has a red (RD) wire connected to it. This wire carries the L1 signal to one of the capacitor's electrical terminals. From that terminal, another red wire leads to the run winding terminal on the compressor. The other side of the capacitor has an orange wire that connects it to the start winding of the compressor.

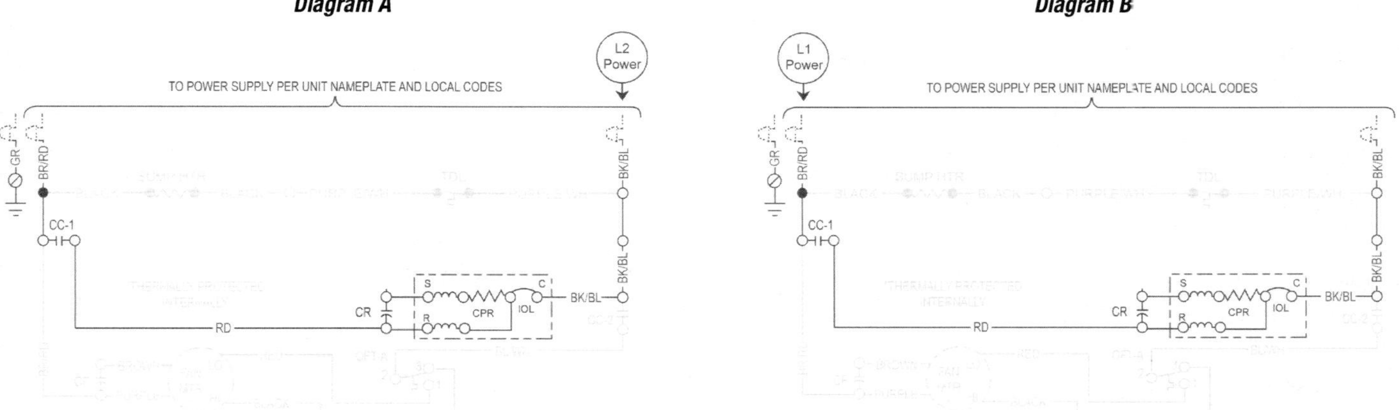

The compressor remains off until the contactor's 24VAC solenoid coil is energized on a call for cooling.

It is important to remember that, when a single pole contactor is used, the voltage to ground at the load side of the CC-1 contact will be 120VAC even when the contactor is not energized. When checking for voltage at the load terminals of the contactor, always check the voltage from both L1 and L2 to confirm there is 240VAC power to the circuit.

On a call for cooling (see diagram below), the thermostat makes a connection between the R and Y terminals. 24VAC is then sent via the low voltage thermostat wire to the condensing unit.

At the condensing unit, the 24VAC signal is received. The yellow wire carries the 24VAC signal to the contactor's 24VAC solenoid coil, which causes the contactor to close its normally open CC contacts.

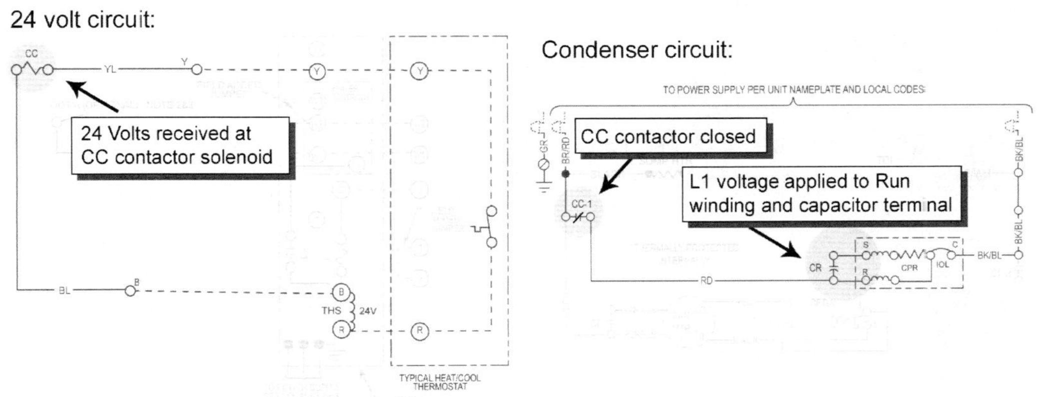

The CC-1 contact completes a circuit from L1 to the run winding terminal and one of the capacitor terminals. At the first one-half cycle of the AC sine wave, the capacitor is charged, and power is supplied to the run winding. The other side of the capacitor completes the start winding circuit to L2. The motor now has 240VAC potential between its common terminal and the run winding.

When the AC sine wave changes polarity in the other half cycle, the potential difference between the plates of the capacitor causes the capacitor to discharge through the start winding and a voltage lag results in the start winding.

The run capacitor charges at the first one-half cycle of the AC sine wave, and then power is applied to the run winding.

The out-of-phase condition between current and voltage in the start winding helps the motor start. Once the motor is running, the run capacitor continues to create a phase displacement that aids in power factor correction during operation.

A voltage is also generated in the start winding at about 75% of full RPM. This voltage is higher than the line voltage. It is called "back electromotive force (EMF)". This voltage will be present at the start terminal of the compressor motor whenever the motor is running.

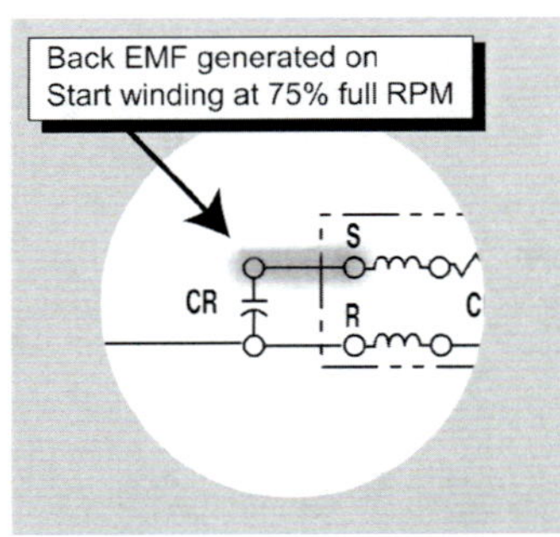

Single-Phase Compressor with Hard Start Kit

Introduction

Hard start kits help single-phase compressors start against potential heavy working loads by generating an out-of-phase current condition between the motor's start and run windings. The circuit operates by temporarily placing a start capacitor in the compressor circuit during the initial start sequence. The circuit removes this capacitor when the motor reaches about 75% of full running RPM. The circuit consists of a high microfarad start capacitor with bleed resistor, and a potential relay with a normally closed set of contacts. Both components are located in the condensing unit's electrical control box.

Pictured here are two different start kits. The one on the left combines the potential relay and start capacitor. The kit on the right has separate units.

Circuit Operation

The wiring diagram below highlights a typical single-phase compressor circuit with hard start kit. The wiring diagram shows that a red (RD) wire carries the L1 signal from the CC-1 contacts to one terminal of a start capacitor. The other terminal of the start capacitor has a brown (BR) wire connected to a normally closed set of contacts labeled CSR-1. An orange (OR) wire on the load side of these contacts carries the voltage to the start winding terminal on the compressor. A bleed resistor discharges the capacitor during off cycle time.

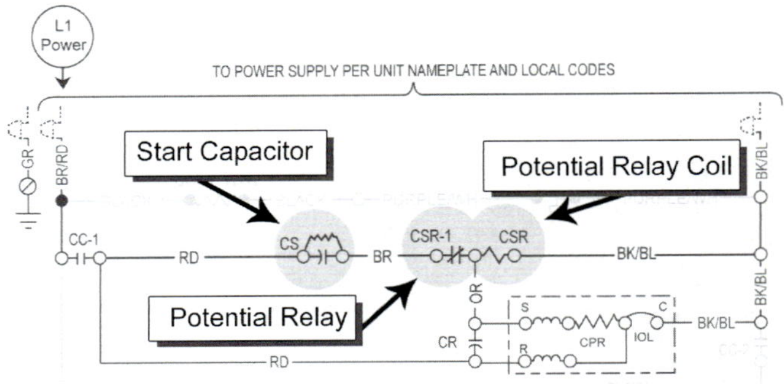

The potential relay coil that opens the CSR-1 contact is wired in series with the start capacitor and CC-1 contacts. The voltage at which the relay coil energizes is higher than 240VAC line voltage, therefore, the relay will not energize when the CC-1 contacts close.

When the CC-1 contacts close, the motor is energized. The normally closed CSR-1 contact allows the start capacitor to create a significant out-of-phase current condition within the motor's internal windings. This out-of-phase condition allows the windings to generate an incredibly strong hold on the motor's rotor. This strong hold on the rotor allows the compressor to overcome the initial starting loads.

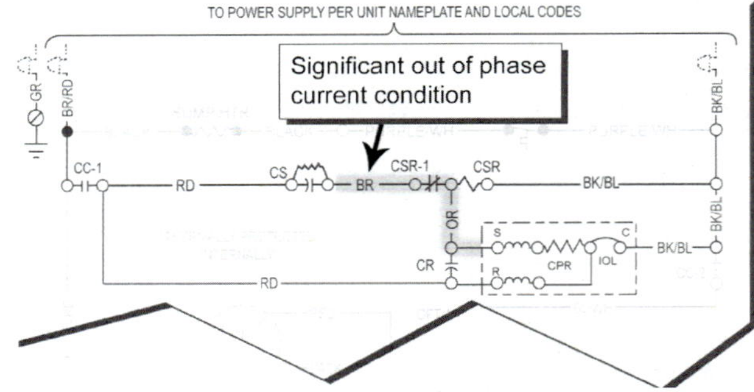

Once the compressor motor near its full RPM, a high counter voltage is generated by the motor's start winding. This voltage travels out of the start winding terminal and places a large voltage on the potential relay's CSR coil via the orange wire. The CSR coil energizes and opens the CSR-1 contacts to remove the start capacitor from the electrical circuit. The relay will remain energized as long as the compressor is running at full RPM.

When the compressor shuts off at the end of a call for operation, the CSR-1 contacts close to place the start capacitor in a position to be active on the next call for cooling.

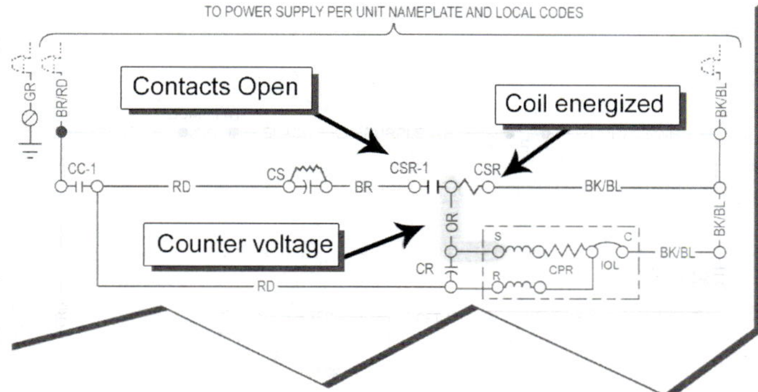

Single-Speed Condenser Fan Circuit

Introduction

The condenser fan circuit energizes the condenser fan motor on a call for cooling operation. The circuit is comprised of a 240VAC PSC single speed motor and run capacitor. The speed of this motor will remain constant regardless of the outdoor air temperature; therefore head pressure problems can be encountered when outdoor air temperature falls below the minimum allowable level. In the case of residential condensing units, the typical minimum allowable run operation for this type of circuit is 55° F.

The single speed condenser fan circuit consists of a 240VAC PSC single speed motor and a run capacitor. The diagram below shows the wiring configuration of the circuit.

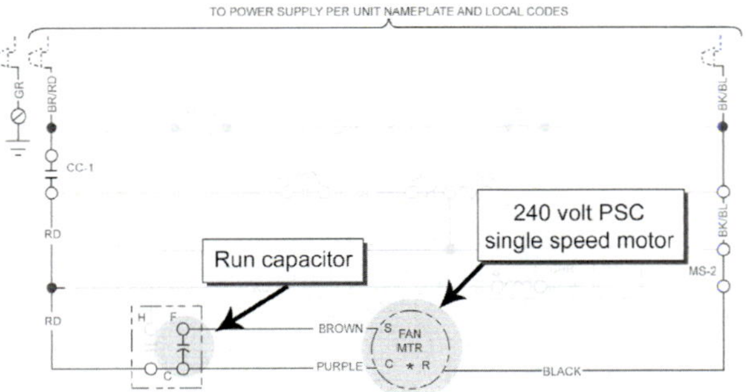

Circuit Operation

The L2 leg of the supply voltage is wired directly to the condenser fan motor's run winding terminal via a connection on the load side terminal of contact L2.

Since the L2 signal is not switched, there will be 120VAC potential to ground from the run terminal of the condenser fan motor whenever line voltage is applied to the condensing unit.

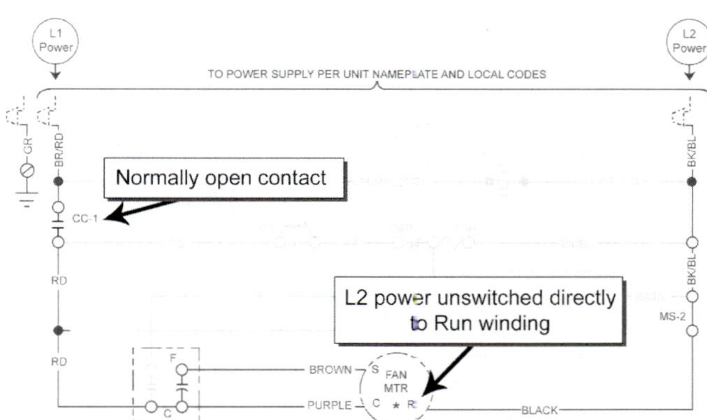

The L1 leg of the supply voltage is delivered to the line terminal of the cooling contactor by the BR/RD (brown and red) wire. A normally open set of contacts labeled CC-1 prevents the L1 voltage from reaching the condenser fan motor when no call for cooling is present.

A red wire (RD) is connected to the load side terminal of the CC-1 contacts. The wire delivers the L1 power to the condenser fan motor's common winding when a call for cooling occurs. When de-energized, some voltage will be present from the start and common windings to ground when there is 120VAC potential applied to the run terminal of the condenser fan motor by the L2 leg.

On a call for cooling, 24VAC is applied to the cooling contactor's 24VAC solenoid coil, which causes the contactor to close its normally open set of contacts. The contactor's CC-1 contact closes and allows the L1 120VAC voltage to pass to the load side terminal.

The red wire carries this voltage to the common winding terminal on the condenser fan motor. With L1 and L2 voltage now at the motor's common and run terminals, the motor starts.

When the call for cooling ends, 24VAC is removed from the cooling contactor's 24VAC solenoid, which causes the contactor to de-energize. The L1 120VAC voltage is removed from the common winding terminal, causing the motor to de-energize.

Single-Speed Condenser Fan Motor Testing

Introduction

In this section, troubleshooting of the single-speed condenser fan circuit will be covered. When there is a problem with the single-speed condenser fan circuit, the problem will likely be in one of four areas: line voltage supply, the contactor, the capacitor, or internal motor problems.

Note: If the compressor has shut off on its IOL due to high pressure conditions, this can be caused by a lack of condenser fan motor operation.

Service Procedure: Checking Condenser Fan Motor Windings

If the no problems have been found at the contactor, the problem is either an open IOL, open motor winding, or bad run capacitor. To determine the problem, disconnect power to the unit and disconnect the wires leading to the condenser fan motor at both the contactor and the condenser fan motor run capacitor. Discharge the condenser fan motor run capacitor.

Tools:
- Digital Multimeter

STEP 1

Using an ohmmeter, check the resistance between the Run and Common wires. A very small resistance should be measured. If infinite resistance is measured, there is an open circuit between the Run and Common terminal.

STEP 2

Next, measure the resistance between the brown wire and the purple wire. This is the Start winding. A resistance that is slightly higher than the Run winding should be measured. If infinite resistance is measured, both of the motor's internal windings have an open circuit to Common. This indicates the motor's internal overload is open. Allow time for the IOL to cool down and reset. If the IOL does not reset, replace the motor.

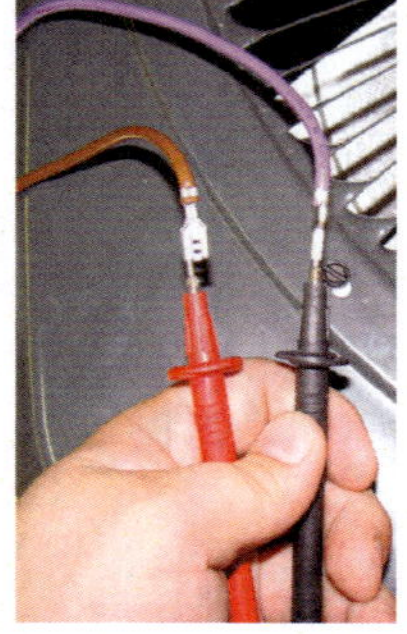

If only one winding has infinite resistance to common, that winding is open. If this is the case, the motor must be replaced.

If the motor windings check out OK, proceed to check the motor for internal grounding.

Service Procedure: Checking for Grounded Motor Windings

To check for internal grounding of the motor windings, set your multimeter to the OHM function.

Tools:
- Digital Multimeter

STEP 1

Place one meter lead to the Common wire and the other meter lead to the motor housing.

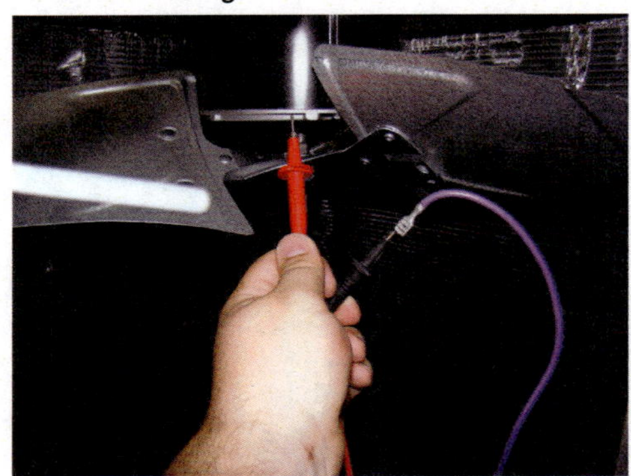

Infinite resistance should be measured. If a measurable resistance is read, the motor is grounded from the Common terminal. Replace the motor. If it is OK, proceed to testing the other windings.

STEP 2

Place one meter lead to the Run winding terminal and the other motor lead to the motor housing. Again, infinite resistance should be measured.

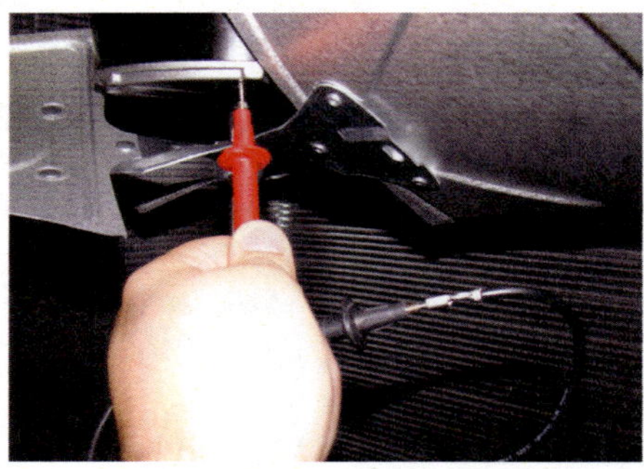

If a measurable resistance is read, the motor is grounded from the Run terminal. Replace the motor. If it is OK, proceed to testing the other winding.

STEP 3

Place one meter lead to the Start wire and the other lead to the motor housing. Again, infinite resistance should be measured. If a measurable resistance is read, the motor is grounded from the Start terminal. Replace the motor.

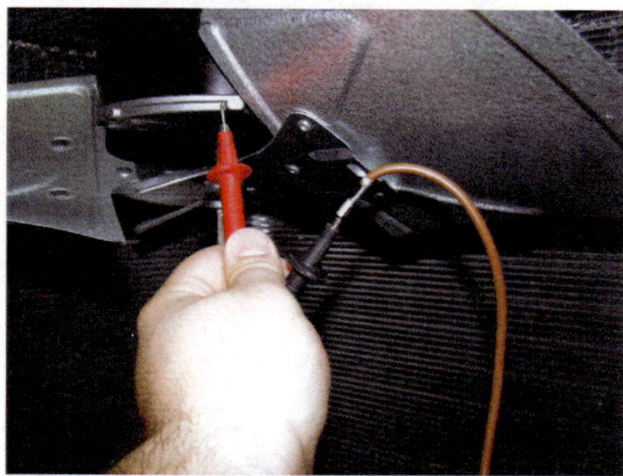

If all of the motor windings check out OK, test the run capacitor.

Service Procedure: Testing The Run Capacitor

Set your multimeter to the OHM function to perform this test. Make sure the capacitor is discharged.

Tools:

- Digital Multimeter

STEP 1

Place one meter lead to each capacitor terminal. The meter should slowly register resistance and then return to infinite ohms reading as the ohmmeter battery charges and discharges the capacitor.

If the meter fails to respond, switch your ohmmeter leads and try again. If the meter still does not show charging and discharging of the capacitor, or if the meter reads a steady resistance, replace the run capacitor. If the meter does respond properly, the capacitor is good.

Note: Most modern digital multimeters have a capacitor testing function, which can be used to measure the actual energy-storing capability of a capacity (microfarads).

Compressors

Continued....

Compressor Introduction

The compressor has two functions in the AC system, to move refrigerant through the system by creating a pressure differential, and to convert low pressure, low temperature refrigerant vapor into a high pressure, high temperature vapor. By raising the temperature of the refrigerant, heat absorbed from the indoor air via the evaporator coil can be easily transferred to the outside air via the condenser coil.

Service technicians must understand how compressor performance is affected by external conditions if they want to become successful in the field.

Suction Pressure and Compressor Capacity

Fundamentally, a compressor is nothing more than a vapor refrigerant pump.

The refrigerant suction vapor that enters the compressor varies in density based on suction pressure. Density is the weight, in pounds, of one cubic foot of a substance. When suction pressure is low, the refrigerant suction vapor is less dense. When the suction pressure is high, the refrigerant suction vapor is more dense. The amount of refrigerant vapor the compressor pumps, in pounds, will change with the suction pressure and corresponding refrigerant density. Compressors are fixed-volume pumps, so changes in compressor capacity are due to the changing density of refrigerant.

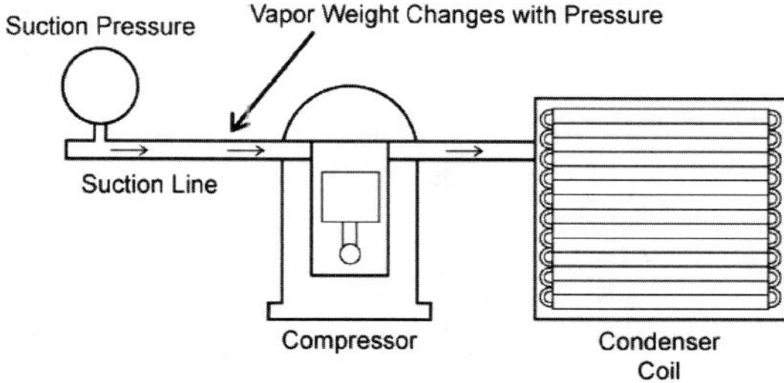

When suction pressure is high, a large amount of refrigerant is being pumped into the condenser coil. The heat transfer of the system will increase since each pound of refrigerant being moved through the condenser coil contains a large amount of heat absorbed into the refrigerant in the evaporator coil.

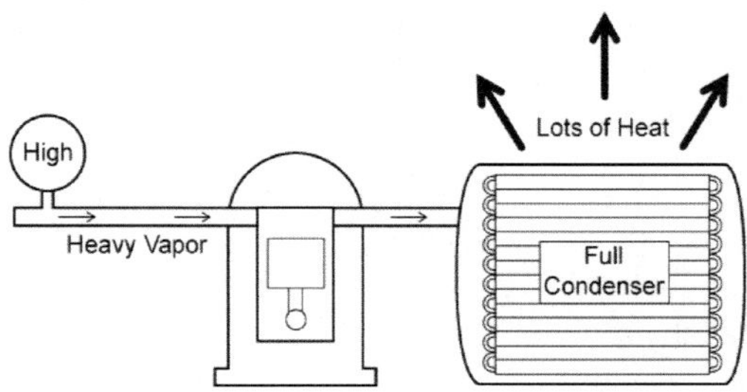

If, for some reason, the suction pressure at the compressor inlet were to fall (such as when the indoor air filter is dirty), the resulting lower pressure will cause the density of the refrigerant vapor entering the compressor to drop. At a lower density, the number of pounds of refrigerant pumped per unit time drops as well. As a result, less heat is being pumped through the condenser coil.

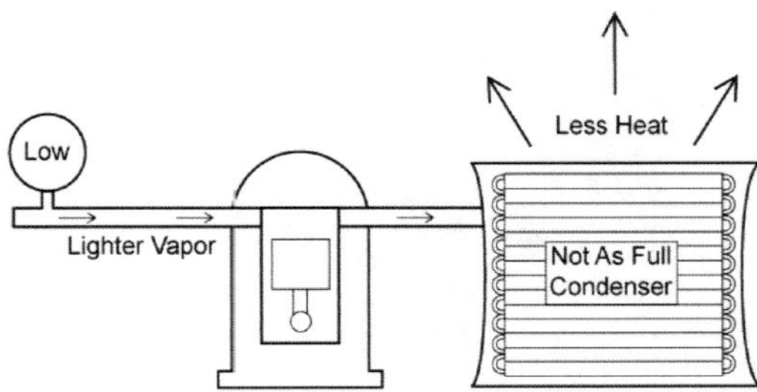

Suction Vapor Density of Refrigerants

The weight property of a refrigerant is called its Suction Vapor Density. Suction vapor density is measured in pounds (#) per cubic foot of vapor. Refrigerant manufacturers publish charts that list the refrigerant suction vapor density at given suction pressures. These charts are available for both saturated refrigerants and superheated suction vapors. For simplicity, we will discuss saturated refrigerant properties in this chapter.

In the example R-22 saturated properties table below, 40°F saturated temperature is highlighted in yellow. This temperature occurs at a gauge pressure near 69 PSIG. From the table, it is determined that vapor entering the compressor at a suction pressure of 69 PSIG will have a density of about 1.521 pounds per cubic foot.

In the same table, 50°F saturated temperature is now highlighted in blue. The corresponding suction pressure at this temperature is near 84 PSIG. From the chart, it can be seen that the density of R-22 is now much greater at almost 1.8 pounds per cubic foot.

Saturation Properties of R-22

Saturation Temp	Gauge Pressure PSIG	Suction Vapor Density (lb/cu.ft.)	Enthalpy Liquid (Btus/lb)	Enthalpy Vapor (Btus/lb)
26	49.95	1.191	17.5	106.93
28	52.39	1.234	18.05	107.11
30	54.9	1.279	18.61	107.28
32	57.47	1.324	19.17	107.46
34	60.12	1.372	19.73	107.63
36	62.84	1.42	20.29	107.8
38	65.64	1.47	20.86	107.97
40	68.51	1.521	21.42	108.14
42	71.46	1.573	21.99	108.31
44	74.48	1.627	22.56	108.47
46	77.58	1.683	23.13	108.63
48	80.77	1.74	23.7	108.8
50	84.03	1.798	24.28	108.95
52	87.38	1.859	24.85	109.11
54	90.81	1.92	25.43	109.26
56	94.32	1.984	26.01	109.42
58	97.93	2.049	26.59	109.56
60	101.62	2.115	27.17	109.71

BTUs in a Pound of Suction Vapor

Refer to the far right column of the example table above. This column, highlighted in red, is called suction vapor enthalpy. It is the rating of how many BTUs are contained in each pound of vapor refrigerant. At 40°F, there are about 108 BTU in each pound of vapor and at 50°F, there are about 109 BTU in each pound of vapor.

Compressor Displacement in CFM

Compressor displacement is a rating of how many cubic feet of vapor the compressor can pump per minute (CFM). If a 10 CFM compressor were pumping R-22 refrigerant with a suction pressure of 85 PSIG (Fig 1), the refrigerant being compressed would have a density of 1.8 pounds per cubic foot. This means a 10 CFM compressor would pump about 18 pounds of refrigerant per minute at 85 PSIG. The same compressor, operating instead at a suction pressure of 68 PSIG (Fig 2), would be compressing vapor with a density of near 1.5 pounds per cubic foot. This compressor would only pump 15 pounds of refrigerant per minute at 68 PSIG. This lower refrigerant flow rate will cause a reduction in system capacity.

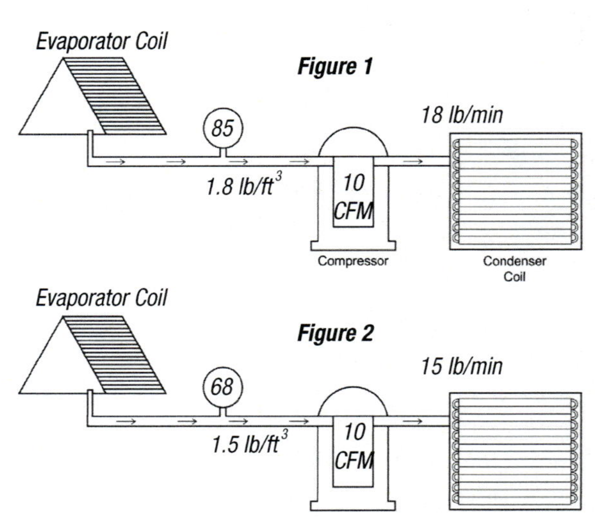

Capacity Loss Relationship to Suction Pressure: R-22 and R-410A

The amount of capacity lost by a drop in suction pressure is a 1% drop for every 1 pound of suction pressure loss for units using R-22. Units using R-410A refrigerant lose 0.6% of capacity for every pound of suction pressure loss.

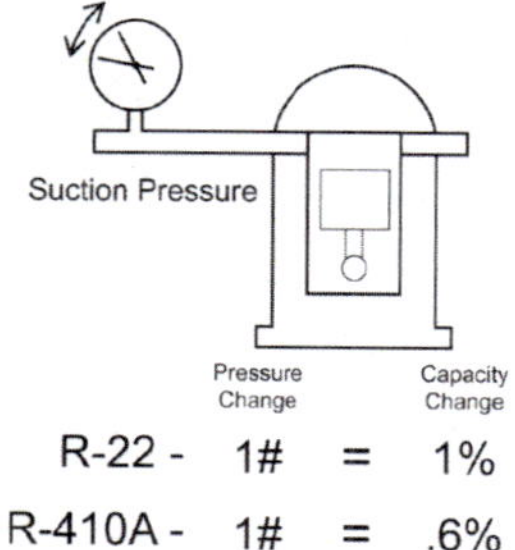

Important note: Compressors should not operate with a liquid/ vapor refrigerant mix entering them. Avoid conditions that will cause the suction pressure to drop to saturated refrigerant levels.

Causes of Low Suction Pressure

There are three main causes of low suction pressure: low heat load at the evaporator coil, suction line piping pressure drop, and lack of refrigerant entering the evaporator circuit. (See Chapter B: Refrigeration Circuit Diagnostics).

Discharge Pressure and its Effect on Compressor Capacity

Discharge vapor pressure also affects the capacity of the compressor. High discharge pressure reduces the capacity of the compressor. This capacity loss is due to the fact that all compressors have an area of clearance that allows liquid oil through the compression process without causing stress to the compressor. This area of clearance is called clearance volume.

At the top of the compression stroke, some hot refrigerant vapor will remain in the clearance volume. This vapor is called clearance volume gas. On the suction stroke, this remaining gas will expand and occupy a portion of the compressor cylinder, effectively reducing the available compressor cylinder space and reducing the amount of refrigerant vapor that will be drawn in from the suction line.

Detail

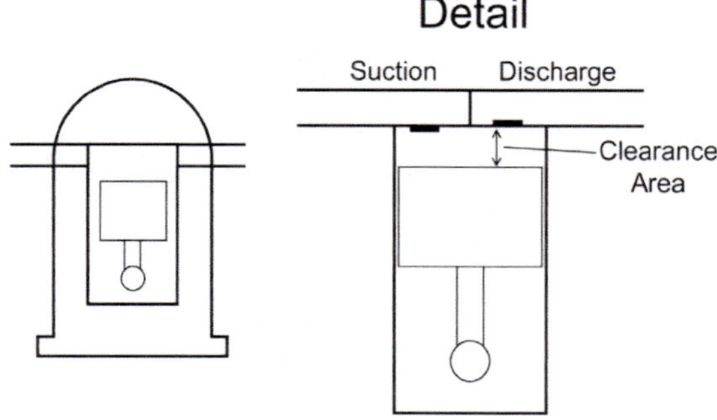

The higher the discharge pressure rises, the denser the clearance gas becomes, and will take up more area in the cylinder during the suction stroke.

When discharge pressure increases, the condensing temperature in the condenser coil increases. The higher condensing temperature will increase the temperature (heat content) of the liquid refrigerant that exits the condenser coil. The additional heat content in the liquid refrigerant is heat the condenser could not reject to the outdoor air, and is now sent to the evaporator coil. The additional heat in the liquid refrigerant reduces the capacity of the system.

Causes of High Discharge Pressure

Causes of high discharge gas pressure include high outdoor air temperature, overcharge, dirty outdoor condenser coil, non condensable gases. (See Chapter B: Refrigeration Circuit Diagnostics).

Performance Data Tables

Published compressor performance information confirms the effect of discharge pressure on compressor capacity. The example performance data table below lists the cooling capacity of a compressor based on the saturation temperature of the evaporator coil, and the condensing temperature of the condenser coil. We will use this table on the following pages, to prove this an example of discharge pressure vs. compressor capacity.

Performance Information

Evap Temp		Condensing Temperature						
		80	90	95	100	110	120	130
35	BTUH	19194	17715	16965	16211	14700	13195	11712
	WATTS	1046	1146	1191	1233	1307	1369	1418
	AMPS	3.62	3.81	3.9	3.98	4.14	4.28	4.4
	LB/HR	232.6	222.6	217.2	211.7	200	187.6	174.5
40	BTUH	21708	20095	19278	18456	16807	15161	13535
	WATTS	1047	1164	1217	1267	1355	1430	1492
	AMPS	3.62	3.84	3.95	4.04	4.23	4.4	4.55
	LB/HR	261.6	251	245.4	239.5	227.2	214.1	200.2
45	BTUH	24381	22632	21745	20854	19062	17273	15500
	WATTS	1037	1172	1234	1292	1395	1485	1560
	AMPS	3.61	3.86	3.98	4.09	4.31	4.51	4.7
	LB/HR	292.1	28.1	275.2	269.1	256.1	242.3	227.7
50	BTUH	27210	25321	24363	23400	21463	19525	17602
	WATTS	1015	1169	1239	1305	1426	1530	1620
	AMPS	3.59	3.87	4	4.13	4.38	4.62	4.84
	LB/HR	324.2	312.7	306.6	300.2	286.6	272.2	256.9
55	BTUH	30192	28158	27128	26091	24005	21916	19838
	WATTS	979	1152	1232	1308	1445	1566	1671
	AMPS	3.55	3.86	4.01	4.16	4.44	4.72	4.98
	LB/HR	357.7	345.8	339.5	332.8	318.7	303.6	287.7

Table courtesy of Tecumseh Cool Products

For the first part of this example (Table 1), we will start with an R-22 system operating with a suction pressure of 69 PSIG. This suction pressure will be found to indicate an evaporator refrigerant saturation temperature of 40°F. The discharge pressure is 200 PSIG, indicating a condenser refrigerant saturation temperature of 100°F. Locate 40° on the far left "Evap Temp" column. Next, find the 100° column under "Condensing Temperature" at the top of the table. Locate the point on the table where the BTUH for 40° intersects the 100° column. Here, the table tells us that this compressor has a capacity of 18,456 BTUH at this suction and discharge pressure combination.

Table 1

Performance Information

Evap Temp	Condensing Temperature						
	80	90	95	100	110	120	130
35				16211			
				1233			
				3.98			
				211.7			
BTUH	21708	20095	19278	18456	16807	15161	13535
40				1267			
				4.04			
				239.5			
45				20854			
				1292			
				4.09			
				269.1			
50				23400			
				1305			
				4.13			
				300.2			
55				26091			
				1308			
				4.16			
				332.8			

For the second part of the example (Table 2), we are going to raise the suction pressure by placing a larger evaporator coil on the compressor. The discharge pressure and saturation temperature will remain the same (200 PSIG/100°F), but the suction pressure is now at 85 PSIG, indicating a 50°F saturation temperature. Locate the point on the chart where the BTUH for 50°F Evap Temp intersects the 100°F Condensing Temperature column. Note the capacity has now gone up to 23,400 BTUH from 18,456 BTUH in the previous part of this example. Because suction vapor density increases with increased pressure, the suction pressure increase from 69 to 85 PSIG increased the capacity due to the increased vapor density entering the compressor cylinder.

Table 2

Performance Information

Evap Temp	Condensing Temperature						
	80	90	95	100	110	120	130
35				16211			
				1233			
				3.98			
				211.7			
40				18456			
				1267			
				4.04			
				239.5			
45				20854			
				1292			
				4.09			
				269.1			
BTUH	27210	25321	24363	23400	21463	19525	17602
50				1305			
				4.13			
				300.2			
55				26091			
				1308			
				4.16			
				332.8			

For the third part of this example (Table 3), we will look at the discharge pressure affect on the compressor. We will keep the suction pressure at 85 PSIG with a saturation temperature of 50°F. We will overcharge the outdoor unit to raise the discharge pressure to 300 PSIG . This will cause the condenser saturation temperature to rise to 130°F. Locate the point where the 130°F Condensing Temperature column intersects the BTUH for 50°F Evap Temp. Note the cooling capacity has been reduced to 17,602 BTUH. This loss of capacity had nothing to do with the suction vapor density. The higher discharge pressure caused an increase in the amount of clearance volume gas, thus reducing the volume of suction vapor the compressor could draw into the compressor cylinder. The net effect is an increase in liquid line heat causing an adverse affect on the performance of the system.

It's likely that at some point, a fellow technician or instructor may have told you (or will tell you) that high suction pressure and low discharge pressure are an indication of an inefficient compressor. In fact, in almost all cases this is the opposite of the truth. Let's use the same table to explain why...

For this example (Table 4), Let's consider a system operating with 93 PSIG suction pressure (55°F evaporator saturation temperature) and a discharge pressure at a very low 147 PSIG (80°F condenser saturation temperature). When we locate the intersect of these temperatures on the chart, we find that the capacity is all the way up to 30,192 BTUH, and the Watts at the lowest on the chart. This means the compressor is operating at its maximum EER rating! Do not condemn a compressor as inefficient when operating at these pressures, as it is actually performing at maximum rated performance.

A Note about Evaporator Saturation Temperature

Cold evaporator coils can better dehumidify air in a home by condensing indoor air water vapor on the surface of the evaporator coil, where it collects in the drain pan and is disposed of via the condensate drain system. This is why residential comfort cooling systems operate at evaporator refrigerant saturation temperatures that average between 35°F and 55°F. Even though a 55°F evaporator temperature might operate at a high efficiency, it may not produce comfort. The coil's relatively high surface temperature will impede its ability to remove humidity from the home's air.

High evaporator coil operating saturation temperatures are due to high suction pressures. Causes of high suction pressure include: very large evaporator coils that absorb large amounts of heat; oversized metering pistons; high levels of indoor heat or air volume; new system start up with large levels of heat in the air, and overcharged piston equipped systems (See Chapter E: Refrigeration Circuit Diagnostics).

Table 3

Performance Information

Evap Temp		80	90	95	100	110	120	130
35								11712
35								1418
35								4.4
35								174.5
40								13535
40								1492
40								4.55
40								200.2
45								15500
45								1560
45								4.7
45								227.7
50	BTUH	27210	25321	24363	23400	21463	19525	17602
50								1620
50								4.84
50								256.9
55								19838
55								1671
55								4.98
55								287.7

Table 4

Performance Information

Evap Temp		80	90	95	100	110	120	130
35	BTUH	19194						
35	WATTS	1046						
35	AMPS	3.62						
35	LB/HR	232.6						
40	BTUH	21708						
40	WATTS	1047						
40	AMPS	3.62						
40	LB/HR	261.6						
45	BTUH	24381						
45	WATTS	1037						
45	AMPS	3.61						
45	LB/HR	292.1						
50	BTUH	27210						
50	WATTS	1015						
50	AMPS	3.59						
50	LB/HR	324.2						
55	BTUH	30192	28158	27128	26091	24005	21916	19838
55	WATTS	979						
55	AMPS	3.55						
55	LB/HR	357.7						

How a Mismatched System Affects Compressor & System Capacity

Simply stated, compressor capacity is greatly reduced when there is a mismatch between the indoor evaporator coil and outdoor condensing unit. System nameplate capacity is achieved when the indoor and outdoor units are properly selected and sized to work with seamlessly with each other.

Older evaporator coils were smaller than today's higher efficiency models. To gain efficiency, modern, high-efficiency air-conditioning systems use larger evaporator coils. The purpose of this is to increase the density of the refrigerant suction vapor by creating higher suction pressures. By raising the density of the refrigerant entering the compressor cylinder, the size of the compressor could be reduced to increase the energy efficiency of the system. Yes, today's 3-ton condensing units do not use a 3-ton compressor. Take a look at the compressor in that 3-ton model, it is likely a 2-½ ton compressor!

A contractor, not realizing that newer systems are matched to larger evaporator coils, may make the mistake of selling a consumer a replacement outdoor unit while leaving the old evaporator coil in place. The old evaporator coil cannot generate a suction pressure high enough to get the compressor to pump at its rated capacity. Remember, for every pound of pressure lost, system capacity is reduced by 1% with R-22 units (0.6% with R-410A units).

When the system is installed, there is yet another problem...the older evaporator coil metering device is sized for the higher discharge pressure of the older, less efficient model. Newer systems run with lower discharge pressures than older models. As a result, the new system will have to be overcharged to get the discharge pressure high enough to properly meter refrigerant through the metering device, which was originally sized for the old unit. The result is abnormally high discharge pressure, which will reduce the capacity of the system and lower the suction pressure. This further reduces the capacity of the system.

The consumer will complain that on hot days the new system does not cool as well as the old one did. The consumer will be absolutely correct.

Compressor Introduction Summary

- Higher suction pressure raises system capacity due to a greater amount of refrigerant being moved by the compressor per unit time.

- Low suction pressure reduces the density of the suction vapor and therefore reduces system capacity.

- There are three primary causes of low suction pressure: low heat load at the evaporator, suction line piping pressure loss, and a lack of refrigerant entering the evaporator circuit.

- All compressors have clearance volume for liquid oil to clear during the compression stroke.

- Hot gas is trapped in the clearance volume. This gas reduces the effective size of the compressor cylinder.

- Avoid system conditions that can cause low suction pressure and high discharge pressure.

- The best system capacity occurs when the suction pressure is high and the discharge pressure is low.

Why Compressors Fail Mechanically

Whether the compressor is a scroll or reciprocating model, causes of mechanical failure are similar for both compressor types. Mechanical failure can typically be isolated to damage from liquid refrigerant entering the compressor, moisture, and high discharge gas temperatures due to abnormal refrigeration circuit operation. The potential for each problem to cause damage can be minimized by proper installation and service practices.

Liquid Migration into the Compressor Oil during Shut Down Periods

Liquid refrigerant entering the compressor will cause lubrication problems. Because liquid refrigerant will migrate to the point of lowest pressure and temperature, which is usually the compressor oil, liquid refrigerant enters the compressor during off cycle periods. When the compressor is started, the liquid refrigerant that has migrated into the compressor oil vaporizes, which causes the refrigerant oil to foam. The oil foam is then carried out of the compressor along with the vapor refrigerant. The oil enters the refrigeration circuit coils and tubing. The compressor is now operating with a lack of oil reaching its internal weight bearing surfaces. Bearings are scored, and the compressor motor must work harder. This additional work results in an increase in the amperage draw of the compressor. Additionally, the bearing damage may cause compressor starting problems.

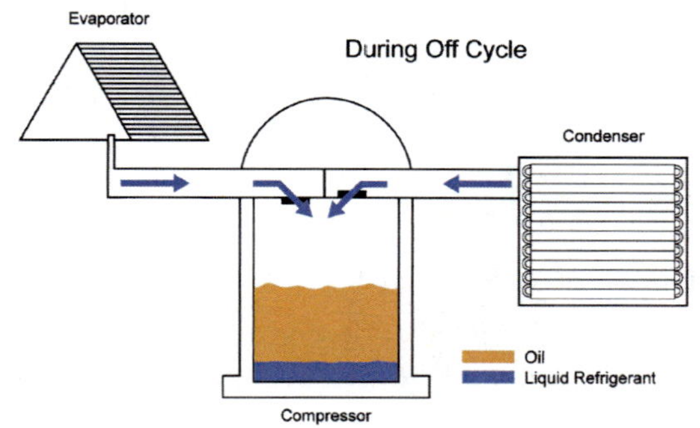

During off cycles, the liquid refrigerant will move from areas of higher pressure, the Evaporator and Condenser, to areas of lower pressure, namely the compressor.

The amount of liquid refrigerant migration increases as the charge level of the system is increased. High-efficiency systems with very large condenser coils, often have larger than expected refrigerant charges. These larger charges have the potential to increase the amount of liquid migration. Long line set applications will also result in higher charge levels.

To minimize liquid migration, a crankcase heater should be installed on the compressor. Crankcase heaters will keep the compressor oil warm and prevent excessive liquid migration from occurring during periods where the system is off. Most air conditioning system manufacturers equip their systems with a factory-installed crankcase heater.

This well type heater should be installed in the heater well under the compressor.

The recommended level of charge requiring a crankcase heater to be installed is surprisingly low. In some cases, scroll compressors -due to their small shell- require a crankcase heater installed when as little as 7 lbs. of refrigerant charge is present the system. To determine the charge level where a crankcase heater is required, consult the manufacturer of the system. If in doubt, install a crankcase heater on the system as a preventative measure.

Liquid Flood Back during System Operation

Liquid refrigerant may leave the evaporator when the system heat load is low, when the system is overcharged, or when the metering device is overfeeding the evaporator. When liquid refrigerant leaves the evaporator coil, the suction line carries it back to the compressor. If liquid refrigerant is present in the suction vapor, the liquid refrigerant will be vaporized by the compressor motor windings. Like in the case of liquid migration, the vaporizing refrigerant will foam the refrigerant oil. The refrigerant oil will travel with the discharge vapor into the condenser coil and through the refrigeration circuit. As the oil travels through the system, the compressor operates without proper lubrication, resulting in potential bearing damage inside of the compressor.

Image Courtesy of Trane
If there is liquid refrigerant present in the suction vapor, it will be vaporized by the compressor motor windings and will foam the refrigerant oil.

Oil Breakdown

Compressor discharge temperature will rise when the indoor heat load is high, when refrigerant circuit restrictions are present, or when the evaporator circuit is lacking refrigerant. As temperature rises, the refrigerant oil in the compressor may begin to break down and form varnish and sludge within the system. In extreme cases, the sludge can circulate in the system and cause mechanical failure of internal compressor components such as suction valves.

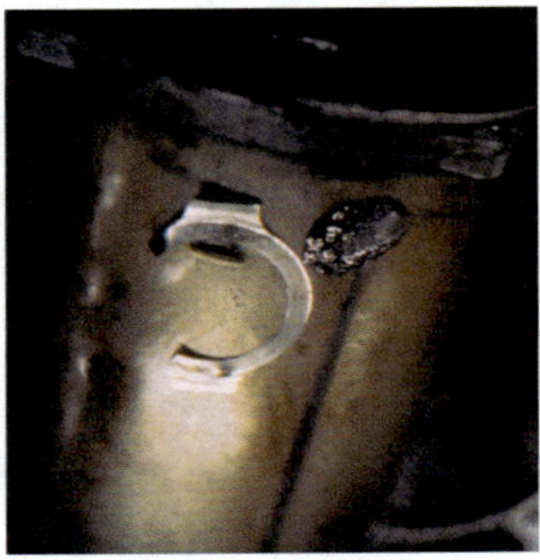

The sludge produced from oil break down can cause damage to internal compressor components such as the broken suction valve shown here.

Image courtesy of Trane

Older scroll compressors feature external temperature limit switches that monitor the discharge temperature of the compressor. If the temperature approaches 300°F, these limit switches will open, stopping compressor operation. Newer scroll models feature an internal temperature hot gas bypass limit that will trip the compressor motor winding overload protector in the event of high discharge temperature. Compressor discharge temperatures should be kept below 250°F to help prevent sludge formation.

Moisture

Moisture in the refrigeration system will react with the refrigerant and oil to form acids. The amount of moisture that will begin to form acid is quite low. For example, with R-410A systems using POE oil, acid may form when there are only 75 Parts Per Million (PPM) of moisture in the system. Some moisture indicating sight glasses will begin to show caution color warnings at this level.

Moisture Meter and Sight Glass

Sight glasses are used to monitor moisture level in refrigerant systems.

If acid forms in the refrigeration circuit, driers can be used to remove some of it. However, once a system has reached a high level of acidity, corrosion will often occur. The acid will corrode copper surfaces, causing copper particles to find their way to surfaces within the compressor, such as the bearings. The hot bearings can cause the copper to bond to their surface. This is called copper plating. The copper plating reduces the effective area for lubricants, causing an increase in friction. The amperage draw of the compressor and the starting load will increase.

Image Courtesy of Trane
Copper plating can occur on weight bearing surfaces if the acidity of the refrigerant oil is too high.

Compressor Mechanical Failure Prevention

To prevent or reduce the potential for mechanical damage to internal compressor components, the following practices should be followed:

- Install crankcase heaters on units when not factory installed. This is especially true for long line set applications and high efficiency systems.
- Measure and adjust air volume to match factory charging chart requirements. Make sure evaporator surfaces are clean.
- Charge for proper superheat levels to prevent liquid floodback. Superheat should also be kept to a minimum to prevent oil breakdown due to excessive compressor discharge temperature.
- Make sure metering piston sizes are correct to prevent abnormally high discharge temperatures.
- Evacuate all systems to 500 microns when opened for installation or service.
- Install and replace driers when installing new equipment or opening existing equipment for service.

Unique Features of Scroll Compressors and Reciprocating Compressors

First of all, it must be pointed out that internal components on hermetically-sealed compressors cannot be acessed. All diagnostic information must be gathered from the motor winding terminals and from the refrigerant gauge ports. From the gauge ports, the system pressures can be monitored and the oil can be tested. Beyond that, the compressor components are beyond reach. Therefore, diagnostic decisions must be made based on the external information gathered at the electrical terminals and pressure ports.

Reciprocating Compressor Process

Reciprocating compressors have one or more pistons that move up and down within one or more cylinders.

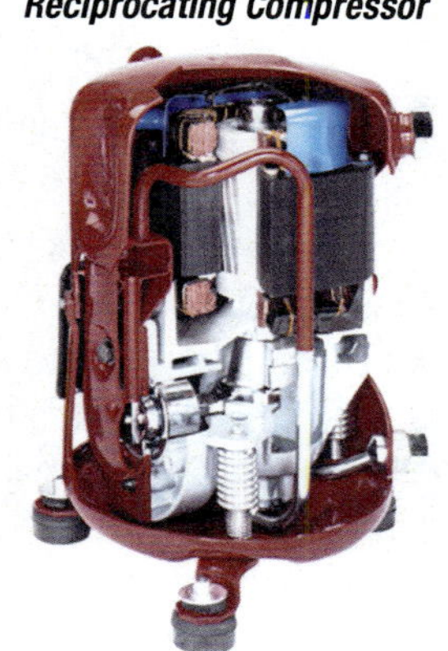

Reciprocating Compressor

At the top of the cylinder are tensioned valves that open and close based on the pressure difference across the valve as well as whether the piston is moving up or down within the cylinder. These valves are called the suction and discharge valves. When the piston travels downward, the suction valve is drawn open to allow refrigerant to enter the cylinder. When the piston travels upward, the suction valve is forced closed and the discharge valve opens to allow the refrigerant to escape into the discharge line of the compressor.

Image courtesy of Trane

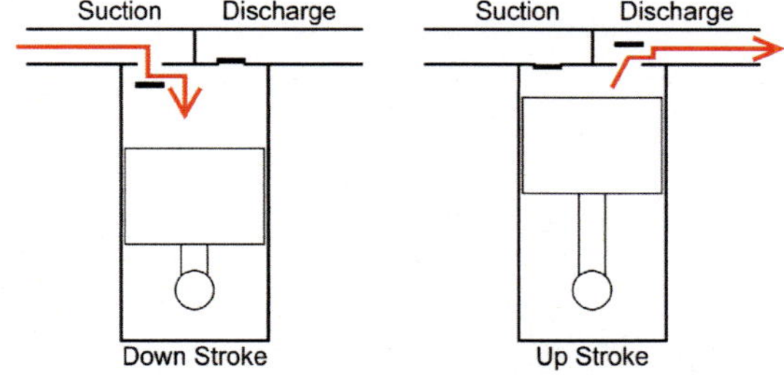

The piston does not travel all the way to the top of the cylinder. There is a small volume of clearance left that allows droplets of liquid to clear the cylinder without causing damage to the compressor due to the fact liquid cannot be compressed. This liquid may be refrigerant or oil. The area of clearance is called the clearance volume.

Scroll Compressor Process

Scroll compressors rotate two nested, helix-type assemblies, called scrolls, that compress the refrigerant vapor trapped between them. One scroll is stationary, while the other orbits or wobbles. When energized, the compressor motor rotates the rotor assembly, which is attached to the scroll. Refrigerant is drawn into the space between the scrolls, where it is compressed, and discharged from the compressor via the discharge line.

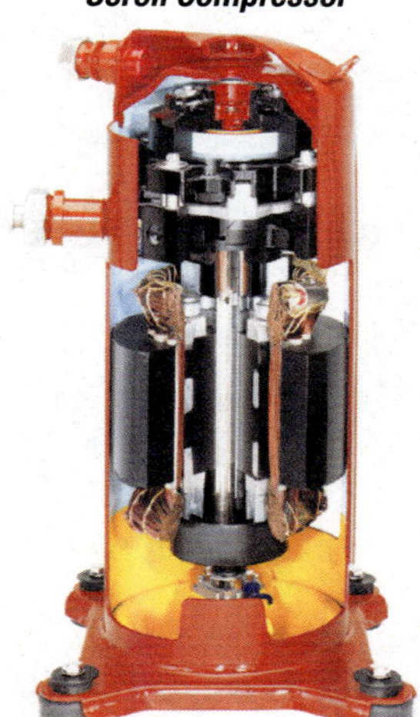

Scroll Compressor

Image courtesy of Trane

Scroll Separation

Most scroll models have an internal safety device that allows the scrolls to separate if excessive liquid refrigerant or oil slugs enter the scrolls. The scrolls will not compress the liquid, and the liquid will pass without damaging the scroll compression assembly. This is called separating. During scroll separation, a loss of pumping efficiency is experienced.

Discharge Check Valve

A discharge line check valve is located inside the compressor shell. This check valve will help prevent hot discharge from circulating back through the scrolls during an off period. The check valve will keep the discharge pressure high, yet internally, the compressor is in an unloaded state. This means the suction and discharge sections of the compressor are at the same pressure.

Low Vacuum Protection

Pulling a compressor into a vacuum can damage the compressor. Starting a compressor in a vacuum may cause the motor to fail electrically, and could cause serious injury if one of the motor terminal pins were to come loose from the compressor shell. To prevent low vacuum operation, modern scrolls have an internal safety feature that will cause the scrolls to equalize pressure if the discharge gas pressure is 10 times higher than the suction pressure. The equalizing of discharge gas to the suction side of the system will open the compressor internal temperature protector and shut the compressor off.

Pressure Test Limits

When leak testing systems with nitrogen, the maximum pressure used should not exceed 150 PSI. If exposed to excessive nitrogen pressure, compressor shells can and have exploded. UL ratings require the compressor shell to be able to withstand a specific amount of pressure based upon the type of refrigerant in the system. This pressure can be pretty low considering many nitrogen regulators can supply 500 PSI. In fact, R-12 compressors by UL rating had to withstand 350 PSI of pressure on the low side of the shell. This pressure is well below that which can be delivered by some nitrogen regulators. The best practice is to install a pressure relief in the nitrogen line between the regulator and air conditioner being serviced. This pressure relief should open at 175 PSIG.

Starting Characteristics of Scroll and Reciprocating Compressors

When a scroll compressor shuts off, the nested scrolls separate. The compressor's high and low side components are at equal pressure during the next call for operation. Therefore, the compressor always starts in an unloaded state.

Reciprocating compressors must be started with refrigerant circuit pressures equalized. If the system is started with differential pressures present, the compressor will likely cycle off on internal overload, as it will have a difficult time overcoming the pressure differential during start operation.

The minimum voltage for scroll and reciprocating compressors should not drop below 90% of the nameplate's minimum voltage as the compressor contactor pulls in and the compressor tries to start. The maximum voltage should not exceed 110% of the nameplate maximum voltage. The starting voltage should be measured at the load side of the contactor as the compressor tries to start, and with all major electrical appliances in the home turned on.

Crankcase Heat Requirements

Crankcase heaters are required when there is a high charge level in the system. If there is significant liquid migration, the compressor may fill with liquid refrigerant as the refrigerant migrates up into the scrolls. Most modern scroll compressors on the market are manufactured by Copeland. Visit the Copeland website to download technical information on maximum charge levels allowed for the scroll compressor model before a crankcase heater is needed.

In general, liquid migration can cause starting problems such as the compressor tripping off on overload, and/or lights dimming during starting attempts. The liquid will eventually clear out of the compressor shell from the heat generated during the starting attempt. The time of the day when this is most likely to occur is in the early morning when the system is started after a long shutdown period. Be aware that long line sets and overcharge conditions can make this condition worse.

High Temperature Limit

A bi-metal temperature limit is located at the top of the discharge area of the compressor. This limit will trip and release hot gas onto the motor overload if the temperature gets too hot in the discharge line area. This limit will open at around 300°F. Earlier models had an external temperature limit located at the top of the compressor

that opened the low voltage circuit to the compressor contactor when an abnormally high temperature was sensed.

High discharge gas temperatures can be caused by high suction vapor superheat levels, excessive compression ratio, dirty outdoor condenser coils, and non-condensable gases in the system.

Internal Motor Overload

All motors used in air conditioning systems have a temperature overload that opens and shuts off the motor when the motor winding temperature is too high. These limits are automatic reset limits that will close when the temperature in the compressor cools. If the limit does not reset, the compressor must be replaced.

In most cases, these limits will open when the system lacks charge or when the motor is overloaded due to the same problems that would cause a high temperature limit trip.

Open Internal Pressure Relief Valve (IPR)

The Internal Pressure Relief (IPR) valve is a spring loaded device that protects the compressor from excessively high pressure difference between the suction pressure and the discharge pressure. Trying to pump refrigerant into the outdoor condenser coil during repairs can cause the IPR valve to open. When the IPR valve opens, the compressor will make a loud internal noise and gauge pressures will equalize. The internal motor overload protection will shut the compressor motor off. The valve is an auto reset device. At times, it may not reset. If this problem occurs, the compressor must be replaced.

Scroll Compressor Noise

Single-phase scroll compressors may run backwards if short cycled for a brief moment. The compressor will eventually shut off on its internal overload. When the motor cools, the compressor will start in the correct direction. No damage to the compressor should occur.

Broken Scroll Assembly

If oil, sludge, or debris enters the space between the scrolls, modern scroll assemblies will separate to protect the compressor. In extreme examples, it may be possible to break the scroll assembly. If this occurrs, the compressor may attempt to run, but the differential between the suction and discharge pressures will be low.

Seized Scroll Assembly

When compressor bearings have been damaged to the point the compressor has seized, the compressor will trip overload protection and shut down during attempted starts. When the scroll assembly has become damaged, the compressor must be replaced.

Compressors

Internal Bearing Damage

It is common for compressors to experience bearing damage. Compressors operating with damaged bearings will operate at amperage levels higher than specified by the compressor manufacturer. The compressor may also vibrate excessively and make a metallic sound as it runs.

Compressor bearing damage is difficult to detect because the running amperage of the compressor must be compared to data charts published by the manufacturer. These charts publish the amp draw of the compressor at combinations of suction and discharge pressure. If the measured amps is 15% higher than the published specifications, the compressor bearings may be damaged. Be aware that, without the published data charts from the compressor manufacturer, this determination cannot be made.

Reciprocating Compressors - Broken Valve Plate

If a reciprocating compressor gets enough liquid trapped in the cylinder to fill the clearance volume between the bottom of the compressor valve plate and the top of the piston, the valve plate may break as a result of the enormous pressure created as the liquid attempts to compress. The suction valve is the most likely component to break. When this valve breaks, the condensing pressure will bleed into the suction side of the system. The suction pressure will be high and the condensing pressure too low.

Image courtesy of Trane
A broken valve plate caused by liquid slugging.

Many technicians are taught to perform a pump-down test. During this test, the system is run with the liquid line service valve closed. The system suction pressure falls as refrigerant is pumped into the condenser coil. When the suction pressure reaches 0 PSIG, the system is shut off and the refrigerant gauges monitored. If the suction pressure rises slightly and then holds, the compressor valves are determined to be OK. Note: this test is not approved by all compressor manufacturers as a valid method of detecting valve damage in a reciprocating compressor.

So what is the proper method for detecting valve plate failure? If the suction pressure is too high, and the condensing pressure too low, the metering device may be overfeeding the evaporator circuit. Check for a properly sized piston. Make sure the indoor coil is properly sized for the outdoor condensing unit. A large evaporator will elevate suction pressure. Pump the system down and see if the

Piston-type metering device.

compressor can hold some of the refrigerant in the condenser. If the pressures quickly equalizes, there might be a broken valve. Contact the manufacturer to obtain proper confirmation of valve damage.

Two-Step Unloading Scroll Compressors

Two-stage systems feature scroll compressors that can reduce capacity from 100% to 67% in order to achieve higher efficiency ratings. These compressor use a standard single-speed PSC motor. Many manufacturers are using two-step compressors in high efficiency systems due to their excellent reliability and performance.

The compressor features an unloading ring that is moved by a solenoid coil located inside the compressor. The ring opens or closes two internal unloading ports inside the compressor.

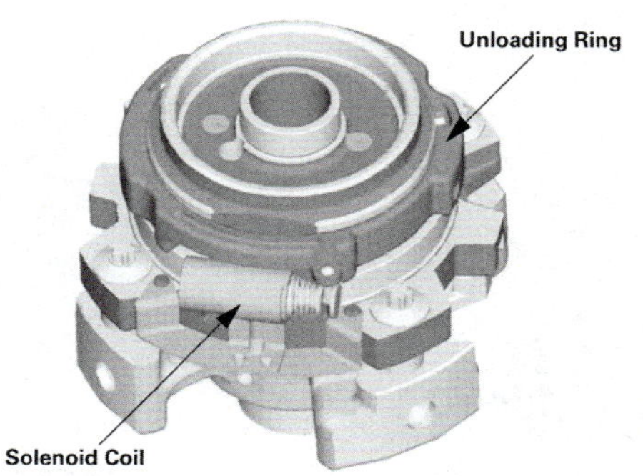

Image Courtesy of Trane

In first stage low capacity operation, the unloading ports inside the compressor are open. The system will operate at 67% capacity in this mode. When the thermostat energizes the Y2 second stage cooling circuit, 24VAC is directed to a bridge rectifier circuit that converts the 24VAC to DC voltage. The voltage output of the bridge rectifier circuit is typically between 15 and 27VDC.

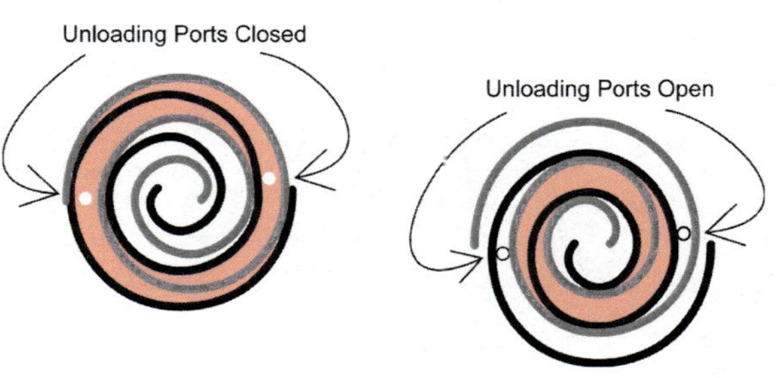

The bridge rectifier circuit is usually found in two configurations, one as an external chip located in the control panel of the air conditioning system, and the other where the bridge rectifier is molded into the plug located on the side of the compressor as shown below.

The bridge rectifier circuit can be found in a molded plug such as shown above.

When 24 VAC is applied to the rectifier circuit, the circuit in the molded plug applies a DC voltage to the compressor solenoid pins. The unloading ring then moves to close the unloading ports, and the compressor goes to 100% cooling capacity. As the compressor loads to 100% capacity, a noticeable increase in compressor amps will be observed.

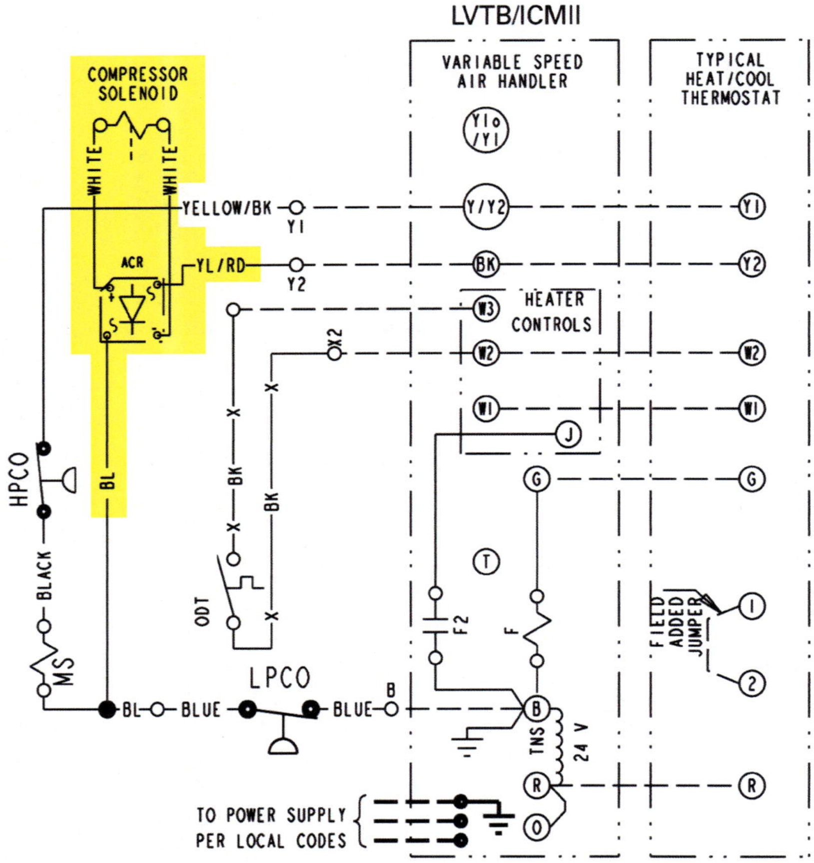

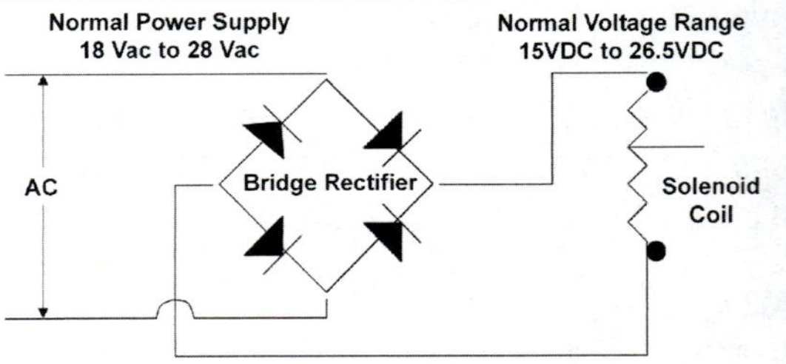

To ensure reliability and prevent failure of the compressor internal solenoid, voltage should never be applied to the unloading solenoid when the system contains no refrigerant, or when the system's low pressure switch has opened. If voltage is applied to the solenoid circuit under these conditions, the solenoid will likely overheat and burn out. With the solenoid burned out, the compressor will only be able to operate at 67% of its rated capacity.

Compressors

Compressor Electrical Operating Characteristics (208/230 Single Phase Reciprocating and Scroll Models)

Compressor Permanent Split Capacitor (PSC) Motors

Other than in high efficiency variable speed air conditioning systems, most residential air conditioning systems use PSC motors inside the compressor shell. Compressor motors used in residential systems are single-speed type. The motor uses an external run capacitor and may also have a start assist device in the electrical circuit.

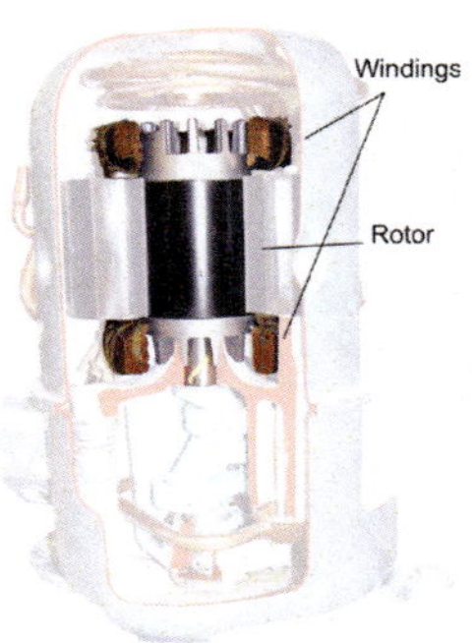

In this picture of a cutaway reciprocating compressor the rotor and copper windings of the PSC motor can be seen. The run capacitor is mounted externally.

How PSC Compressor Motors Work

PSC motors have two motor windings, a START winding, and a RUN winding. The windings are located around the compressor motor rotor. When electrical power is applied to the motor, a circular rotating magnetic field is created by the windings that pull and push on the rotor, causing the rotor to rotate. The rotor will continue to rotate as it is attracted to, and repelled by, the magnetized coils of wire. This motion will continue until power is disconnected.

PSC Motor Run Capacitor

A run capacitor is an energy storage device that will charge and discharge with changes in the electrical polarity of the applied voltage. When a capacitor is applied to the motor winding circuit, it creates a time delay where the applied voltage to the winding must first charge and then discharge through the capacitor. PSC motors use run capacitors to create the best possible magnetic hold on the motor rotor by changing the timing of the magnetic field generated by the start winding.

A continuous duty run capacitor is wired in series with the start winding. The capacitor stores and discharges power, and will change the timing of the start winding's magnetic field. This is done to shift the electrical field to the optimum position in relationship to the run winding magnetic field. With both magnetic fields in proper place, the motor operates at its most efficient power performance level.

This continuous duty run capacitor is wired in series with the Start Winding in order to create the best magnetic hold possible on the motor rotor.

Capacitors are rated by microfarad range (storage capacity) and voltage range. Each PSC motor model will have a matched run capacitor rating to provide maximum performance. If a capacitor is replaced with one of incorrect microfarad range, the position of the magnetic field will be affected, and motor performance will decrease. The voltage rating of the capacitor should always be equal to, or greater than, what is specified for the motor.

Start Assist Devices

Potential Relay and Start Capacitor

A hard start kit helps a single-phase compressor start against potentially heavy work loads. A hard start kit consists of a potential relay and a start capacitor. The required start kit potential relay and start capacitor microfarad requirements are specified by the compressor manufacturer.

Shown here (right) is a potential relay and start capacitor of a start kit, as well as a combo start kit (left) that has both components in one unit.

The hard start kit places an intermittent duty start capacitor in the compressor motor winding circuit for a very brief moment. The additional capacitor generates a stronger magnetic hold on the motor rotor. This stronger hold helps the rotor turn quicker. It is important to know that this device gets the rotor started quicker, but does not limit the locked rotor amps at the initial moment when power is applied to the compressor motor.

The start capacitor must be removed very quickly from the motor circuit or it will fail electrically. To remove the capacitor from the circuit, a special relay called a potential relay is used. A potential relay has a set of normally closed contacts that place the capacitor into the electrical circuit when the compressor is not running. The relay has a coil, that is energized by back voltage that is generated when the compressor motor rotor begins to turn.

This back voltage will energize the potential relay coil when the motor speed reaches approximately 75% of the motor's rated RPM. When the relay coil is energized, the normally closed contacts open to remove the start capacitor from the circuit.

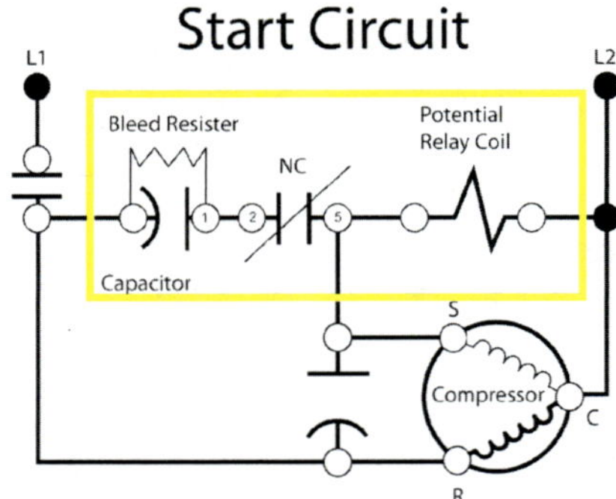

The voltage that is generated by the motor to energize the relay coil is called a pick up voltage. The potential relay is rated by the pick up voltage level that will energize the relay. If the wrong relay is used, and the motor is not capable of generating enough pick up voltage quickly, the capacitor will remain in the circuit too long and will be destroyed.

A start kit can be used to reduce dimming light complaints in homes. Be aware that in many cases, the start kit will not eliminate dimming lights, but will reduce the amount of time it occurs and potentially reduce the intensity of light dimming.

Adding a start kit should not be the first thing done when a compressor experiences start problems. First, check the run capacitor and confirm the run capacitor's microfarad level is within 10% of the rating on the capacitor's label. Make sure the system is not trying to start against unequalized refrigerant pressures. Next, the line voltage at the condensing unit should be measured as the compressor tries to start. All major electrical appliances in the home should be on, and the voltage measured at the load side of the contactor as the contactor pulls in. The voltage should be a within 10% of the rated voltage. If the voltage is not within the specified limits, troubleshoot the line voltage (see Chapter F: High Voltage Circuit).

If the voltage and run capacitor check out good, the compressor may have bearing damage (see Internal Bearing Damage, page G13). If there is no bearing damage found, add a start kit to the compressor.

If the compressor still experiencing starting problems, the compressor may need to be replaced.

PTC Start Assist Device

A Positive Temperature Coefficient (PTC) start assist device uses a special element that will allow electrical current flow for a brief moment, and then stop current flow.

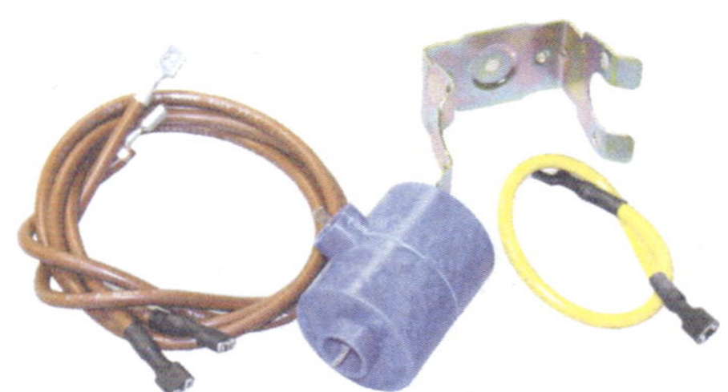

A PTC device shown before installation.

The electrical resistance of the PTC device will rapidly rise as the electrical current flows through it. It is sometimes wired directly to the compressor run capacitor. In this configuration, there is a direct electrical circuit to the compressor start winding during compressor start. This applied voltage to the start winding increases motor torque. The device quickly opens and removes the applied voltage to the start winding. The run capacitor is then placed back into the compressor motor winding circuit.

Sometimes PTC devices are used in place of a potential relay as the switching device to add and remove a start capacitor from the compressor start circuit.

A disadvantage of this type of start device, when compared to a potential relay device, is that if the system is short cycled, and starting is attempted against non-equalized system pressures, the PTC device will still be hot and will not be an active start component when it is needed.

Fusite Plugs & Safety

The compressor motor windings are connected to a plug, called a fusite plug. The fusite plug separates the internal electrical components of the compressor from the outside of the compressor shell. There are three terminal pins located on a fusite plug. The pins are the connection between the motor windings and the wiring connections external to the compressor shell.

Fusite Plug

The pins are: common (C) connection between the motor start and run windings; the run winding (R); and the start winding (S). The resistance of the compressor motor windings can be checked at the opposite end of the wires that are connected to these pins. For safety reasons, it is not recommended that motor winding resistance checks be made directly at these pins.

If an internal short circuit were to occur, the terminal pins on the fusite plug could violently eject out of the fusite plug. This is called terminal venting. If power is present, and an electrical short blows out a pin, the bullet-like force of a pin ejecting from the compressor shell is in itself capable of causing serious physical harm or death. Additionally, refrigerant oil and refrigerant will erupt through the hole and may catch fire, creating a deadly stream of high velocity flaming oil. This is called terminal venting with ignition. If there is any indication of electrical charring or burning on one of the fusite plug pins, exercise extreme caution to ensure the pin is not loose and about to eject the refrigerant charge and pin.

Terminal venting can cause serious injury and/or death.

To reduce the potential for terminal venting, never reset a condensing unit circuit breaker, or replace an open/blown line voltage fuse, without first performing a motor grounding check. Also, never start or pump a residential compressor into a vacuum as internal arcing may occur inside the compressor shell and lead to terminal venting.

At no time should the protective cover be removed from the compressor motor terminal area when power is present to the condensing unit.

If a compressor is energized, and sizzling or popping noises are heard coming from the compressor, get away from the compressor immediately! Water may be present in the compressor. This water may generate enough steam to cause an explosion and/or the release of scalding steam. Shut off power and allow the compressor to cool completely before approaching.

Checking Motor Winding Resistances and Winding Condition

When measuring compressor motor winding resistance, power to the condensing unit must be turned off.

DISCONNECT POWER TO THE OUTDOOR UNIT. FAILURE TO FOLLOW THIS INSTRUCTION CAN RESULT IN PERSONAL INJURY OR DEATH.

The resistance of the compressor motor windings will typically be low. The start winding (S) has greater resistance than the run winding (R). In the picture below left, the resistance of the motor windings is being read between the common (C) and the R winding leads. The R winding has a resistance of about 1.6 OHMS. In the lower right picture, the resistance of the S winding is read from between the C and the S winding leads. This reading has a resistance of about 2.2 OHMS. All compressors will have different motor winding resistance readings, but in all cases, the S winding resistance to C should be higher than the R winding resistance to C.

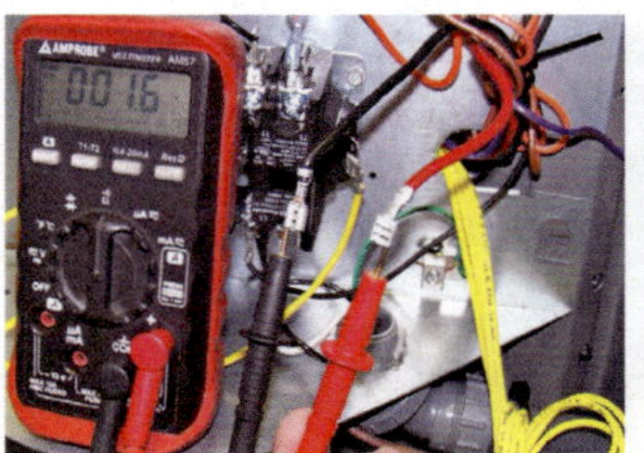 

Always perform this check on the compressor winding wire leads only. Do not perform check directly on compressor terminals. Terminal venting may cause serious injury or death.

Checking Internal Motor Winding Overload Protector Switch

When measuring compressor motor winding resistance, power to the condensing unit must be turned off.

DISCONNECT POWER TO THE OUTDOOR UNIT. FAILURE TO FOLLOW THIS INSTRUCTION CAN RESULT IN PERSONAL INJURY OR DEATH.

Inside the compressor, an internal motor overload protection switch will open when excessive temperature occurs in the motor windings. This switch is wired in series between the connection point for the run (R) and start (S) windings, and the common (C) pin on the fusite plug.

If the overload switch is in an open position, the motor will not start when power is applied to the compressor. When checking the resistance of the motor windings when the internal overload switch is open, infinite resistance will be measured between the R and C terminals, as well as between the S and C terminals.

If the overload switch has opened, allow time for the compressor to cool. This switch is an automatic reset type that should close after the motor has cooled. If the switch fails to reset, the compressor must be replaced.

To be sure the motor windings have not burned out (opened), measure the resistance between the R and S winding leads. A resistance should be measured if the windings have continuity. If infinite resistance is measured, one of the windings is open. An open winding cannot be repaired and the compressor would need to be replaced.

Checking the resistance of the motor windings.

 Always perform this check on the compressor winding wire leads only. Do not perform check directly on compressor terminals. Terminal venting may cause serious injury or death.

Checking for a Grounded Motor Winding

When measuring for a grounded motor winding, power to the condensing unit must be turned off.

DISCONNECT POWER TO THE OUTDOOR UNIT. FAILURE TO FOLLOW THIS INSTRUCTION CAN RESULT IN PERSONAL INJURY OR DEATH.

The compressor motor windings can develop a short circuit to ground. This short circuit is caused by a motor wiring coming in direct contact with a ground-connected metal portion of the compressor. This will result in excessive electrical current flow, causing either the circuit breaker or fuse to open and disconnect power from the condensing unit.

To test for a grounded compressor motor, measure the resistance from each wire connected to the fusite plug to the discharge line of the compressor. (Remember, measure the resistance at the opposite end of the wires connected to the fusite plug, not directly at the plug pins.)

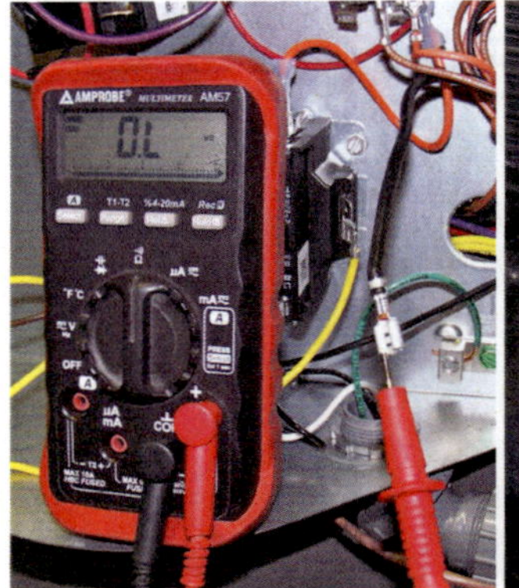

Testing for a grounded compressor

Always perform this check on the compressor winding wire leads only. Do not perform check directly on compressor terminals. Terminal venting may cause serious injury or death.

Check the resistance between the compressor discharge line and the common wire (C), the run wire (R), and the start wire (S). There should be measurable resistance somewhere above 1,000,000 OHMS. If the resistance is below this level, contact the manufacturer of the condensing unit to determine if resistances below 1,000,000 OHMS is acceptable. In some older compressors, manufacturers have published acceptable OHMS to ground as low as 500,000 OHMS. If a compressor motor is shorted to ground, the compressor must be replaced.

Compressor Starting Problems

When a compressor has trouble starting, the problem may be as simple as a bad run capacitor, or as complicated as internal bearing damage. The first step is to determine if anything in the refrigeration system is contributing to the compressor's inability or difficulty in starting. Refrigeration conditions that can cause starting problems include:

- Long line sets resulting in high system charge requirements.
- Overcharge.
- Liquid refrigerant entering the compressor.
- Restrictions causing unequalized refrigerant pressures during start operation. (Most new scroll compressors start unloaded internally even when gauges appear unequalized.)

If it appears there are no refrigerant circuit problems present, electrical tests must be done on the compressor and associated components.

Service Procedure: Finding the Cause of a Compressor Starting Problem

Tools:
- Refrigeration Gauges
- Multimeter
- Ammeter

STEP 1

Make sure the compressor is not trying to start against unequalized refrigerant pressures. Also, The compressor may be full of liquid refrigerant that migrated to the compressor shell during off cycle. Make sure the compressor does

not need a crankcase heater. If the charge level of the system is high due to long refrigerant lines, the compressor may require a crankcase heater. Call the manufacturer to confirm.

STEP 2

Disconnect power to the outdoor unit and discharge the run capacitor. Remove the wires from the run capacitor. Perform a run capacitor test to confirm the microfarad level is within 10% of capacitor's design rating. If the run capacitor is defective, replace it with one of equal

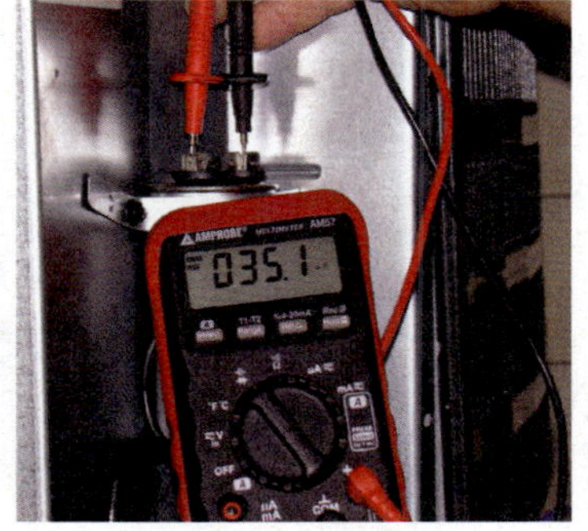

microfarad value. The voltage rating of the replacement capacitor should be equal to, or greater than the one being replaced.

If the capacitor is OK, continue to the next step.

STEP 3

Measure the line voltage at the load side of the compressor contactor as the compressor tries to start.

Voltage should not fall below 90% of the system's rated voltage. (Make sure all major electrical appliances in the home are on when performing this measurement.)

If voltage drops below 90% of the system's rated voltage, check for a voltage drop from the home's electrical panel to the outdoor disconnect at the condensing unit.

If there is a voltage drop, confirm proper wire sizing and make corrections.

If the voltage is OK, check the voltage drop from the outdoor unit disconnect switch to the line side of the compressor contactor.

If there is excessive voltage drop, fix the problem.

If it is OK, check for a voltage drop across the compressor contactor. If voltage drop is present across the contacts of the contactor, clean or replace the contactor.

STEP 4

If steps 1 through 3 check out OK, disconnect power to the condensing unit and make sure all capacitors are discharged. Find the wires that lead to the compressor motor fusite plug terminals: run (R), start (C) and common (C). Take a resistance reading between each motor windings at the end of these wires. If a winding is open, replace the compressor. If they are OK, continue.

STEP 5

Install a hard start kit on the compressor. If the compressor runs, measure the amperage of the compressor motor on the common (C) wire to the compressor motor and note the refrigerant pressures. Contact the system manufacturer to confirm what the amperage should be. If it is 15% higher than it should be, bearing damage may be present and the compressor may need to be replaced.

Amperage should read within normal parameters

STEP 6

If the system starts and runs with normal amperage, compressor is OK.

When the Circuit Breaker Trips or a Line Fuse Opens

If a condensing unit is tripping the circuit breaker or opening a line fuse, there is excessive electrical current to the outdoor unit, or the fuse/breaker size is incorrect. If the outdoor unit is under a heavy heat load (such as when it is very hot outside, with a high indoor air temperature), the compressor may overload and operate with excessive amperage. A dirty condenser coil will also cause excessive condensing pressure and higher than normal amperage draw. If there is no obvious external reason for the circuit breaker or fuse to open, the high current may be caused by a direct short to ground in the outdoor condensing unit. This short can come from any electrical component in the high voltage circuit, including the crankcase heater, wiring, contactor, capacitors, condenser fan motor, and compressor.

In any instance where the circuit breaker or line voltage fuses have opened, the compressor must be checked for a short to ground before the circuit breaker or fuse is reset/replaced. Failure to follow this instruction may result in injury or death.

Service Procedure: Testing for a Grounded Compressor

Tools:

- Multimeter

STEP 1

Disconnect power to the outdoor condensing unit and discharge all capacitors.

STEP 2

Set electrical multimeter to read OHMS at the highest resistance scale.

Measure the resistance between the compressor discharge line and each wire leading to the motor fusite plug terminals: run (R), start (S), and common (C).

 *Always perform this check on the compressor winding wire leads only. Do not perform check directly on compressor terminals. Terminal venting may cause serious injury or death.*

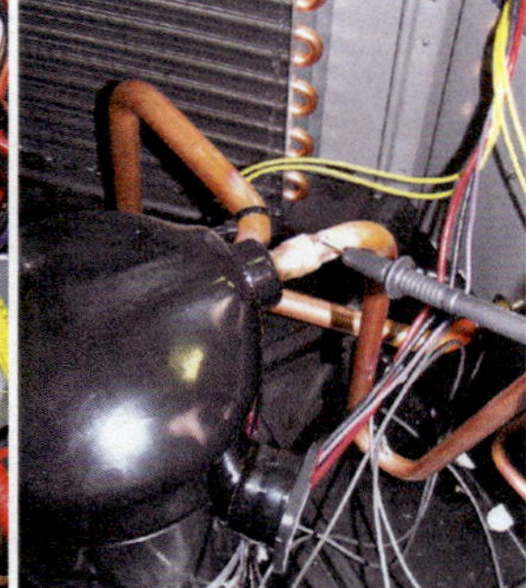

Infinite resistance should be measured if the compressor has no conductive path between the motor and the discharge line. There should be measurable resistance somewhere above 1,000,000 OHMS. Resistance below 1,000,000 OHMS is considered shorted by most compressor manufacturers. If low resistance is measured, the compressor is shorted to ground and must be replaced.

STEP 4

If the compressor checks out OK, check other components for a short circuit to ground.

For three-phase motors, follow the same procedure as above, but check resistance from motor terminals 1, 2, and 3 to ground.

Troubleshooting a Compressor Start Circuit

A compressor start circuit consists of a potential relay and start capacitor. When there is a malfunction of the start circuit, the start capacitor is usually destroyed (relief plug opens on top of capacitor). The capacitor fails because it's an intermittent duty capacitor that should only be in the electrical circuit for a brief moment.

Causes of start circuit failure include low line voltage. This causes the compressor pick up voltage to be too low to energize the potential relay, thus the capacitor remains in the circuit for too long.

When a compressor start circuit has failed, check to ensure that starting voltage at the compressor contactor does not fall below 90% of the system's rated voltage. Make sure all major electrical appliances in the home are on to simulate the lowest voltage possible at the condensing unit. If voltage is low, find and correct the cause.

 **Service Procedure: Testing the Compressor Start Relay**

Tools:
- Screwdriver with Insulated Handle
- Multimeter

STEP 1

Disconnect power to the unit and discharge the start capacitor using a screwdriver with an insulated handle. Remove the potential relay from the electrical circuit.

STEP 2

Testing the potential relay is simple. There are three terminals on the relay. These terminals are labeled 1, 2, and 5. Terminals 1 and 2 are the normally closed contacts. Terminal 2 to 5 is the internal solenoid coil of the relay. Measure the resistance between terminal 1 and 2. The resistance should be 0 OHMS.

Next, turn the ohmmeter to the diode function test so a beep tone indicates a short circuit. Place the ohmmeter leads on relay terminals 1 and 2. The ohmmeter should make a steady tone. Holding the meter leads against the relay terminals, shake the relay.

As the shaking causes the internal contacts to snap open, the beep tone will break. This indicates the relay contacts are not welded closed. If the relay contacts do not open, replace the relay. If the relay contacts open, proceed to the next step.

STEP 3

Measure the resistance between relay terminals 2 and 5. There should be resistance measured. If there is either no resistance, or infinite resistance measured, replace the relay. The solenoid coil inside the relay has opened. If the relay coil checks out OK, the relay is not at fault.

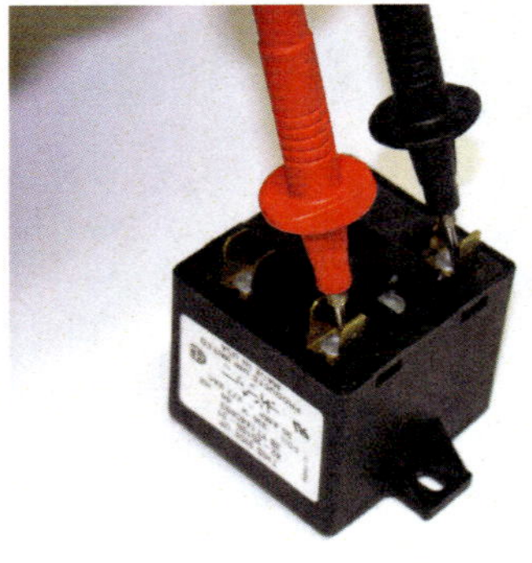

Testing a Capacitor for Proper Operation

Disconnect power to the condensing unit. Discharge capacitors before removing from the electrical circuit. Start capacitors should be discharged before handling. There is a bleed resistor soldered across most start capacitor terminals. This resistance slowly discharges the capacitor. Do not assume the capacitor is discharged by the bleed resistor. Discharge it using a screwdriver with an insulated handle.

Service Procedure: Testing a Start Capacitor

Tools:
- Screwdriver with Insulated Handle
- Multimeter

STEP 1

Inspect the capacitor case for a vented top or cracks in the case. If the capacitor is vented or cracked, replace it with one of equal microfarad and voltage rating. If the capacitor shows no visible sign of failure, proceed to the next step.

STEP 2

Set multimeter to OHM setting. Place meter leads on each capacitor terminal. The battery in the meter should charge the capacitor and the ohm reading on the meter should fluctuate up and down.

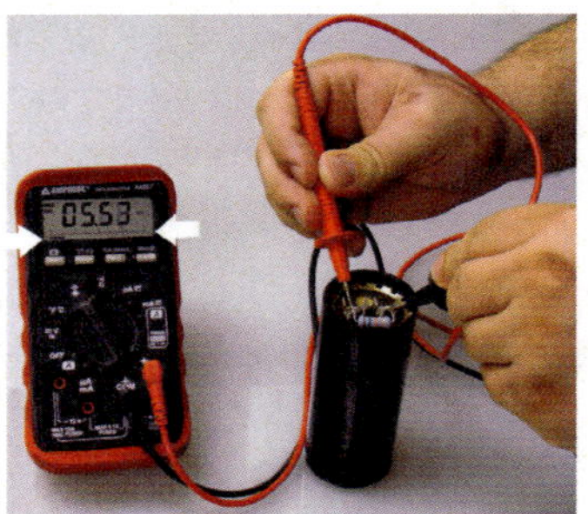

*The meter should end up reading the value of the bleed resistor. **If the meter does not change reading, replace the start capacitor with one of equal microfarad ratings.** Voltage rating may be equal or greater than the original rating.*

Service Procedure: Testing a Run Capacitor for Proper Microfarad Rating

DO NOT PERFORM THIS TEST ON A START CAPACITOR!

Tools:
- Screwdriver with Insulated Handle
- Multimeter

STEP 1

Disconnect power to the outdoor unit. Discharge the run capacitor.

STEP 2

Set the multimeters to the capacitance measuring function, and place the meter leads across the discharged capacitor. The multimeter will display the measured microfarad level of the run capacitor.

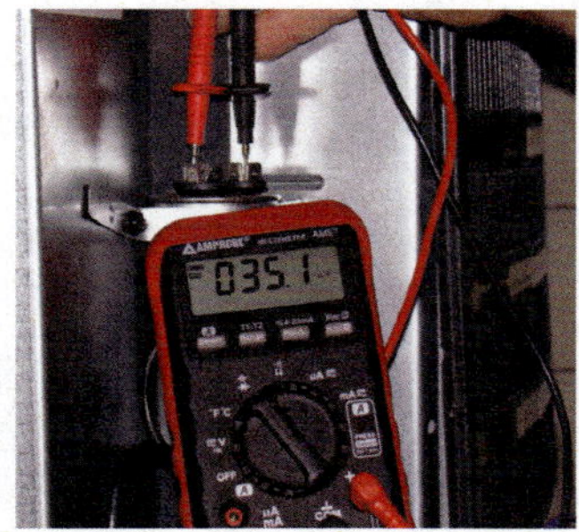

If the measured microfarad value exceeds a 10% difference from the value specified on the capacitor label, replace it with one of equal microfarad rating. Voltage rating may be equal or greater than the original rating.

Service Procedure: Run Capacitor "Power On" Test

If a multimeter with a microfarad reading function is unavailable, a simple test rig can be constructed from a power cord and a 5 amp automotive type fuse. Power will be applied to the rig while attached to the capacitor.

DO NOT PERFORM THIS TEST ON A START CAPACITOR!

BEFORE PERFORMING THIS TEST, MEASURE RESISTANCE FROM EACH CAPACITOR LEAD TO THE CAPACITOR CASE. IF A SHORT CIRCUIT IS FOUND, DO NOT PERFORM THIS TEST. INJURY OR DEATH COULD RESULT.

Tools:
- Screwdriver with Insulated Handle
- Multimeter

STEP 1

Make a wire & fuse rig as shown below for the capacitor check. Attach the rig to the discharged capacitor. and plug the wire rig into a wall outlet. With power applied to the capacitor, measure the amp draw to the capacitor and multiply by 2650.

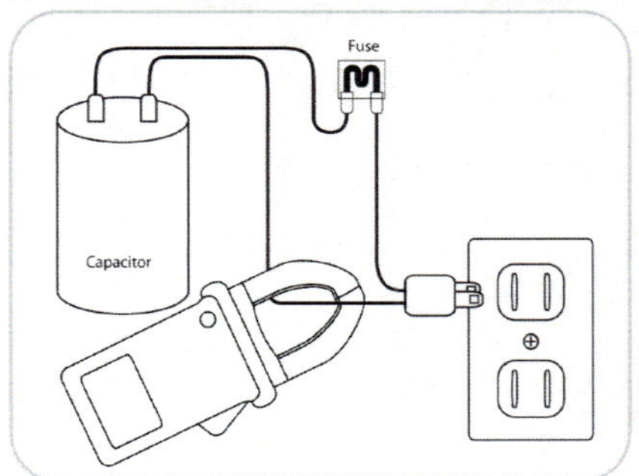

STEP 2

Measure the voltage applied to the capacitor.

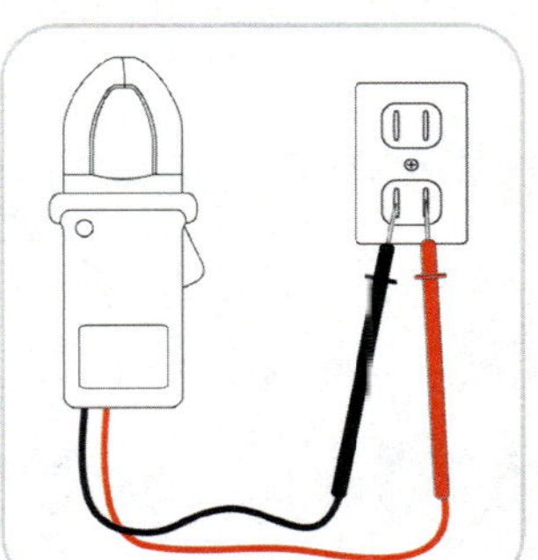

STEP 3

Apply this formula to find the microfarad rating of the capacitor:

$$\text{Microfarad} = \frac{\text{Amps} \times 2650}{\text{Applied voltage to capacitor}}$$

Example:
Amps = 4
Voltage Applied = 110 VAC

$$\frac{4 \times 2650}{110} = \frac{5300}{110} = \textbf{48 MFD}$$

Troubleshooting a Two-Step Compressor to Determine if it is Loading and Unloading

A two-step compressor uses a standard single speed PSC motor, so testing of the motor is identical to any single phase compressor motor test. However, the unloading solenoid should be tested to ensure it is working properly.

Service Procedure:
Testing a Two-Step Compressor Unloading Solenoid

Tools:

- Ammeter
- Multimeter

STEP 1

Run the system in first stage cooling and allow pressures to stabilize. Measure the running amps to the common lead of the compressor motor.

STEP 2

Call for second stage cooling and observe the amp reading. When the compressor loads to second stage, the amp draw should increase by around 25%.

The increase in amps indicates the compressor has loaded to 100% capacity. If the amps do not change, proceed to the next step.

STEP 3

With the continued call for second stage cooling, check to ensure 24VAC is present to the molded plug or rectifier chip. If 24VAC is not present, find the cause and make corrections. If 24VAC is present, proceed to the next step.

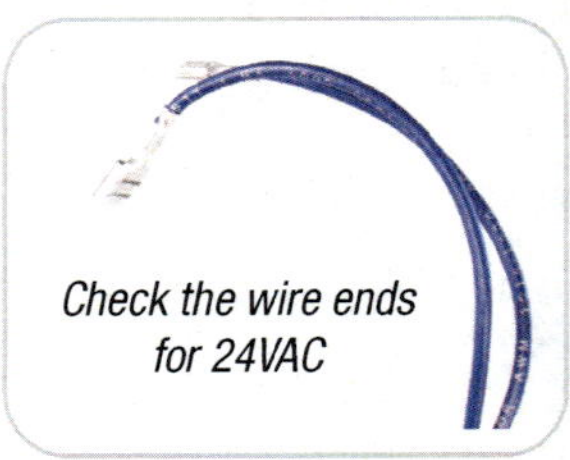

Check the wire ends for 24VAC

STEP 4

End the call for cooling, disconnect line voltage to the condensing unit, and remove the molded solenoid plug from the compressor. Restore line voltage and the call for second stage cooling to the unit.

Set voltmeter to read DC Volts and measure the voltage at the plug solenoid terminals.

15-27 VDC should be present with 24VAC applied to the plug or rectifier. If there is no output voltage at the plug, the internal bridge rectifier circuit has failed. Replace the plug. (Replace the rectifier chip on models with an external chip).

If 15-27 VDC is measured at the terminals on the plug, the compressor is failing to load and unload due to the internal solenoid being burned out, or the compressor may have an internal mechanical failure. In either case, the compressor must be replaced.

NOTE: The compressor unloader solenoid can burn out or get damaged by applying voltage to the solenoid when there is inadequate charge in the system. Systems using the two-step compressor have a low pressure switch that opens the 24VAC power circuit when there is low refrigerant pressure. Jumper bypassing this switch to get the system to operate without low pressure will likely cause the unloader solenoid to be destroyed.

Compressor Valve Plate Damage

If a compressor has tried to compress liquid or sludge, one of the compressor valves may have broken. Typically, the suction valve is damaged. When a compressor operates with a broken valve, the compressor cannot operate at proper refrigeration pressures. The system will have higher than normal suction pressure, and lower than normal discharge pressure.

To confirm a failed compressor valve plate, a pump down is usually performed by closing the liquid line service valve while the compressor is running. With the service valve closed, the compressor will try to pump all of the refrigerant charge into the condenser coil. The refrigerant can't escape the condenser coil with the liquid line service valve closed unless the refrigerant circulates back through the compressor valve plate assembly. If the refrigerant circulates back through the compressor valve plate, the pressures will equalize. The equalizing pressures indicate a leakage occurring through the valves of the compressor.

At no time during a pump down test should a reciprocating residential compressor be allowed to operate below 0 PSIG. Also, never perform this test on a scroll compressor, as a scroll compressor does not have valves. The scroll compressor may become damaged.

Note: This test is not universally accepted by all compressor manufacturers, as some compressors are not designed to hold a pump down.

Tools:

- Refrigeration Gauges

STEP 1

Connect refrigeration gauges to the service ports and run the system.

STEP 2

With the compressor running, close the liquid line service valve. The pressures will begin to fall. When the suction pressure reaches 1-2 PSIG, shut power off to the condensing unit.

DO NOT PUMP THE SYSTEM INTO A VACUUM!

STEP 3

Observe the suction gauge. The gauge pressure will typically rise slightly, and then stop. This is an indication that the valves are holding back the refrigerant charge in the condenser coil.

If the pressures begin to equalize, this is an indication of leakage of refrigerant through the valve plate. Remember, however, that some compressors are not designed to hold a pump down.

STEP 4

If there is indication of a broken or damaged valve plate, contact the compressor manufacturer to determine if the compressor should be replaced.

If the compressor passes the test, open the liquid line service valve and allow pressures to equalize before restarting the system.

> NOTE: A small rise in suction pressure may occur after the compressor is shut down due to some refrigerant remaining in the suction side of the system. This small rise in pressure does not indicate bad compressor valves.

Single or Three-Phase Scroll Compressor Mechanical Failure

When a higher than normal suction pressure and a lower than normal liquid pressure are present, many technicians begin to assume that the compressor is failing. In fact, there are other causes of high suction pressure and low liquid pressure. These causes include:

- Bypassing or oversized metering devices
- Leaking heat pump indoor check valve assemblies
- Heat pump reversing valve leakage

Compressor valve plate leakage is typically the primary suspected cause of high suction pressure and low liquid pressure, however, scroll compressors do not have valve plates, so the same presumption cannot be made.

To cause elevated suction pressure and lowered liquid pressure, scroll compressors would have to allow some of the discharge gas to leak back into the suction side of the system. This condition is highly unlikely to occur in scroll designs. Many scroll compressors feature a discharge line check valve that prevents discharge gas from re-entering the compressor shell. However, this discharge line check valve is not leak proof.

A symptom that may indicate mechanical damage to a scroll compressor is the compressor amp draw. If there is additional drag due to mechanical friction in the compressor body, it can be detected with an ammeter. The measured amperage will be higher than the published amps at the combination of indoor air heat load and outdoor air temperature. Unfortunately, many systems will not have this data available at the job site. Additionally, many systems will not have the correct indoor air volume to reference the required amps. Air volume should be checked and corrected as necessary before performing the following procedure (See Chapter J: Air Volume).

Service Procedure: Testing a Scroll Compressor for Proper Pumping Capability

Tools:
- Refrigeration Gauges

STEP 1

Attach refrigeration gauges to the service ports and run the system.

STEP 2

Assuming a proper charge level, the suction pressure should drop and the discharge pressure rise. If the pressures do not reach proper levels, check for an improperly sized metering piston or properly operating expansion valve. If these are OK, and the pressures do not reach normal levels, the compressor may be bad.

STEP 3

Make sure the compressor is running in the proper direction. (three phase models). If the compressor is running backwards, it will be excessively noisy and will not pump.

3 Phase Models:

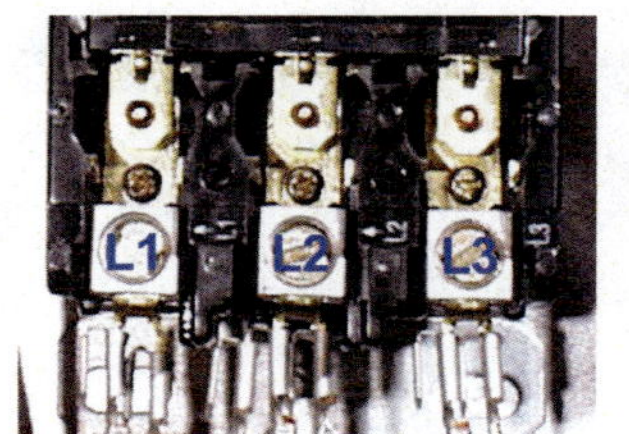
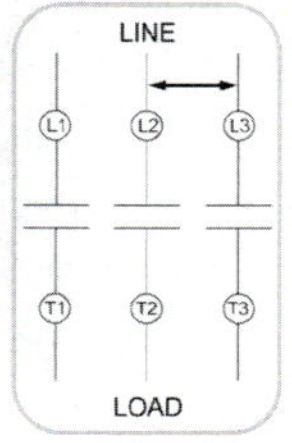

To change direction, switch any two incoming line power phases (L1,L2,L3).

SHUT OFF LINE VOLTAGE TO THE UNIT BEFORE REVERSING THE POWER LEADS OR INJURY/DEATH MAY OCCUR.

STEP 4

Obtain performance tables from the compressor manufacturer's website. Compare the operating pressures against compressor amp draw charts. If the compressor is not operating within range specified by the compressor manufacturer, replace the compressor.

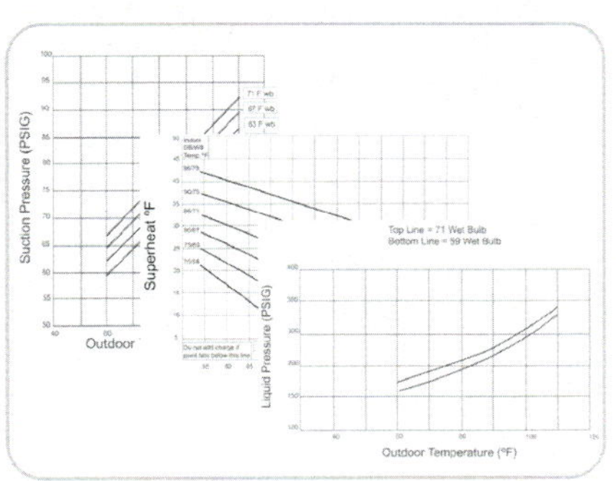

Compressor Bearing Damage

Compressor bearings can be damaged by liquid refrigerant diluting oil, by copper plating, and by overheating. Compressors that are operating with damaged bearings will operate at higher than specified amps for the operating refrigerant pressures. Comparing operating amps against performance tables provided by the compressor manufacturer is a valid method of detecting bearing damage.

Compressors operating with damaged bearings may have symptoms such as starting problems, excessive vibration, and metallic sounding operation.

Service Procedure: Testing for Damaged Bearings

Tools:
- Refrigeration Gauges
- Ammeter

STEP 1

Attach refrigerant gauges to the service ports and run the system.

STEP 2

Measure compressor amps at the common (C) compressor motor lead.

Amperage should read within normal parameters

Compare the measured operating amps against performance data table information for the specific compressor. If the amps are 15% greater than those specified by the data table, the compressor may have damaged bearings. Confirm diagnosis with the compressor manufacturer. (Performance tables may be obtained at compressor manufacturer's website).

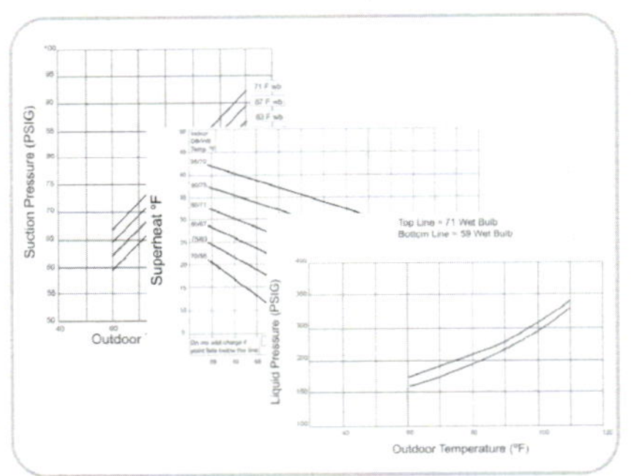

Seized Compressor (Single-Phase)

Determining if a compressor is seized requires eliminating all other possible causes before condemning the compressor as seized.

Symptoms

A potentially seized compressor will try to start, hum, and then open its internal overload. The compressor will draw Locked Rotor Amps (LRA) as it tries to start. Electrical problems must be eliminated as a potential cause before condemning the compressor. Electrical problems that can cause a compressor to fail to start will make a compressor act as if it is seized. These problems include:

- Low line voltage
- Pitted contactor
- A broken/bad wire connection between the compressor and run capacitor or contactor
- A failed run capacitor
- Unequalized system pressures when the compressor tries to start

Before beginning the electrical test sequence, make sure the compressor is not trying to start against unequalized system pressures. If the compressor is trying to start against unequalized pressures, check for refrigeration circuit restrictions or short cycling conditions.

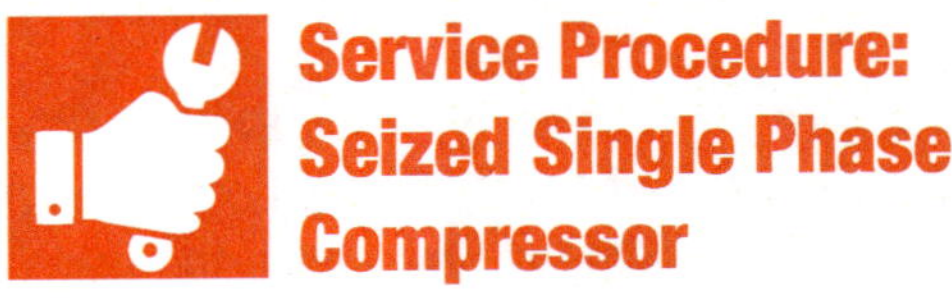

Service Procedure: Seized Single Phase Compressor

Tools:

- Ammeter
- Multimeter
- Ohmmeter/Capacity Tester

STEP 1

Remove power to the outdoor condensing unit. Place an ammeter on the common (C) motor winding lead of the compressor. (Typically the BLACK WIRE coming from the compressor motor terminal cover.)

Restore power to the condensing unit and call for cooling. Confirm the compressor drawing LRA. If the compressor is trying to start, it should hum and then shut off on its internal overload. If the compressor does not try to start, the internal overload may be open. Allow time for it to reset. Once confirmed that the compressor is trying to start, proceed to step 2.

STEP 2

Measure the voltage at the compressor contactor while it is under a load. Confirm the correct line voltage level at the load side terminals of the compressor contactor.

Typically the voltage must be within 10% of the unit nameplate voltage rating. If the line voltage is too low at the load terminals, check the voltage to the line terminals of the contactor.

If the voltage is low at these terminals, there is a problem with the electrical supply to the condensing unit. If the line voltage is normal, but is low at the load terminals, check voltage across the contactor points with the contactor energized.

0VAC should be measured. If voltage is measured across the electrical points of the contactor, there is either pitting or an obstruction at the contactor. Correct the problem and retry starting the compressor. If the voltage level at the contactor is within tolerance, proceed to step 3.

STEP 3

Disconnect power to the condensing unit and test the run capacitor (see run capacitor test procedures on page G23). If the run capacitor has failed, replace it with one of equal value. If the capacitor is OK, proceed to step 4.

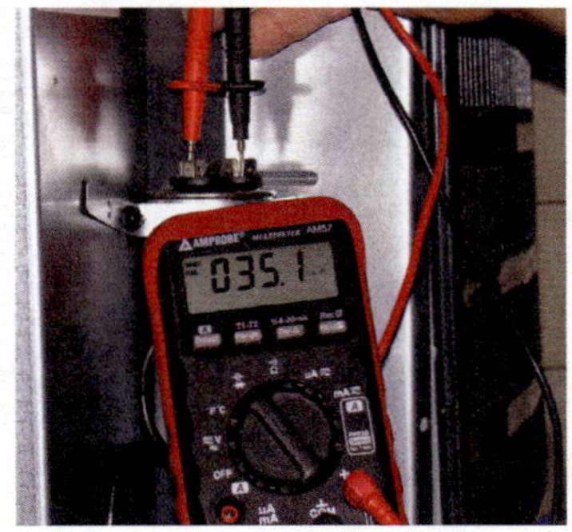

STEP 4

With power to the unit still off, perform a motor winding test to check for an open run or start winding. (See "Single-Phase Motor Winding Test", page G30) If either winding is open, replace the compressor. If the windings are OK, proceed to step 5.

STEP 5

Check for a grounded compressor motor. (See "Testing for a Grounded Compressor", page G20) If any terminal is shorted to ground, replace the compressor. If the motor is not shorted, proceed to step 6.

STEP 6

If everything checks out good electrically, and the compressor is not trying to start against unequalized pressures, add a hard start kit. If the compressor still fails to start, replace the compressor.

Single-Phase Motor Windings

This test procedure will determine if the PSC compressor motor has failed due to an open motor winding. This test will also detect a possible open Internal Overload (IOL).

Symptoms

A compressor with an open winding will fail to start. The compressor may do nothing, or may hum and then trip off on its internal overload. If the IOL is open, the compressor will not try to start.

Service Procedure: Single-Phase Motor Winding Test

Tools:

- Multimeter

STEP 1

Remove power to the outdoor condensing unit and discharge all capacitors.

STEP 2

Set your ohmmeter to read low ohms. (Down to ½ of an ohm may be read.) Place one ohmmeter lead to the compressor Start (S) winding wire lead and the other ohmmeter lead to the Common (C) wire lead.

If a low resistance is measured, the S winding is not open. Next measure the resistance between the compressor Run (R) winding wire lead and the C wire lead. Again, a small resistance should be measured. If so, the compressor does not have an open winding.

If infinite resistance is measured between C and either the S or R winding leads, there is an open winding. Replace the compressor. If infinite resistance is measured between C and <u>both</u> the S and R winding leads, proceed to the next step.

STEP 3

To determine if the problem is an open winding or an open IOL, check the resistance between the R winding wire lead and the S winding wire lead.

If resistance is measured, the IOL is open. Allow up to four hours for the IOL to re-close. If the IOL does not re-close, replace the compressor.

If infinite resistance is measured between the R and S windings there is an open winding. Replace the compressor.

Three-Phase Motor Windings

This test procedure will determine if a three-phase compressor motor has failed due to an open motor winding. This test will also detect a possible open Internal Overload (IOL).

Symptoms

A three-phase compressor with an open winding may fail to start. In some cases, the motor may run on one phase. The compressor may do nothing, or may hum as it tries to start and then trip off on its internal overload. If the IOL is open, the compressor will not try to start.

Service Procedure: Three-Phase Motor Winding Test

Tools:

- Multimeter

STEP 1

Remove power to the outdoor condensing unit and disconnect the three wires to the compressor 1, 2, and 3 motor terminals from the load side connections of the compressor contactor.

(The load side terminals are the terminals that connect the compressor motor to the contactor. NOT THE INCOMING LINE VOLTAGE.)

STEP 2

Set your ohmmeter to read low ohms. (Down to 1/2 of an ohm may be read.) Check resistance between all combinations of motor terminals. For example, between terminals 1 to 2, 1 to 3, and 2 to 3.

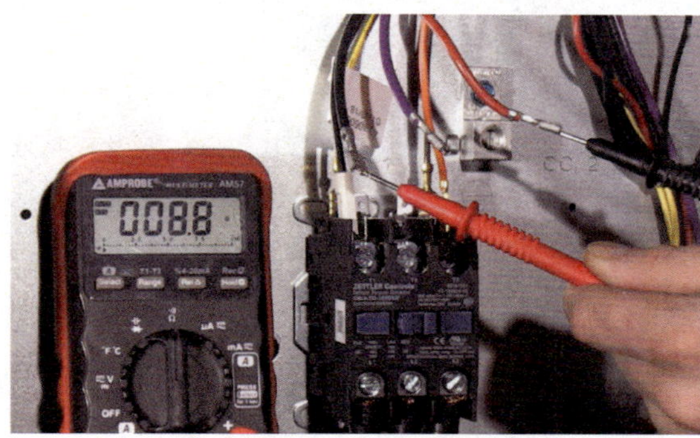

Equal resistance should be measures between all terminal combinations. If infinite resistance is measured between any terminals, the IOL may be open. Allow time for the motor to cool down and the IOL to re-close. This may take up to four hours. If the IOL does not re-close, replace the compressor.

If infinite resistance is measured between any combination of 2 terminals, yet have resistance between the other terminals, the compressor has an open winding. Replace the compressor.

NOTE: some three phase compressors have separate terminals for the IOL. Check the schematic drawing to determine which electrical configuration being worked with.

Three-Phase Voltage Imbalances

When three-phase power is supplied to a commercial building, electricians take off power from different phase legs to power 230VAC single-phase equipment. If the same leg is used for the majority of the single phase equipment, that leg may drop its voltage too far below the other legs. In practice, a maximum voltage imbalance of 2% is all that is allowed. This is due to the fact that large voltage imbalances generate excessive heat in a three phase motor winding, such as those found in expensive three phase compressors.

We can determine the voltage imbalance present at a job site by performing the following test.

Service Procedure: Calculating Voltage Imbalance

Tools:

- Calculator
- Multimeter

Step 1

Run the condensing unit and measure the voltage being supplied to the compressor contactor on all three legs. Make sure the compressor is running.

> *Example:*
> **Leg 1 = 235 Leg 2 = 227 Leg 3 = 229**

STEP 2

Add the three measured voltages and divide by three. This is the average voltage.

> *Example:*
> **(L1)235 + (L2) 227 + (L3) 229 = 691**
> **691 ÷ 3 = 230**

STEP 3

Identify the leg with the largest difference from the average.

> *Example:*
>
> | L1 = 235 | L2 = 227 | L3 = 229 |
> |----------|----------|----------|
> | -230 | -230 | -230 |
> | (5) | -3 | -1 |

STEP 4

Divide the largest difference by the average voltage and multiply the result by 100.

> *Example:*
> **5 ÷ 230 = .0217 × 100 = 2.17%**

STEP 5

The result is the percentage of difference from the average. Note that a maximum percentage of 2% is all that is allowed.

> *Example:*
> **L1 = 2.17%**
>
> *This imbalance creates an increase in the temperature of the winding in the phase where the L1 phase is connected.*

STEP 6

To correct the problem, try moving the lines forward on the compressor contactor lugs, such as L1 to L2, L2 to L3, and L3 goes to L1, and then recheck the voltages. If the problem is not corrected, move the lines forward one more position and retest. If the problem is still not corrected inform the building owner to have an electrician analyze the buildings load distribution.

Failure to correct this type of problem will create the potential for premature motor failure at the compressor and indoor blower motor.

ECM Blower Motors

Some information in this chapter is provided courtesy of Regal-Beloit Corporation

Overview of ECM Motor Technology

What is ECM technology? ECM (Electronically Commutated Motor) technology is based on a brushless DC permanent magnet design that is inherently more efficient than the shaded-pole and permanent-split-capacitor (PSC) motors commonly found in air handlers, furnaces, heat pumps, air conditioners and refrigeration applications throughout the HVACR industry. By combining electronic controls with brushless DC motors, ECMs can maintain efficiency across a wide range of operating speeds. Plus, the electronic controls make the ECM programmable, allowing for advanced characteristics that are impossible to create using conventional motor technologies. The benefit of all of this technology is increased electrical efficiency and the ability to program a more precise operation of the motor, over a wide range of HVAC system performance needs, to enhance consumer comfort.

PSC Motor

Early HVAC literature listed these motors as ICM (Integrated Control Motor), meaning that a control was integrated or used in conjunction with a motor to control its operation. This was later changed to ECM (Electronically Commutated Motor) as they are typically referred to today. The definition of commutate is to reverse the direction of an alternating electric current (the means by which all electric motors rotate). In an ECM, this process is controlled electronically by a microprocessor and electronic controls, which provide the ability to program and control the speed and/or torque of the motor.

ECM Motor

The GE ECM™ motor, currently used by most residential HVAC systems is a brushless DC, three-phase motor with a permanent magnet rotor. Motor phases are sequentially energized by the electronic control, powered from a single-phase supply. These motors are actually made of two components, a motor control (control module) and a motor, sometimes called a motor module.

Motor Control

The motor control is the brains of the device, where single phase (1Ø) 120VAC or 240VAC 60 cycle (Hertz/frequency) power is connected. The control then converts AC power to DC power to operate the internal electronics, thus the name DC motor. The microprocessor in the motor control is programmed to then convert DC power (by means of electronic controls) to a three phase (3Ø) signal to drive the motor, thus the name Three Phase Motor. It also has the added ability to control the frequency (which controls the speed in revolutions per minute) and the amount of torque (current/power) it delivers to the motor.

Motor

The motor is essentially a three-phase motor with a permanent magnet rotor. The permanent magnet rotor contributes to the electrical efficiency of the ECM and also to its sensor-less ability to control the rpm (revolutions per minute) and commutation (when to alternate the cycle). Typical DC motors require brushes to provide the commutation function. This is where the motor gets the name brushless DC motor.

ECM Motor Diagnostics

When servicing ECM blowers, do not automatically assume that the ECM motor has failed. Follow these step to test the ECM motor and ECM motor control before replacing these components.

> **WARNING:** Working on the motor with power connected may result in electrical shock or other conditions that may cause personal injury, death or property damage.
>
> **WARNING:** On Models 2.0/2.3/2.5 always disconnect power from the HVAC system and wait at least 5 minutes before opening the motor, i.e. removing the two bolts from the motor control (end bell) and disconnecting the 3-pin plug to the motor. This to allow the capacitors to dissipate for safety.
>
> **WARNING:** Always disconnect the power from the HVAC system before removing or replacing connectors, servicing the motor, removing the high voltage plug, and before reconnecting.
>
> **WARNING:** Disconnect AC power from the system and make sure the blower wheel has come to a complete stop.
>
> **WARNING:** Do not operate motor without blower wheel attached. Such operation will cause motor to oscillate up and down.
>
> **WARNING:** The failed module must be replaced with the correct direct replacement module from the manufacturer. USING THE WRONG MODULE VOIDS ALL PRODUCT WARRANTIES AND MAY PRODUCE UNEXPECTED RESULTS.

Before troubleshooting the ECM motor, check these system basics:

- Confirm that the correct thermostat input and ONLY the correct input voltage is present at the main control board on the furnace/air handler. Loose or broken low-voltage wires are also potential problem areas and can cause intermittent problems. Use the manufacturer's guide to confirm proper demands (heat or cool), especially on multi-stage systems. Use Sequence of Operation charts and thermostat wiring diagrams to confirm proper wiring and operation.

- Check the setting of the jumper pins or DIP switches on the manufacturer's control board. Do not assume they are correct; use the manufacturer's guide to select the proper airflow, delays, and profiles. Always disconnect the main power to the unit when making these adjustments.

- Check all terminal/plug connections both at the furnace/air handler control board and at the motor. Always disconnect power to the system before disconnecting and reconnecting plugs. Look for loose plugs and/or loose pin connections in the plug, and for burnt, bent or loose pins or seats.

- Confirm there are no limits, rollouts, or safeties tripped. Also check for any fault codes present on the furnace/air handler control boards. If fault codes are present, follow the manufacturer's recommendations to resolve the problem.

- It is normal for the motor to rock back and forth on start up. Do not replace the motor if this is the only symptom identified.

Service Procedure: Motor Not Running

Tools:
- Multimeter
- TECMate PRO

STEP 1

Check for proper high voltage and ground at the 5-pin connector at the motor. These are dual voltage motors capable of operating in 120 or 240 volt systems. Input voltage within +/- 10% of the nominal 120VAC or 240VAC is acceptable. Correct any voltage issues before continuing.

STEP 2

If the motor is a 2.0/2.3 motor and has proper high voltage and ground at the 5-pin connector, go to procedure "GE ECM™ TECMate PRO ECM Motor Troubleshooting". If the motor is a 2.5 motor and has proper high voltage and ground, reference the equipment manufacturer's manuals for further check procedures.

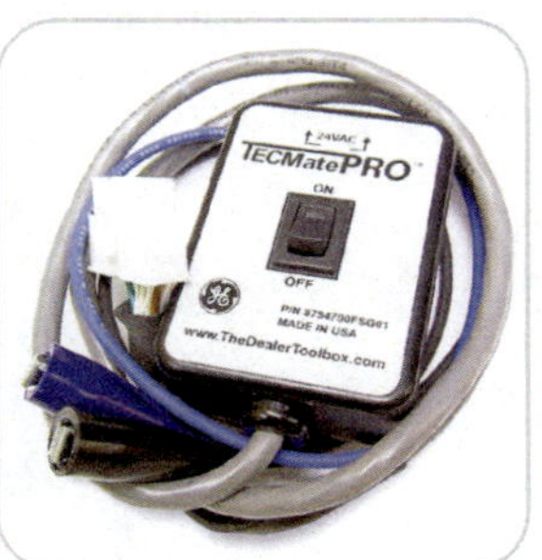

Service Procedure: Motor Running Poorly

Perform this procedure if the system is excessively noisy, does not appear to change speeds in response to a heat or cool demand, or is having symptoms during the cycle such as a tripping limit or freezing coil.

Tools:

- Multimeter
- TECMate PRO

STEP 1

Wait for programmed delays to time out. If delays are too long, then reset them using the manufacturer's charts.

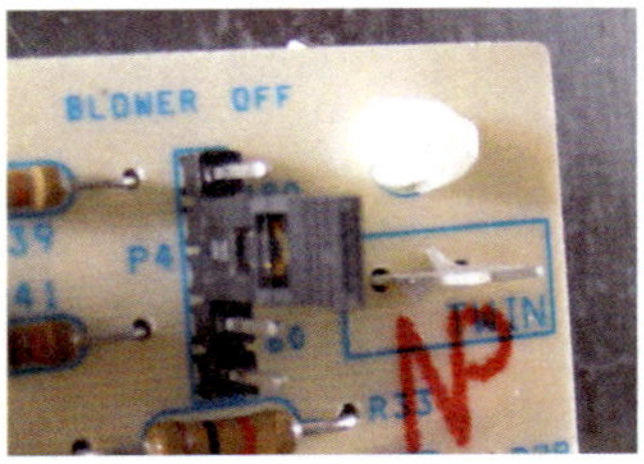

STEP 2

Ensure the airflow settings are correct for the installed system using the manufacturer's charts. Remember that the change in airflow between continuous fan and low stages of operation may be very slight depending on the size of the system.

STEP 3

Remove the filter and check that all of the dampers, registers, and grills are open and free flowing. If removing the filter corrects the problem, replace with a clean or less restrictive filter.

STEP 4

Check and clean the blower wheel, secondary heat exchanger (if applicable), and evaporator coil (if applicable).

STEP 5

Check the external static pressure (See Chapter J: Air Volume). If it is higher than the manufacturer's recommendations, correct the airflow restriction.

STEP 6

If the motor does not shut off at the end of the cycle, check the delay times and wait for the delay to time out. Also, make sure there is no call for continuous fan" on the G terminal. The motor may take a while to come to a complete stop with the selected delays and normal ramp down time.

ECM Blower Motors

STEP 7

Check for proper high voltage and ground at the 5-pin connector at the motor. These are dual voltage motors capable of operating with 120VAC or 240VAC systems. Input voltage within +/- 10% of the nominal 120VAC or 240VAC is acceptable. Correct any voltage issues before continuing.

STEP 8

If the motor is a 2.0/2.3 motor and has proper high voltage and ground at the 5-pin connector, go to procedure "GE ECM™ TECMate PRO ECM Motor Troubleshooting". If the motor is a 2.5 motor and has proper high voltage and ground, reference the equipment manufacturer's manuals for further check procedures.

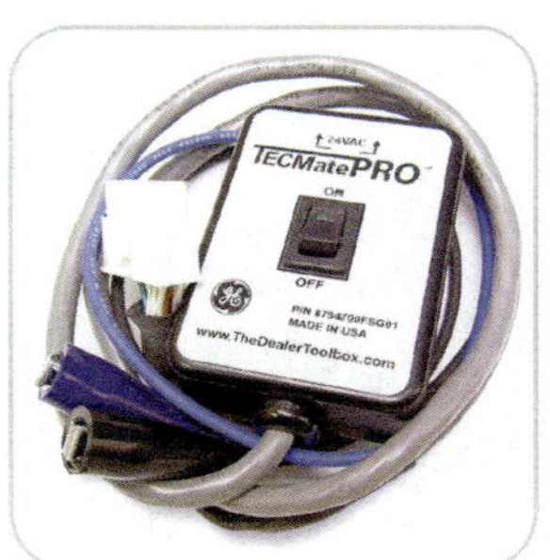

GE ECM™ TECMate PRO
ECM Motor Troubleshooting
(for 2.0/2.3 motors only)

The GE ECM™ TECMate PRO is designed to assist field service technicians by troubleshooting GE ECM™ 2.0 and 2.3 motors independently of the HVAC system. Analysis of field returns shows that quite often the GE ECM™ is misdiagnosed as faulty. This is due in large part because of the unavailability of effective

GE ECM™ TECMate PRO

troubleshooting tools for the GE ECM™ motor. The TECMate PRO is designed to isolate a motor failure from other HVAC system controller failures. GE ECM™ motors are used in one of two modes: Thermostat Mode and Variable Speed Mode. Thermostat Mode is controlled by a 24VAC signal usually from a thermostat, whereas Variable Speed Mode is controlled by a Pulse Width Modulating (PWM) signal. In either mode, the TECMate PRO is capable of identifying a motor failure versus other HVAC controller failures.

Tools:

- TECMate PRO
- Socket Wrench/Sockets
- Multimeter

STEP 1

Disconnect power from the furnace, air handler, or system being serviced.

STEP 2

Remove the 16-pin connector from the motor, and connect the 16-pin connector from the TECMate PRO to the motor. Do not disconnect the 5-pin AC power connector from the motor.

STEP 3

Connect the two alligator clips from the TECMate PRO to 24VAC.

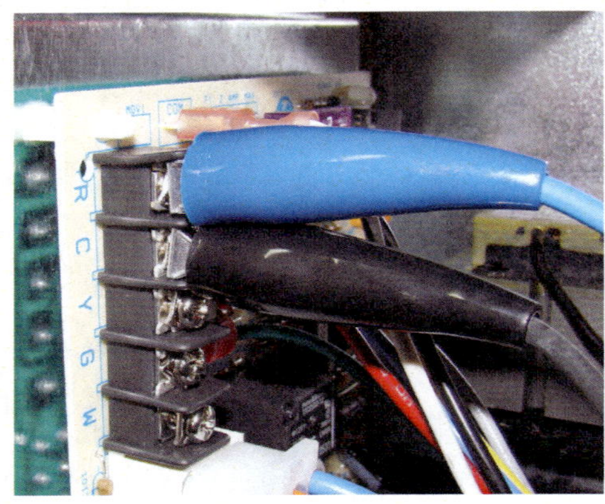

STEP 4

Place the switch on the TECMate PRO in the off position.

STEP 5

Restore power to the system.

STEP 6

Place the switch in the On position and observe the motor. If the motor starts, the motor is good.

STEP 7

If the motor passes the test, then the fault is with some other component in the HVAC system.

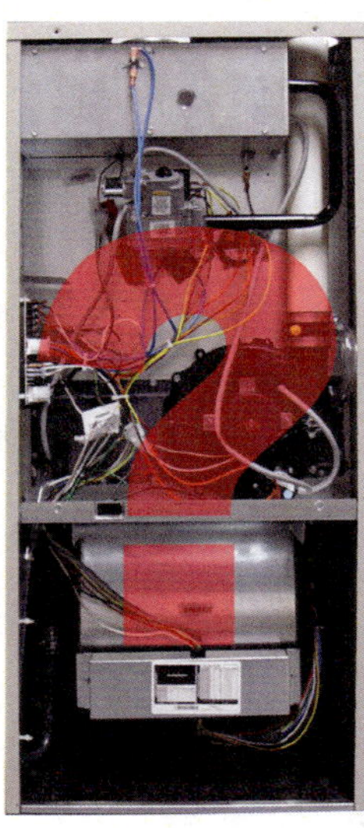

STEP 8

If the motor does not start, then proceed to replace the motor control. NOTE: Before replacing the control, test to insure the motor is not damaged: (1) check insulation, and (2) check for mechanical integrity by rotating shaft by hand. The following steps describe these checks.

STEP 9

Remove the control by locating and removing the two bolts from the back of the control. The control module is now free of mechanical attachment to the motor end shield, but is still connected by a plug and three wires inside the control.

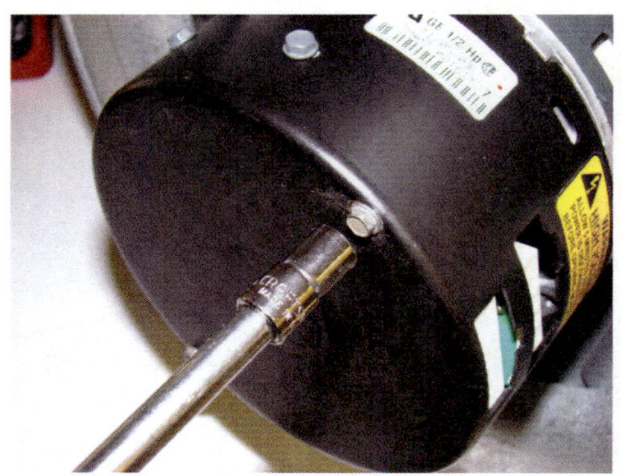

STEP 10

Carefully remove and rotate the control to gain access to the plug at the control end of the wires. Release the plug latch tab and gently pull the plug out of the connector socket in the control. DO NOT PULL ON THE WIRES. GRIP THE PLUG ONLY. The control module is now completely detached from the motor.

STEP 11

Use an ohmmeter to verify that the resistance from each connector pin (in the motor plug just removed) to the motor shell is greater than 100K ohms (measure to unpainted motor end plate).

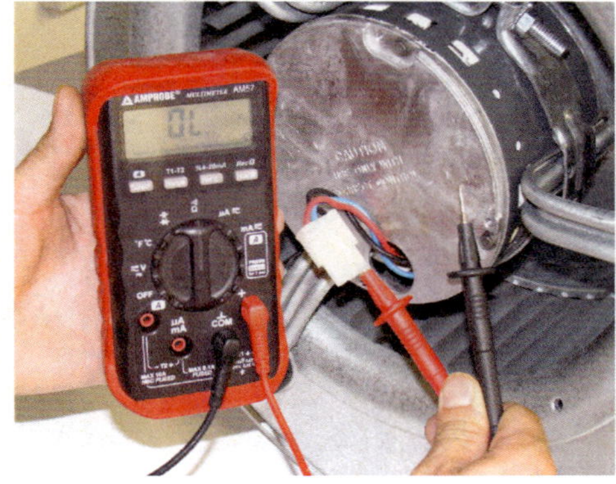

STEP 12

Use an ohmmeter to verify that the resistance from each connector pin to the other two connector pins, is similar and less than 20K ohms.

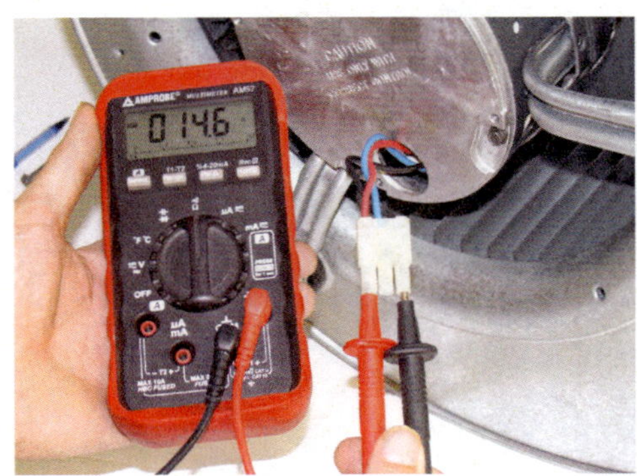

STEP 13

Rotate the motor shaft to test for rubbing and/or mechanical defect.

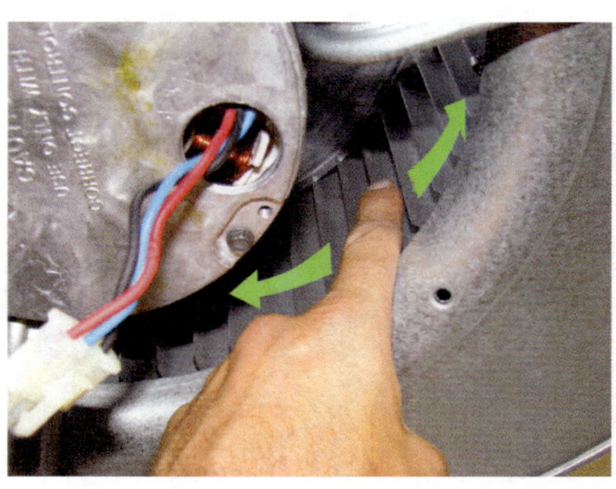

STEP 14

If any connector pin fails this test and/or the shaft doesn't spin easily, the motor and control is defective and must be replaced. Contact your dealer for a replacement motor. If the motor passes this test, but failed the Thermostat and Variable Speed Mode tested by the TECMate PRO, then only the control needs to be replaced. Contact your authorized dealer for a replacement control.

Service Procedure: Replacing the ECM Control Module

Tools:

- Screwdrivers
- Socket Wrench/Sockets

STEP 1

After the system power has been off for 5 minutes, unplug the 16-pin connector and the 5-pin connector from the motor control.

STEP 2

Remove the blower assembly from the HVAC system and remove the two hex-head screws from the back of the control.

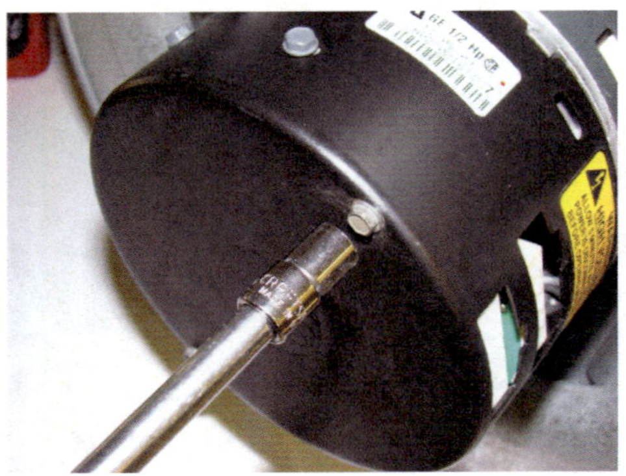

STEP 3

Unplug the 3-pin connector from the inside of the control by squeezing the latch and gently pulling on the connector.

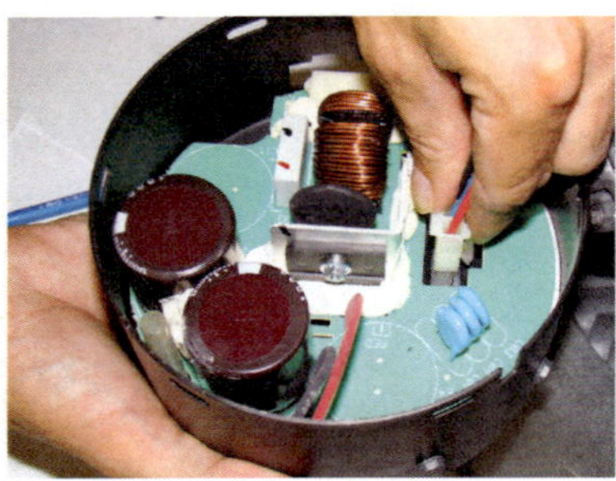

STEP 4

Insert the 3-pin connector into the new control module. A slight click will be heard when inserted properly. This connector is keyed for proper connection direction.

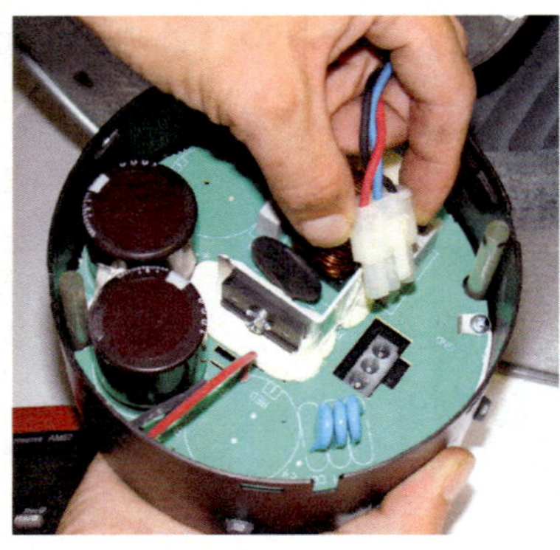

STEP 5

Orient the new control to the motor's endshield with connectors facing down, insert bolts, and tighten. If replacing an ECM 2.0 control with an ECM 2.3 control, insert plastic tab into perimeter of replacement control and align tab with mating hole in the endshield. Use the new shorter bolts provided to ensure a secure attachment. Orient the control connectors to the endshield between 4 and 8 o'clock positions

STEP 6

Reinstall the blower/motor assembly into the HVAC system by following the manufacturer's guidelines.

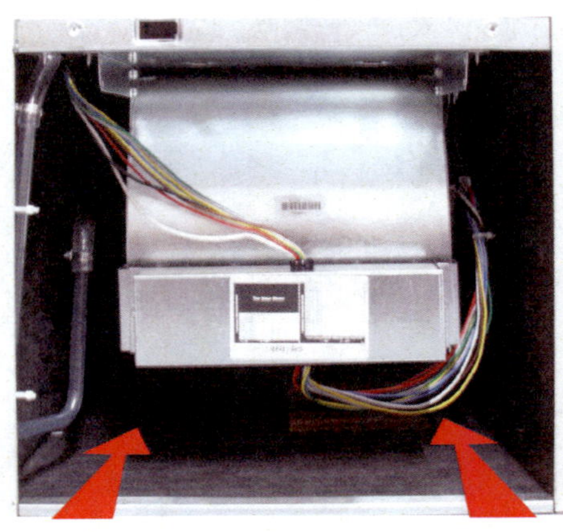

STEP 7

Plug the 16-pin connector and 5-pin connector back into the motor. The connectors are keyed. Observe proper orientation.

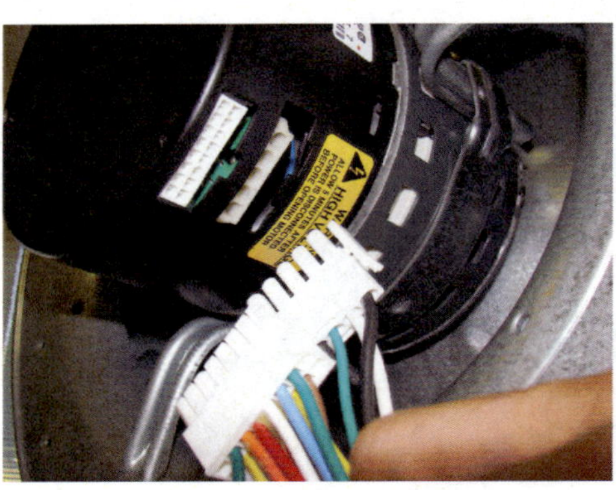

STEP 8

Be certain to form a drip loop so that water cannot enter the motor by draining down the cables.

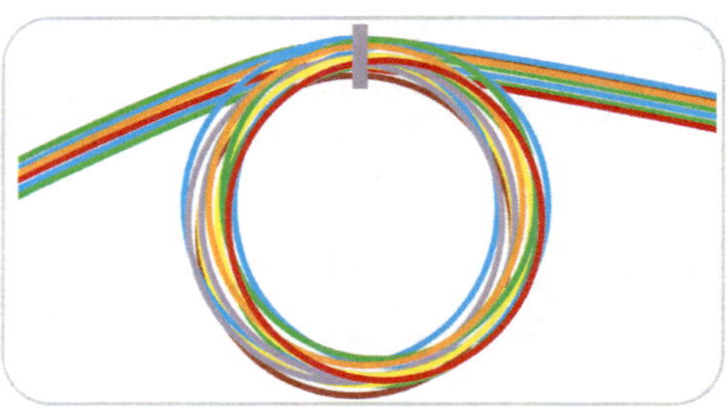

Service Procedure: Replacing the ECM Motor Module

Tools:
- Screwdrivers
- Socket Wrench/Sockets

STEP 1

After the system power has been off for 5 minutes, unplug the 16-pin connector and the 5-pin connector from the motor control.

STEP 2

Remove the blower assembly from the HVAC system and remove the two hex-head screws from the back of the control.

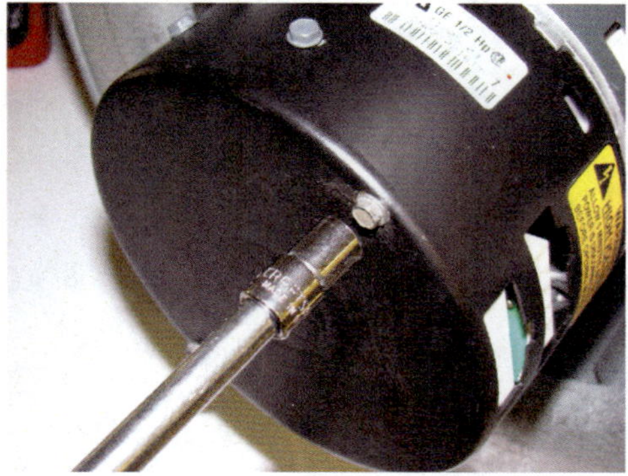

STEP 3

Unplug the 3-pin connector from the inside of the control by squeezing the latch and gently pulling on the connector.

STEP 4

Loosen the bolt securing the blower wheel to the motor shaft.

STEP 5

Loosen the bolt securing the belly band around the motor.

STEP 6

Carefully remove the motor from the belly band and blower wheel.

STEP 7

Slide the new motor module into the belly band and through the blower wheel. Ensure the belly band is not covering any shell holes.

STEP 8

The motor module does not have a specific orientation, however, the motor control does. Make sure the three-wire plug will reach the motor control when it is oriented properly before tightening the belly band.

STEP 9

Tighten the belly band and secure the blower wheel to the motor shaft.

STEP 10

After installing the new motor module, reinstall the motor control as described previously in "Replacing the ECM Control Module"

Final Checks

☐ Check all wiring and connections, especially those removed while servicing.

Ensure the system is set up as follows:

☐ Verify the condensate drain is not plugged or clogged.

☐ Reconnect power to the HVAC system and verify that the motor is working properly.

☐ Check and plug leaks in return ducts and equipment cabinet.

☐ Verify that the system is running quietly and smoothly, in all modes (heating, cooling, and continuous fan) and all stages (if applicable).

☐ Return all thermostat settings to the customer's preference.

If this is a repeat failure, it is important to check the following:

☐ If any evidence of moisture, correct the issue.

☐ If the area is subject to increased lightning strike activity, the use of additional transient voltage protection may be helpful.

PSC Blower Motors

PSC Blower Introduction

Permanent Split Capacitor (PSC) blower motors are the most common type of motor found in residential HVAC systems. These motors are used to power compressors, condenser fan assemblies, and indoor air blower assemblies. The motors used in residential systems are fractional horsepower motors.

Indoor fan assemblies feature multi-speed PSC motors that are directly coupled to the blower wheel. These motors are called direct drive motors and consist of the motor and matching run capacitor.

Operating Characteristics of a PSC Motor

Permanent split capacitor motors have two internal motor windings, a run winding, and an auxiliary start winding. The two windings, when energized, create a powerful magnetic field that rotates around the rotor of the motor. The magnetic field alternately pushes and pulls the motor rotor in a circular motion to turn the motor shaft.

The run capacitor discharges voltage in time with the voltage coming into the motor. When the AC sinewave changes polarity, the capacitor charges and discharges. The discharged voltage generates the best alignment of magnetic fields to increase motor power. The run capacitor is selected by the motor manufacturer to have a microfarad rating range that will get optimum performance from the motor.

Run Capacitor

In the illustration below, a typical PSC blower motor circuit is shown. The motor has various speed taps that can be energized to change blower speeds. Be aware that when the blower is running, or when voltage is applied to the motor, all of these taps will have voltage present! Voltage cannot be applied to more than one speed tap at a time. If this happens, the motor windings will fail.

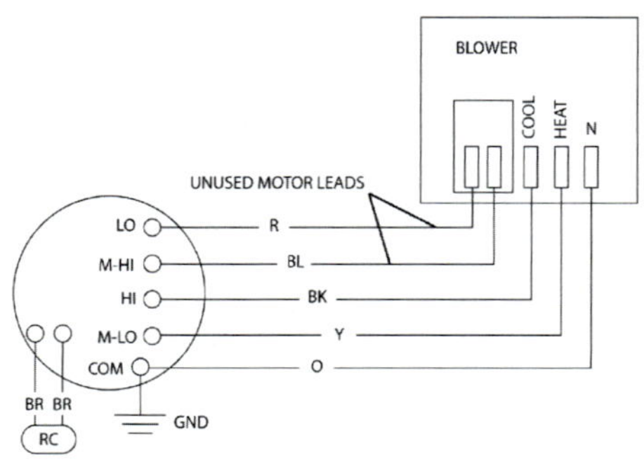

PSC Motor Failures

Bearing Seizure

PSC motors can suffer bearing seizure if excessive strain is placed on the motor due to issues like blower wheel problems or excessive dirt buildup. If the motor has a bearing problem, the bearings are not replaceable. The motor must be replaced.

To test for a motor bearing problem, turn power off to the furnace or air handler. Reach into the blower housing and spin the blower wheel. The wheel should freely spin. If the motor does not spin, make sure the blower wheel is not dragging on the blower housing. If it is, repair the problem.

If no blower wheel issues are found, yet the motor will not spin, or makes grinding noises as it turns, replace the motor.

Electrical Failure

PSC motors can have an electrical failure of the motor windings. One of the windings could be broken (open), or the internal overload can open and fail to reset. The motor could have an electrical winding short to ground. The run capacitor can also fail, which will cause the blower to run slow or trip its internal overload protection.

Tools:
- Multimeter

Check for proper line voltage to the motor as it tries to start. If voltage is incorrect, find the cause of the problem. It is likely to be in the furnace control board. If proper voltage is present at the motor, yet the motor hums and goes off on internal overload, remove power to the furnace.

PSC Motors

Service Procedure: Run Capacitor Check

Tools:
- Multimeter

STEP 1

Remove power to the furnace. Properly discharge the capacitor.

STEP 2

Remove the two motor wires that attach to the run capacitor. Set multimeter to the Capacitor Test mode. In this mode, the measured microfarad rating of the capacitor will be displayed on the meter.

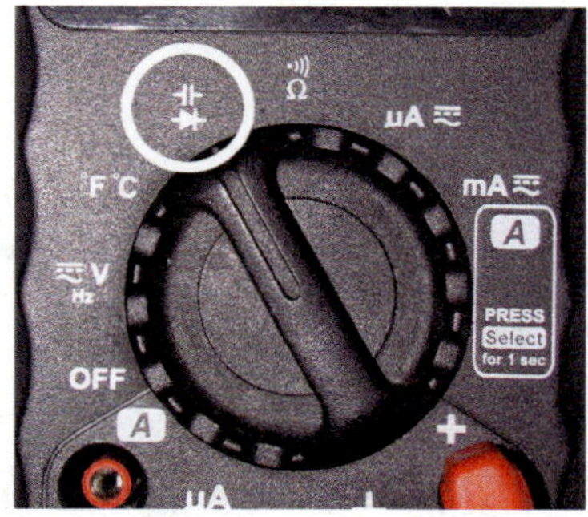

STEP 3

Place a meter lead on each capacitor terminal. Read the microfarad level of the capacitor. It should be within 10% of the rating printed on the capacitor. If it is, the capacitor is good. If it is out of range, replace it with one of equal rating and try running the motor. If the reading is within 10% of microfarad rating, the capacitor is good.

Service Procedure: Checking for an Open Motor Winding
(Open Internal Overload)

Tools:
- Multimeter

STEP 1

Remove power to the furnace and properly discharge the run capacitor. Remove the wires from the run capacitor.

STEP 2

Set the multimeter to read OHMS. See the motor label to find the wiring diagram. Identify the motor's two line voltage wires. In this case, they are the orange wire and the black wire. The black wire is the high speed motor winding. The orange wire is the common connection of the motor. Read the resistance from the common motor winding wire to all of the motor speed tap wires.

The resistance will increase as the speed taps are measured from high to low. The high speed tap will have lower resistance than the low speed tap.

If there is an infinite resistance read through any winding, replace the motor. If the motor windings all check out with proper resistance, go to the next step.

If there is infinite resistance to all motor winding speed taps, the internal overload inside the motor may have opened. If the motor is hot, allow time for the motor to cool down and recheck. If the motor is cool, yet infinite resistance is measured, replace the motor.

Find the two capacitor wires. Measure the resistance from each wire to the common wire identified in STEP 2.

One wire will read a 0 resistance. The other wire should measure resistance. This is the start winding. If infinite resistance is measured on the start winding, the winding is open. Replace the motor.

Service Procedure: Checking for an Electrically Grounded Motor

Tools:
- Multimeter

STEP 1
Remove power to the furnace and properly discharge the run capacitor. Remove the wires from the run capacitor.

STEP 2
Measure the resistance between each motor wire to a bare metal portion the motor frame.

Infinite resistance should be measured from each lead to the motor frame. If there is measurable resistance from any motor speed wire to the frame, replace the motor.

What to Check When a Problem Occurs

Motor hums, tries to start, then goes off on internal overload:
- Improper line voltage
- Mechanical failure of the motor or binding blower wheel
- Bad run capacitor
- Motor winding open
- Grounded motor

Motor runs slow:
- Improper voltage to the motor
- Excessive duct static pressure
- Bad run capacitor

Motor does not try to start, proper voltage is applied to motor:
- Open motor winding
- Open internal motor overload protector. (Allow motor to cool down)

Motor trips circuit breaker to air handler/furnace:
- Motor winding short to ground
- Capacitor short to ground

PSC Blower Motor Troubleshooting Summary
- Check for proper voltage to the motor first.
- Check run capacitor for proper microfarad range.
- Read the motor schematic to find the common and run winding taps.
- Check all wires for grounding to the motor frame.

Air Volume

Air Volume Measurement
Introduction

Many service technicians and installers have never measured how much air the blower assembly is moving. An HVAC system that is serviced without having had an air volume calculation done is likely operating with an improper charge, unknown operating characteristics, unknown efficiency, uncertain capacity, and uncertain reliability.

Here is a simple technical explanation of why an air conditioner's air volume level must be known:

The evaporator coil sits in the air stream at the indoor blower assembly. Refrigerant circulates through the evaporator and absorbs heat from the air. If the air volume is low, the coil will absorb less heat than it should. Since heat levels change pressure and superheat, the system will operate at a pressure level that is too low. The natural tendency of the technician servicing the system will be to add refrigerant in an attempt to raise pressure. The system will now operate with excessive charge and a lack of heat.

A system charged without enough air volume and an overcharge will include problems including compressor starting problems, capacity complaints, premature failure, poor efficiency, and poor reliability.

Air Volume Requirement

Air conditioning system performance is engineered for a specific amount of indoor air passing across the surface of the evaporator coil. The amount of air is measured in cubic feet per minute (CFM). Older generation air conditioning systems typically require 400 CFM of indoor air for every 1 ton of cooling capacity. For example, a 2-ton unit would need 800 CFM of indoor air, since 2 tons times 400 equals 800. It is important to know that air has weight and every pound of air contains a certain number of Btus in it. When the system is designed, the amount of air is set and the heat in the air is measured. This is important to understand as a system can only absorb as much heat as is available. There are two ways the amount of heat fluctuates; by the number of pounds of air flowing, and by the amount of heat in the air. If the heat in the air falls, the system capacity falls. If the number of pounds of air falls, the system capacity falls.

By properly configuring the air volume, the system will always receive the proper amount of air. However, as the temperature in the home changes, the amount of heat in each pound of air will fluctuate. This is not a problem as the amount of heat in the air is measured in direct relationship to wet bulb temperature of the air. Manufacturer charging charts always plot wet bulb temperature as a requirement because it tells them how much heat is in each pound of air. The charts ASSUME that indoor air volume has been properly measured and adjusted so the evaporator coil always receives the proper amount of heat based on the wet bulb temperature of the return air.

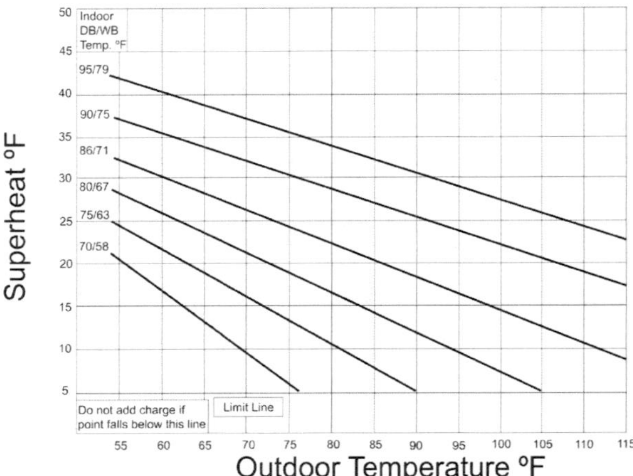

In summary, capacity changes with the amount of heat that is absorbed at the evaporator coil. Manufacturers have established a relationship between indoor air wet bulb temperature and the amount of heat energy contained in a pound of air. By configuring the indoor blower to deliver the specified amount of indoor air CFM, the system will have a fixed constant that allows for wet bulb temperature to be referenced to system suction pressure. If wet bulb is measured and used to plot required operating pressures, but the air volume is incorrect, the charging chart is useless.

Measuring air volume takes about 10 minutes. If not performed, it is likely that the system will not function properly.

There are two methods of air volume measurement that can be easily performed by a qualified service technician. The two methods are Temperature Rise Method, and the External Static Pressure Method. Both methods of air volume measurement will be discussed.

Temperature Rise Method

In the temperature rise method, the heating section of the system is operated with the indoor fan speed set to cooling speed. The temperature of the air entering and leaving the air handler or furnace is measured to find the temperature rise. The BTUH Output of the heating system must also be determined to perform this method. There are two separate calculations for determining BTUH output, one for gas furnaces, and the other for electric heat air handlers. If the system does not have a heat option, this method will not work.

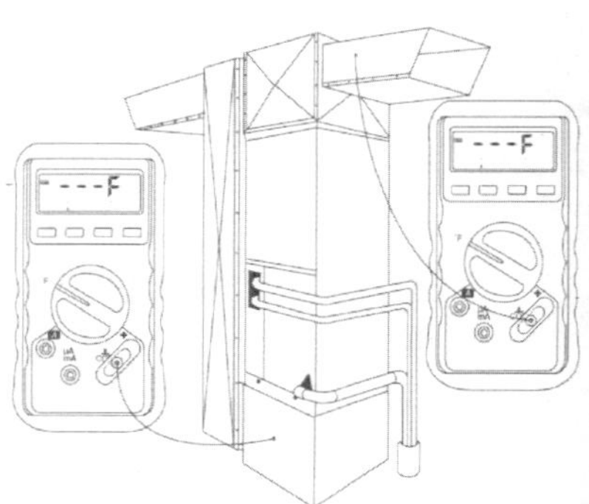

Temperature Rise Method

External Static Pressure Method

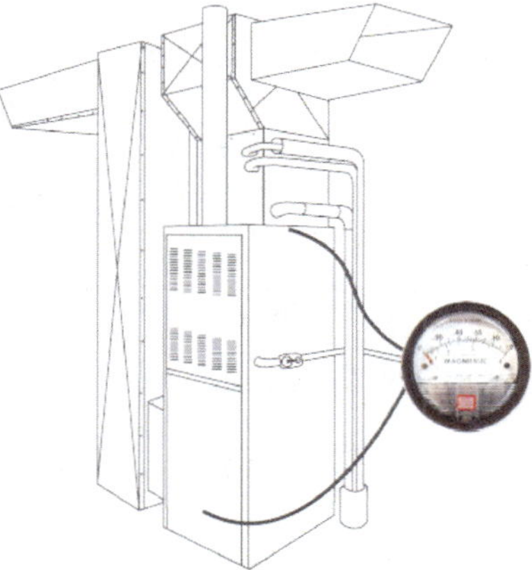

In the external static pressure method, a manometer is used to measure the air pressure at the inlet and outlet of the indoor blower assembly. The measured air pressure is compared to charts provided by the equipment manufacturer to determine how much air volume is flowing in cubic feet per minute (CFM).

External Static Pressure Method

Determining BTUH Output

Gas Furnace BTUH Output

In this series of steps, it is necessary to know the heat content level of the natural gas being delivered to the home. The heat contained in a cubic foot of natural gas can range from just below or above 1050 BTU per cubic foot. To obtain the specific heat content of the natural gas in your area, contact the local gas provider.

Furnace Nameplate

If it is not possible to obtain the heat content of the gas from the gas provider, the furnace nameplate BTUH input should be used to determine the BTUH output of the furnace. Be aware that calculated air volume will not be precise (Deviation will be due to fluctuations in heat content of the fuel).

If the furnace nameplate lists BTUH Input, use the following formula:

BTUH Input x Furnace Efficiency = BTUH Output

Furnace Nameplate Example

CATEGORY 1 FORCED AIR FURNACE FOR INDOOR INSTALLATION ONLY IN A BUILDING
CONSTRUCTED ON-SITE
INSTALLATION
ELECTRIC 115 V.   60 HZ   1 PH,   MAXIMUM TOTAL INPUT 8.5 AMPS
FACTORY EQUIPPED FOR NATURAL GAS
EQUIPPED WITH NO. 45 DRILL SIZE ORIFICE
HOURLY INPUT RATING MANIFOLD PRESSURE   NATURAL GAS PROPANE
BTUH/H   (KW)   INCHES W.C. PO/ (KPA)   INCHES W.C. (KPA)
80,000   (23.44)   3.5   (87)
MAXIMUM PERMISSIBLE GAS SUPPLY PRESSURE TO FURNACE
MINIMUM GAS SUPPLY PRESSURE FOR PURPOSES OF INPUT ADJUSTMENT
LOW INPUT
LIMIT SETTING
DESIGN MAXIMUM

In this example Input BTUH 80,000 BTUH times .80 efficiency = 64,000 BTUH OUTPUT (Estimated)

Clocking the Gas Meter

This method is the most accurate way of determining the BTUH Output of a gas fired furnace. The furnace will be operated to determine how many cubic feet of natural gas enter the furnace burners. The input will then be multiplied by the efficiency of the furnace to determine BTUH Output.

The heat content of a cubic foot of gas should be found by contacting the local gas company. It may be difficult to obtain this information from the local gas company as the heat content of 1 cubic foot of natural gas fluctuates from day to day. If the heat content of the fuel can be determined by making a phone call, it is well worth the effort as the accurate BTUH Input of the gas furnace can be found.

If the actual heat content of the gas cannot be obtained, use a default value of 1050 BTU/cubic foot of natural gas for calculations. Be aware, there will be some inaccuracy in the calculations performed in this procedure if using the default value.

When clocking the gas meter, all other gas appliances in the home should be off. Keep any pilot lights lit. The furnace should be operating at proper manifold pressure as stated on the furnace nameplate. The furnace should be calling for heat with all stages of capacity firing.

Service Procedure: Determine BTUH Output by Clocking the Gas Meter

Tools:
- Gas Pressure Manometer
- Watch/Timing Device

STEP 1

Find the BTUH content of a cubic foot of natural gas. This information is available from your local gas company. The default average used in generating air volume tables is 1050 BTU per Cubic Foot.

STEP 2

Shut off all gas appliances in the home. Do not shut off the pilot lights to any appliances. The only gas appliance that should be operating in this test is the gas furnace.

Set the blower to operate at cooling speed during heat mode operation. This can be accomplished by simply setting the fan switch on the thermostat to the ON position.

STEP 3

Install a gas pressure manometer onto the furnace manifold pressure tap.

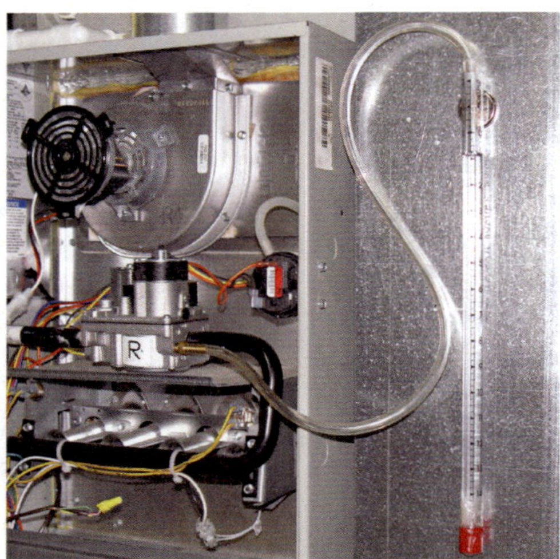

STEP 4

Call for heat. Make sure all furnace stages are calling for heat and the burners are operating at high fire rate. Check manifold pressure versus the nameplate requirement. Make adjustments as needed.

CATEGORY 1 FORCED AIR FURNACE FOR INDOOR INSTALLATION ONLY
IN A BUILDING
CONSTRUCTED ON-SITE
INSTALLATION
ELECTRIC 115 V.   60 HZ   1 PH,   MAXIMUM TOTAL INPUT 8.5 AMPS
FACTORY EQUIPPED FOR NATURAL GAS
EQUIPPED WITH NO. 45 DRILL SIZE ORIFICE

HOURLY INPUT RATING	MANIFOLD PRESSURE	NATURAL GAS PROPANE
BTUH/H (KW)	INCHES W.C. PO/ (KPA)	INCHES W.C. (KPA)
80,000 (23.44)	3.5 (87)	

MAXIMUM PERMISSIBLE GAS SUPPLY PRESSURE TO FURNACE
MINIMUM GAS SUPPLY PRESSURE FOR PURPOSES OF INPUT ADJUST-
MENT
LOW INPUT
LIMIT SETTING
DESIGN MAXIMUM

STEP 5

With the furnace operating at high fire rate, go to the home's gas meter. Time the number of seconds it takes for the gas meter small scale to complete one revolution on the dial.

STEP 6

Determine the scale of the dial on the gas meter. The scale needs to be found so that total cubic feet of natural gas entering the home can be found using the chart below. Plot time measured on the chart to determine cubic feet of gas being burned by the furnace per hour.

GAS FLOW IN CUBIC FEET PER HOUR (2 CUBIC FOOT DIAL)							
Sec.	Flo	Sec.	Flo	Sec.	Flo	Sec.	Flo
8	900	29	248	50	144	82	88
9	800	30	240	51	141	84	86
10	720	31	232	52	138	86	84
11	655	32	225	53	136	88	82
12	600	33	218	54	133	90	80
13	555	34	212	55	131	92	78
14	514	35	206	56	129	94	76
15	480	36	200	57	126	96	75
16	450	37	195	58	124	98	73
17	424	38	189	59	122	100	72
18	400	39	185	60	120	104	69
19	379	40	180	62	116	108	67
20	360	41	176	64	112	112	64
21	343	42	172	66	109	116	62
22	327	43	167	68	106	120	60
23	313	44	164	70	103	124	58
24	300	45	160	72	100	128	56
25	288	46	157	74	97	132	54
26	277	47	153	76	95	136	53
27	267	48	150	78	92	140	51
28	257	49	147	80	90	144	50

For 1 Cu. Ft. Dial:
Gas Flow CFH = Chart Flow Reading divided by 2

For ½ Cu. Ft. Dial:
Gas flow CFH = Chart Flow Reading divided by 4

Example: If it took 36 seconds for 1 revolution of the gas meter dial and the dial was a 1 cubic foot dial, the cubic foot used is 200 divided by 2 = 100 cubic feet of fuel per hour.

STEP 7

Multiply the cubic feet of gas burned per hour by the heat content value obtained from the gas supplier. (1050 BTU default average) The answer is BTUH INPUT to the furnace.

Example: Lets say we timed 36 seconds per revolution on a 1 Cu. Ft. dial, the furnace is an 80% furnace:

36 seconds = 100 cubic feet per hour

100 x 1050 = 105,000 BTUH Input

STEP 8

Multiply BTUH Input obtained in step 7 by furnace efficiency to determine BTUH Output.

Example: Assume an 80% efficient furnace:

105,000 BTUH Input x .80 =

84,000 BTUH Output

Electric Heat BTUH Output

To calculate the airflow CFM for electric heat air handler systems, the heating output of the electric heaters in the air handler in BTUH output per hour must be measured. It is easy to determine the heating output of the single phase electric heaters.

The formula for finding BTUH Output is:

BTUH Output = Volts x Amps x 3.413

Service Procedure: Determining Electric Heat BTUH Output

Tools:

- Multimeter
- Ammeter

STEP 1

Set the blower fan speed for cool mode speed. The blower must run at cooling mode speed during heat mode operation. Run the air handler in the heating mode of operation by calling for heat. Make sure to energize all of the heaters if the heater section has more than 1 stage of capacity.

STEP 2

When the heaters are all on, measure the voltage to the heaters.

Example: 200VAC

STEP 3

Measure the amps being drawn by the heaters.

Example: 50 Amps

STEP 4

Apply your readings to the formula:

BTUH Output = Volts x Amps x 3.413

Example:

200VAC x 50A = 10,000

10,000 x 3.413 =

34,130 BTUH Output

Formula for Three Phase Electric heater BTUH Output

Although not found in residential applications, a three-phase heater may be encountered. The formula for BTUH output is slightly different than single phase heaters:

Volts x Amps x 3.413 x 1.73 = BTUH OUTPUT

Temperature Rise

The temperature rise that occurs across the gas furnace or air handler will be measured with the unit operating in heating mode. The indoor blower motor must be running at its cooling speed during this portion of the procedure. Place the blower speed tap to operate at cooling speed during heat mode operation by switching motor leads at the indoor furnace or air handler control board terminals for fan speed selection.

It is recommended to use the same thermometer for measuring supply air and return air temperatures to avoid any deviances between thermometers. The supply air temperature should be taken out of direct sight of heat exchangers (gas furnace) or heating elements (electric heat air handler). The temperature difference between the supply air and return air temperatures is called the Delta T.

Service Procedure: Determining Temperature Rise

Tools:
- Thermometer

STEP 1

Configure the furnace to run at cooling speed in the heating mode. With all stages of heat calling, measure the return air and the supply air temperature. Take the supply air temperature reading out of direct sight of heat exchangers.

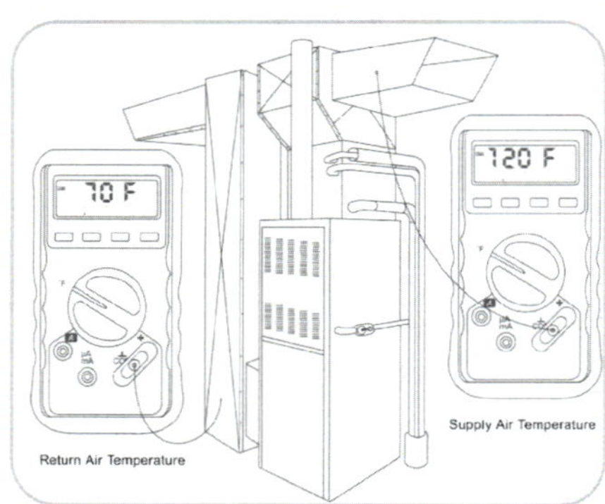

Example: Supply Air Temp: 120°F, Return Air Temp: 70°F

STEP 2

Subtract the return air temperature from the supply air temperature. The result is the temperature rise, also called the Delta T.

Example:
120°F - 70°F = 50°F
Delta T = 50

Calculating Air Volume CFM: Temperature Rise Method

With all information gathered properly - BTUH output & temperature rise (ΔT) - the indoor air volume in CFM may be determined.

The altitude correction factor must also be found to complete the information gathering needed to find CFM. This table is shown below. It lists altitude correction factors for various altitudes. The acronym MSL stands for mean sea level.

Altitude MSL	Correction Factor
<1000	1.08
1000	1.04
2000	1
3000	0.97
4000	0.93
5000	0.9
6000	0.86

The formula for finding CFM for both gas furnace and electric heat air handler systems is:

$$CFM = \frac{BTUH\ Output}{\Delta T \times Altitude\ Correction\ Factor}$$

Service Procedure: Calculating CFM Using the Temperature Rise Method

Tools:
- Calculator

Step 1
Get furnace or air handler BTUH output

Example: Gas furnace BTUH output = 84,000 BTUH

Step 2
Get Furnace or Air Handler Temperature Rise (ΔT)

Example: Gas furnace ΔT = 50

Step 3
Get Furnace or Air Handler Altitude Correction Factor

Example: Gas furnace altitude = 600 ft MSL
Altitude Correction Factor = 1.08

Step 4
Apply BTUH, ΔT, and Altitude Correction Factor to formula:

Example:
$$CFM = \frac{84,000}{50 \times 1.08} = \frac{84,000}{54} = 1,556\ CFM$$

If the Air Volume is Too High or Too Low

It is highly unlikely that the measured air volume will exactly match the 400 CFM per 1 ton of cooling capacity requirement. Acceptable ranges of air volume are between 350 CFM - 450 CFM per ton of cooling capacity.

If the air volume is too high, select the next lowest fan speed and recalculate temperature rise to find the new CFM. If the air volume is too low, increase the fan speed and repeat the temperature rise portion of the equation and find new CFM.

If the fan speed taps are at their highest setting and air volume is too low, diagnose the cause of low heat load on the evaporator coil. (See Chapter E: Refrigeration Circuit Diagnostics.)

Air Volume CFM: Temperature Rise Method Summary

- The formula for both methods is the same.
- The method of determining BTUH Output is different.
- Clocking the meter and just multiplying BTUH input times efficiency to determine CFM for gas furnaces is not an exact measurement.
- Take temperature measurements on the supply air side out of direct sight of heaters.

External Static Pressure

Introduction

External Static Pressure (ESP) is a measurement of the air pressure at the inlet and outlet of the indoor air blower assembly. This pressure is similar to the pressure an inflated balloon has pushing against its internal surface. An air blower has this same type of pressure at its inlet and outlet.

Special probes called static pressure probes are placed into the ducting airstream at the inlet and outlet of the air handler or furnace. These probes have holes drilled into them that are used to sense the static pressure of air in that area.

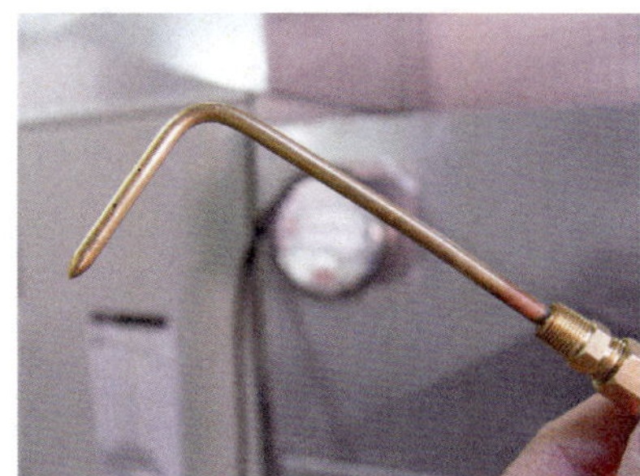

Static Pressure Probe

The static pressure probes are connected to a differential pressure meter, such as a magnehelic gauge. The magnehelic gauge has two ports, a high pressure port, and a low pressure port. The scale on the gauge should read a maximum of 1" of water column. Higher scales cannot be used for air pressure readings common in residential homes.

Magnehelic Gauge

The amount of static pressure at the fan inlet and outlet will have a direct impact upon how much air volume the fan can deliver. As the static pressure across the fan goes up, the ability to move air goes down.

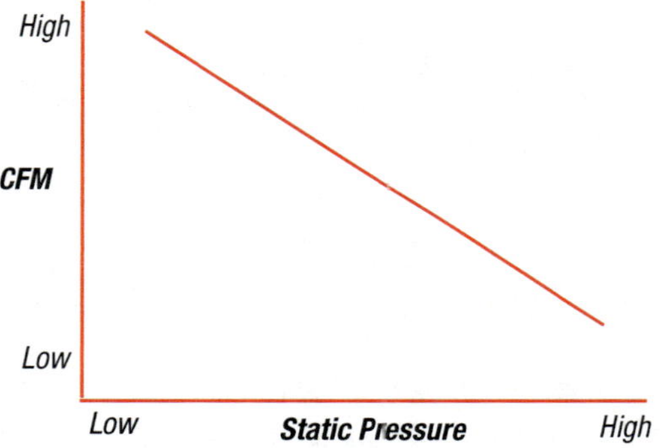

As this graph shows, the higher the static pressure, the less air will be able to be moved through the air handler. And the opposite holds true as well. The less static pressure that is present, the greater the ability to move air through the air handler.

Anything that has a resistance to air movement will cause a pressure drop. The total pressure drop in a supply duct can be measured at the outlet of the blower fan. The static pressure read at this point is the total static dropped from the furthest component in the ducting. When a static pressure probe is placed into a blower door at the inlet to the fan, the total static pressure losses of the return air duct components are measured.

When these two static pressures are added together, the **total external static** pressure is known.

To measure the total external static pressure, the probe that connects to the HIGH PORT of the pressure gauge is inserted into the supply air side of the fan blower. The probe that connects to the LOW PORT of the pressure gauge is inserted into the return air inlet to the fan. The gauge will now read the total external static pressure.

To measure return air ducting static pressure, leave the return air probe in its place and remove the supply air probe, leaving it exposed to atmosphere. The gauge now reads only return air static pressure.

Add the two readings together to get total external static.

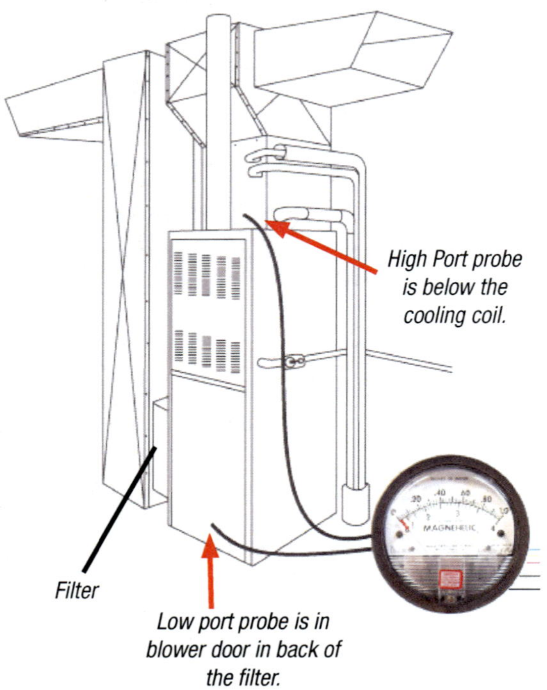

Gas furnace configured to measure total external static pressure

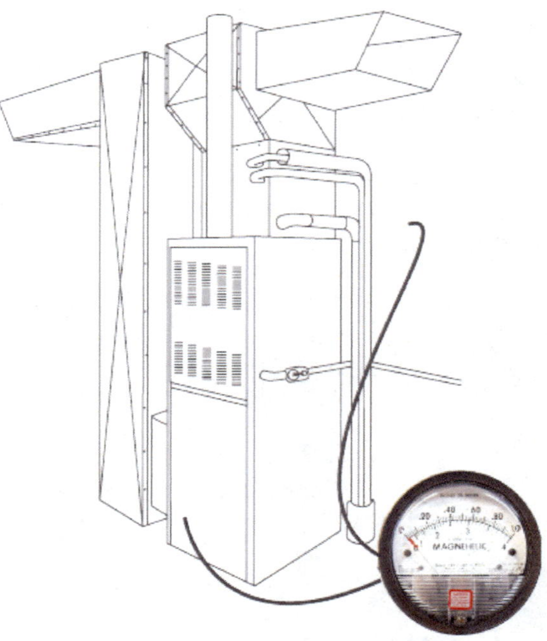

Measuring return air static pressure

Reading Each Duct Static Pressure Independently

The pressure gauge can be used to find the static pressure of the supply ducting or the return ducting individually. To measure the supply ducting static pressure, remove the probe from the return air and leave it exposed to atmosphere. Leave the supply air probe in its place. The gauge will read only supply air static pressure.

Checking the return and supply static pressure individually is useful when trying to determine which duct may be causing high static pressure and low air volume levels. One or both of the ducting systems may have an excessive pressure drop that is hurting blower air delivery performance.

Measuring supply air static pressure

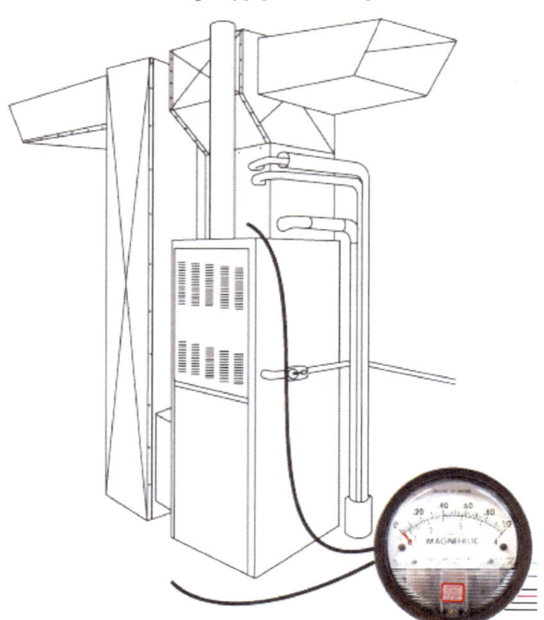

Using Static Pressure to Find Air Volume

Special tables are published by equipment manufacturers that show how much air the system can deliver in CFM based upon the total external static pressure measured.

The chart below is typical of charts found on gas furnaces. Notice the blower is capable of up to 0.9 inches external static pressure. This is a very high range. Typical total external static readings should range anywhere from about 0.15 to 0.65 ESP The reason this table goes so high is because the system uses a powerful ECM type blower motor. Notice the steady air volume produced over a wide range of static pressures. This would not be possible with a PSC blower system.

*DY080R9V3W - FURNACE COOLING AIRFLOW (CFM) AND POWER (WATTS) VS. EXTERNAL STATIC PRESSURE WITH FILTER												
OUTDOOR UNIT SIZE (TONS)	AIRFLOW SETTING	DIP SWITCH SETTING						EXTERNAL STATIC PRESSURE				
		SW 1	SW 2	SW 3	SW 4			0.1	0.3	0.5	0.7	0.9
3.0	LOW (350 CFM/TON)	ON	OFF	OFF	ON	CFM		1050	1050	1050	1010	800
						WATTS		235	295	340	350	290
	NORMAL (400 CFM/TON)	ON	OFF	OFF	OFF	CFM		1200	1200	1200	1040	840
						WATTS		335	385	410	365	310
	HIGH (450 CFM/TON)	ON	OFF	ON	OFF	CFM		1350	1350	1210	1070	900
						WATTS		455	480	435	390	345

NOTES: * First letter may be "A" or "T"
1. At continuous fan setting: Heating or Cooling airflows are approximately 50% of selected cooling value.
2. LOW airflow (350 cfm/ton) is COMFORT & HUMID CLIMATE setting;
   NORMAL airflow (400 cfm/ton) is typical setting;
   HIGH airflow (450 cfm/ton) is DRY CLIMATE setting.

For static pressure readings to be accurate, the static pressure probes must be placed into the system at the proper positions. If placed incorrectly, the pressure drop of components such as return air filters will increase the total external reading and cause a misinterpretation of the information that is gathered from the static pressure tables.

Air Performance Charts: External Static Pressure

In the example below of a CFM versus external static pressure chart, an air handling unit with a PSC blower motor is plotted. The chart requires the user to find the voltage to the blower motor, and the speed tap the blower is operating with. Static pressure is then measured and plotted to find CFM. There are footnotes at the bottom of the chart that indicate, for example, whether the filter is included in the reading. This is important to note as it will affect where the low pressure port probe is placed into the ducting. If the filter is included, the probe is placed upstream of the filter. If the filter is not included, the probe is placed downstream of the filter. Making a mistake here will make the readings invalid, so be careful.

AIR FLOW PERFORMANCE TWE024P13,0A,FA												
EXTERNAL STATIC PRESSURE (INCHES OF WATER)												
	VERTICAL (See Notes)						HORIZONTAL (See Notes)					
	230 VOLTS			208 VOLTS			230 VOLTS			208 VOLTS		
CFM	HI	MED	LO	HI	MED	LO	HI	MED	LO	HI	MED	LO
300						0.65						
350						0.63						
400			0.61			0.6					0.66	0.64
450			0.59		0.55	0.54			0.69		0.64	0.59
500		0.61	0.55		0.53	0.48			0.63		0.61	0.52
550	0.6	0.57	0.5	0.71	0.5	0.4		0.63	0.56	0.6	0.55	0.43
600	0.58	0.52	0.44	0.65	0.45	0.31	0.6	0.57	0.48	0.59	0.49	0.34
650	0.55	0.47	0.36	0.58	0.38	0.2	0.58	0.51	0.4	0.56	0.41	0.22
700	0.52	0.4	0.27	0.51	0.29	0.08	0.55	0.43	0.3	0.53	0.32	0.1
750	0.47	0.32	0.17	0.44	0.18		0.51	0.35	0.19	0.48	0.21	
800	0.42	0.24	0.05	0.37	0.06		0.46	0.26	0.08	0.43	0.09	
850	0.36	0.14		0.3			0.4	0.16		0.36		
900	0.3	0.04		0.23			0.33	0.05		0.28		
950	0.22			0.16			0.25			0.19		
1000	0.14			0.09			0.16			0.09		
1050	0.05			0.01			0.06					

Courtesy of American Standard

NOTES:
Vertical: With filter, no horizontal drip tray   Horizontal: As shipped but without filter.
Small apex baffle   Subtract 0.05" W.G. for Horizontal left.
Subtract 0.06" W.G. for downflow.

Tools:

- Magnehelic Gauge
- Static Pressure Probes
- Drill

STEP 1

Make an access hole between the top of the furnace and the bottom of the cooling coil. DO NOT DRILL INTO THE CONDENSATE PAN OF THE COOLING COIL. If creating a hole is not possible, try using the hole where the main air limit is mounted. Insert the static pressure probe that is connected to the high port of the pressure gauge into this hole. Position the probe so the tip is facing into the air stream.

STEP 2

Drill a hole into the blower door of the furnace in a spot where no damage will occur. Insert the low port probe into this hole and position the tip to face into the flow of return air.

STEP 3

Call for cooling operation. When the blower motor comes up to speed, read the external static pressure. Compare to factory charts to determine CFM.

STEP 4

If necessary, change the blower speed taps to get the desired CFM. If air volume is not high enough and the highest speed tap is being used, the cause of low air volume must be found. Check static pressure at the return and at the supply independently. If one of them has a high reading, find the cause of the air restriction.

Service Procedure:
Measuring Air Handler ESP

Tools:

- Magnehelic Gauge
- Static Pressure Probes
- Drill

STEP 1

Make an access hole at the discharge supply air outlet area of the air handler. Insert the positive pressure probe into this hole. When inserting the probe, place the tip of the probe to face into the air stream.

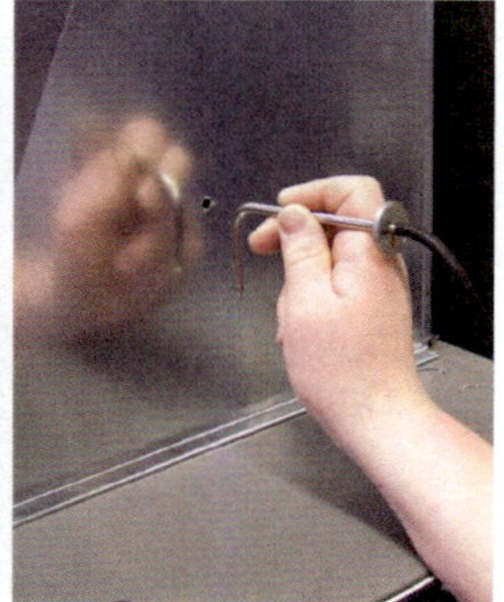

STEP 2

Determine if the filter is included in the air performance data chart. If the filter is included, drill a hole where return air duct attaches to the air handler. If the filter is not included, drill a hole into the blower door of the air handler in a spot where no damage will occur and where the probe when inserted will not touch anything electrical. Insert the low port probe into this hole and position the tip to face the direction of return air.

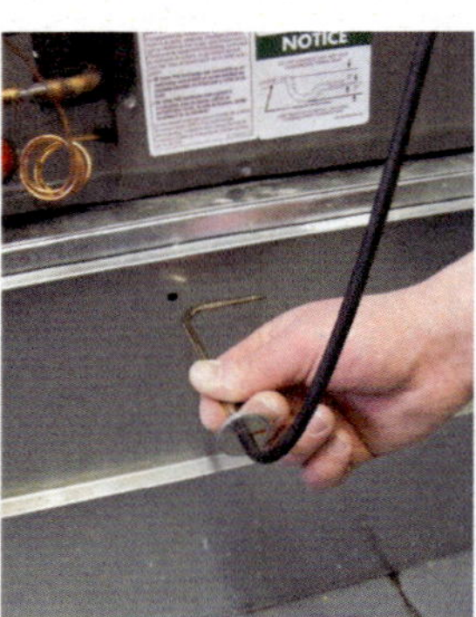

NOTE: Air performance data charts, when including a filter in the data, are specifying a factory installed filter, not a high efficiency add on filter! If a high efficiency filter has been added to the air handler, remove the factory installed filter and measure return air static via a hole in the air handler door.

STEP 3

Call for cooling operation. When the blower motor comes up to speed, read the external static pressure on the gauge. Compare to factory charts to determine CFM.

STEP 4

If necessary, change the blower speed taps to get the desired CFM. If air volume is not high enough and the highest speed tap is being used, the cause of low air volume must be found. Check static pressure at the return and at the supply independently. If one of them has a high reading, find the cause of the air restriction.

When Measured External Static Pressure is Excessively High

What is too high? What is considered too high will depend upon what type of motor is in the air handling component. If it is a variable speed ECM type motor, typically around 0.9" external static pressure is the limit. If the blower is a PSC type blower, above 0.50" is typically too high if the blower is not oversized for the cooling system. In other words, if a 3-ton system has a 4-ton air handler, the air handler can usually deliver 1200 CFM for a 3-ton system at external static pressure that is above .50". If the system is a gas furnace with a 3-ton blower, 1200 CFM will usually be achieved at 0.50" external static pressure. Anything above that will typically cause air volume to be below required levels.

When Excessive External Static Pressure is Present

Check the static pressure in the supply duct by removing the probe from the return air. Next, check the static pressure in the return air by inserting the return air probe back into place and removing the supply air probe.

The duct with the higher of the two readings should be investigated for air volume restrictions and proper duct dimension sizing. Make duct corrections as needed. It may be necessary to remove components from the duct system such as high efficiency filters.

Some high efficiency air filters can drop over 0.2" of static pressure. This is almost half of the available pressure from a gas furnace using a PSC blower motor.

Finding Static Pressure Loss Across Duct Components

When trying to determine how much a duct component, a filter for example, is contributing to the total static pressure loss, a pressure drop can be taken across the component. The high port probe is placed in front of the component in the direction of entering air into the component. The low port probe is positioned where the air leaves the component. The pressure drop is displayed on the pressure gauge.